Minicars, Maglevs, and Mopeds

Minicars, Maglevs, and Mopeds

Modern Modes of Transportation around the World

SELIMA SULTANA AND
JOE WEBER, EDITORS

 ABC-CLIO™

An Imprint of ABC-CLIO, LLC
Santa Barbara, California • Denver, Colorado

Copyright © 2016 by ABC-CLIO, LLC

Library of Congress Cataloging-in-Publication Data

Names: Sultana, Selima, 1965- editor. | Weber, Joe, 1970- editor.
Title: Minicars, maglevs, and mopeds : modern modes of transportation around the world / Selima Sultana and Joe Weber, editors.
Description: Santa Barbara, California : ABC-CLIO, an Imprint of ABC-CLIO, LLC, [2016] | Includes bibliographical references and index.
Identifiers: LCCN 2015039284 | ISBN 9781440834943 (acid-free paper) | ISBN 9781440834950 (ebook)
Subjects: LCSH: Transportation engineering—Encyclopedias. | Vehicles—Encyclopedias. | Roads—Encyclopedias. | Transportation—Encyclopedias.
Classification: LCC TA1009 .M56 2016 | DDC 388.03—dc23
LC record available at http://lccn.loc.gov/2015039284

ISBN: 978-1-4408-3494-3
EISBN: 978-1-4408-3495-0

20 19 18 17 16   1 2 3 4 5

This book is also available as an eBook.

ABC-CLIO
An Imprint of ABC-CLIO, LLC

ABC-CLIO, LLC
130 Cremona Drive, P.O. Box 1911
Santa Barbara, California 93116-1911
www.abc-clio.com

This book is printed on acid-free paper ∞

Manufactured in the United States of America

Contents

Guide to Related Topics

Aerial Vehicles

Aerial Tramways
Flying Cars
Schwebebahn-Wuppertal (Germany)
Ski Lifts

Animals and Animal-Powered Modes

Camel Trains (Mali)
Camels
Dog Sleds
Elephants
Grand Canyon Mule Train (United
    States)
Llamas
Ox/Bullock Carts

Automobiles, Trucks, and Buses

Ambulances
Automobiles
Autonomous Vehicles
Biofuel Vehicles
Biofuels
Buses
Car Sharing
Car2Go
Chicken Buses (Central America)
Double-Decker Buses
Electric Vehicles
Flying Cars
Hydrogen Vehicles
Jeepneys (Philippines)
Lyft (United States)

Mammy Wagons (West Africa)
Massachusetts Bay Transportation
    Authority (MBTA) (United States)
Minicars
Online Electric Vehicle (South Korea)
Plug-In Battery Electric Vehicles
    (BEVs)
Plug-In Hybrids
Slugging (United States)
Smart Cars
Sport Utility Vehicles (SUVs)
TransMilenio, Bogotá (Colombia)
Uber
Yellow Cab, New York City (United
    States)

Bridges

Cable-Stayed Bridges
Covered Bridges
Great Belt Fixed Link (Denmark)
Suspension Bridges

Cable Cars, Trolleys, and Streetcars

Cable Cars (United States)
San Diego Trolleys (United States)
San Francisco Cable Cars (United
    States)
Streetcars (United States)

Cycles

Bicycle Lanes
Bicycle Libraries

Tunnels

Preface

Everyone needs and uses some form of transportation to get around. Much of the time people give little thought to this fact. People rarely realize just how differently this is accomplished in various parts of the world. Unlike many books on transportation that focus on congestion, planning, or aspects of travel behaviors, this book is about the different forms of land transportation that currently exist worldwide. There presently is no other single reference book that covers such a wide variety of current transport modes, technology, and routes used around the world. This book is the first of its kind to provide complete coverage of major land-transport modes used in different parts of the world. The fruition of this book is an extensive team effort between the authors and the contributors. This project started with a listing of every mode that we could think of, and then colleagues and others were invited to contribute to the work. Many of the contributors suggested entries that we never had heard of before. We learned a lot from reading these entries and think that readers will be amazed by all the different forms of transport that people use for their daily mobility in various parts of the world.

This single-volume reference book presents more than 150 entries that provide readers with an understanding of the world's current major forms of transportation in the context of cultures, economies, politics, and environments of a place. Readers will realize that there is no single best way of traveling, and that the possibilities for travelling are almost endless. Many historical land-transport modes such as Roman chariots, stagecoaches, and covered wagons are not included in this book, though they still can be seen on display in museums or in use in re-enactments. Pipelines also are very important forms of transport for certain commodities but are not discussed herein. Instead, this book is limited to those forms of transport in common use that the reader might encounter and hope to use for travel.

We are indebted to our many contributors and the expertise they brought to the project. We are especially very grateful to Michael Kuby at Arizona State University and Shih-Lung Shaw at the University of Tennessee for their encouragement to pursue this project and for the numerous entries they suggested for this book. Stan Thompson, a retired planner in North Carolina, provided assistance in writing the entry on "hydrail."

Other colleagues helped us to connect with their PhD students. We are especially grateful to Mark Horner and Darren M. Scott for this. One can only imagine the work required to communicate with the contributors for a project like this, and we acknowledge and appreciate the assistance of Dylan Coolbaugh in this capacity.

A number of undergraduate students at the University of North Carolina at Greensboro contributed to this book by gathering information on many entries, and we specifically thank Karen Carr, Carly Everhart, Caleb Hanke, Angela Hasz, Brian O'Keefe, and Matthew Walker. We extend special and heartfelt thanks to Julie Dunbar, the editor at ABC-CLIO who gave us the idea for this project and worked endlessly with us throughout the production of this book.

We hope that readers interested in transportation around the world will find this book useful and entertaining, and that it will inspire them to seek out new modes for their travels.

Selima Sultana and Joe Weber
Editors

Introduction

Transportation is one of the world's oldest and most important activities. Everybody requires some means of transport to get around—to get food or water, go to work or school, seek medical care, visit with friends and family members, and even to receive a product purchased online. For some, this could involve traveling at hundreds of miles per hour to a distant city by train, or a drive in one's car, a ride on an animal, or even walking down the stairs. These are all important modes of transport around the world, and are among a nearly infinite list of different varieties and subtypes of modes. We all have some experience of land transport, and for most of the world's population this is the only kind of transport known. There is a long history of ships and travel by sea, and travel by air enables people to cross continents and oceans quickly, but land transport by far was—and remains—the most important way of traveling in the world.

Walking is the principal transportation mode that has allowed humans to sustain or build settlements since the start of human existence and it remains the primary form of transport used in many parts of the world. The first animal used for transportation might have been the dog, more than 10,000 years ago; many others followed, including oxen, horses, donkeys, elephants, llamas, and camels. The wheel first was used for transport around 4000 BCE, ultimately giving rise to railroads, bicycles, automobiles, and many other types of vehicles. Today there are more than 1 billion cars used worldwide, and 147 countries have some sort of railroad. There are countless variants of these and other land-transport modes, and this book discusses more than 150 of them. Each method included is its entry, and entries are listed in alphabetical order with connections to other mode entries. Some entries are about vehicles (or animals), some are about roads or rail lines, and some are about transport routes or networks. A few entries also discuss new forms of propulsion systems for road and rail vehicles. Some are for passenger travel, some are for freight, and some can serve both. Many of these modes serve in part—or entirely—as tourist attractions, including San Francisco's famous cable cars and India's Darjeeling Himalayan Railway.

Many of the modes used are urban, enabling people to move about in the world's cities. About half the world's population lives in cities; and in developed countries this proportion is much greater. These cities are concentrations of people and buildings and require advanced transport systems to move people on streets, underground, or overhead. Many cities rely on trains to move people, but as readers will discover, there are several important types of urban trains, and the way that they operate differs from place to place. For example, there are many entries about urban rail systems; and each shows differences in terminology, history, equipment, how fares are collected, and how the network is arranged. Some rail systems operate above ground, some below ground, some as long trains, and some as single cars. Some modes covered by the book's entries are almost entirely rural—such as dog sleds or ski lifts. In rural areas and developing countries there still are many places where modern and old, animals and machines, still mix. The variety of entries on railroads around the world is amazing.

Since the mechanization of land travel in the late 18th century, travel modes have become faster. Supersonic land vehicles now exist, although—except for trains—increased speeds have not proven useful. Faster speeds, however, have created what is called "space-time convergence," in which travel times between places decreases, bringing those places closer together. Despite the idea of a shrinking world, however, in many ways the opposite is true. The planet's population is growing, as are the world's urban areas. The need for mobility will continue to grow, and new technologies will change how we travel.

Not all modes are fast, however. Those relying on muscle power have biological limits that have not greatly changed in millennia. Speed is not always valued. Increasingly the challenge is to develop modes of travel that use renewable fuels, or at least those fuels not likely to be facing shortages in the near future. Electric cars, fuel cell forklifts, and similar vehicles are part of this trend. Creating travel modes that create less air pollution also is an important goal. Vehicles that don't rely on fossil fuels have an advantage in this regard. These goals are often lumped together and called "sustainable," with the idea that they can continue in service in the future without running out of fuel or causing unnecessary pollution. Sustainable or low-carbon modes of transportation are not a new concept, as many of the modes in the book need only food and water to continue operating. These animal transport modes even are capable of reproducing themselves, something still beyond the capability of mechanical systems. Whether technologically based sustainable modes of transport can be mass-produced and distributed quickly and inexpensively enough to be of significance in the world remains to be seen.

In addition to vehicles, many entries cover particular transport routes or networks. The Interstate Highway System in the United States, the National Trunk Highway System in China, and the National Highways Development Program in India are three of the world's largest and most important highway systems. One important way that land transport differs from air and sea travel is that there are many separate highway and railroad networks around the world, built to different specifications. The sheer variety of different types of railroads around the world, and the varying track gauges, speeds, and forms of propulsion are evidence of this. The Channel and Seikan tunnels and Great Belt Bridges are expensive structures that connect formerly separate areas by using a single transport system. What networks will be connected next? This remains to be seen.

This book explores the most up-to-date surface transport modes; transportation technologies; and some of the most famous streets, rail systems, and highways in use on every continent—which blend transportation geography with culture, politics, economics, and the environment of a place. The entries are written by experts from around the world and are intended to be accessible to general readers without technical backgrounds. Each entry incorporates cross-references that enable readers to easily find related entries, making the book ideal for conducting specific research or projects. Additionally, sidebars explore future innovations in the world of transportation. This book conveys the messages that there are diverse ways to transport people and goods given the context of a place, culture, and environment; that there is no single best way of traveling; and that the possibilities for making our way are almost endless.

Chronology of Modern Modes
of Surface Transportation (1775–2015)

1775

The Wilderness Road—the first long-distance road in the United States—was built by widening an American Indian trail. This was built to expand settlement to Kentucky for European settlers.

1784–1797

The first mail coach was built by John Palmer and in 1784 started carrying mail in Britain between Bristol and London. The mail coach became so successful that by 1797 there were 42 routes covering a total of 16 towns in Britain. The departure of mail coaches was a popular event for spectators every evening in London.

1800–1850

The Industrial Revolution set the stage for massive changes in transportation systems. The development of the steam engine (an external combustion engine that could convert thermal energy into mechanical energy) marked the beginning of a new era of mechanization in land transportation. The adaptation of the steam engine to locomotion engine gave birth to rail transport in Britain in 1814, and this became a dominant mode of land transportation in London by 1836. The first railroad in the United States—the Baltimore & Ohio—was built in 1830. In 1812, the first omnibus came into use in Bordeaux, France. In 1829, London saw the first bus that used a steam engine. In the United States, a new highway—known as the National Road or the Cumberland Road—was completed in 1818 and stretched from Cumberland, Maryland, to the Ohio River. By 1833 this road was expanded to Columbus, Ohio. The route still survives as US 40. In 1835, the first time passenger elevator was installed in a factory in England.

1850–1900

The first metro, the railway that carries a great number of passengers within the city, appeared in 1863 in London, which is today known as the Metropolitan Line of London or London Underground or the "Tube". Soon after the London metro began operation, metro railways were built worldwide.

In 1867, bicycles first appeared in Paris. The Alpine Tunnel was built in 1871 to link France and Italy. Electric engines became available in the 1880s, which created another form of land transport known as "tramways" or "streetcars." In the United States, ambulances became available to the general public in 1865 and first were operated by the Commercial Hospital of Cincinnati.

The idea of the funicular design—a form of transportation that combines the aspects of an elevator and a railroad to propel cars on rails up and down steep slopes—came from the Swiss engineer Carl Roman Abt in 1867. The world's most famous and oldest funicular train, the Tünel, was operating in Turkey by 1875. The cable car system was initiated in San Francisco in 1871 by engineer Andrew Hallidie. Meanwhile, the first and the most well-known mountain railway—the Darjeeling Himalayan Railroad (DHR)—was built in India in 1879.

This period also witnessed the birth of the automobile in Germany (1885). In the United States, the first "modern" subway system was built in Boston, Massachusetts, in 1897, followed by the New York Subway System in 1904. The Alaska Railroad was built in 1900 to connect seaports with the interior gold mining area in Alaska.

1900–1950

The birth of the automobile in the late 19th century was followed by the construction of world's first limited-access road system—the Autobahn—in Germany. Construction of the Autobahn began around 1913 and was completed after World War II. In 1914, the first recorded intercity bus transport began in the United States. Construction of the Autostrade—an Italian major highway system—began in the 1920s. The scenic Blue Ridge Parkway in the United States was built in 1935 to enable motorists to enjoy the natural beauty of the Appalachian Mountains. The Alaska Highway was built in 1942 to connect the United States to the territory of Alaska.

1950–2000

During the period from 1950 to 2000, increased speed became a primary goal for transportation. The moped (a hybrid bicycle with an engine and pedals) was introduced in 1952 to increase bicycle speed. The advancement of transportation technology during this period introduced a new type of railway—the high-speed train. The first high-speed Shinkansen train was in operation in Japan between Tokyo and Osaka in 1964, with an average speed of 150 miles per hour. France launched its own high-speed rail, the Train à Grande Vitesse (TGV), in 1981; the train's top speed was 220 miles per hour. After the TGV was introduced, high-speed trains began operation in many countries in Europe and Asia.

In the United States, construction of the Interstate Highway System began in 1956. In Japan, development of intelligent transportation systems was initiated in the 1960s—but it took 30 years for the United States to pursue this line of research. From the 1950s on, the love of automobiles increased worldwide and the growing economies of many countries enabled them to start building highways similar to the U.S. Interstate Highway System. The nationwide freeway system in China is an example and has been under construction since the 1990s.

The development of new technology made improving the connectivity of the world's transport networks an important goal. The world's longest undersea tunnel—the Seikan Tunnel—was built in 1988 to connect the two main islands of Japan: Honshu and Hokkaido. The world's longest undersea railway, the Channel Tunnel, was built in 1994 to connect the United Kingdom and France.

Since the 1990s, the issue of transportation emissions and global warming has become an important topic and has raised awareness of the negative environmental impacts of automobiles. Many automobile companies started investing in hybrid vehicle technology. Hybrid vehicles such as the Toyota Prius and Honda Insight represent a huge breakthrough in automotive technology. At the same time, alternative transportation modes (such as walking and bicycling) received renewed interest in the developed world, which made greenways popular as a way of creating safe and enjoyable routes for nonmotorized travel.

2000–2015

At the beginning of the 21st century, the issue of transportation emissions resulting in global warming continues to illustrate the need to explore alternative fuels such as biofuels. The use of ethanol and biodiesel (biofuels) could offer significant environmental improvements. Since 2000, the world's first high-capacity bus–rapid transit (BRT) system, the TransMilenio, has been serving the city of Bogota and using biofuels.

In 2002, the Segway was introduced by inventor Dean Kamen. Since 2004, Korean high-speed rail (KTX) has been in operation between Seoul and Busan. In 2004, the first commercial high-speed maglev train began operation in Shanghai, China, between outskirts of Pudong, Shanghai and Shanghai Pudong International Airport. In 2005, the "walking school bus" concept was introduced in the United States to reembrace walking as a mode of transportation for school-age children.

Hybrids have gained popularity in the late 20th century and companies that produce hybrids have been introducing better batteries. Other innovations also abound—electric vehicles, plug-in battery electric vehicles (BEVs), plug-in hybrids, hydrogen vehicles, online electric vehicles (OLEVs), solar roads, and hydrail. In 2008, plug-in hybrids came to market. In 2010, the online electric vehicle—an alternative transport technology with great potential for reducing pollution—was invented in South Korea. Also in 2010, the first solar road was built in the United States. Since 2014 the "SolaRoad" has been in operation in the Netherlands. In 2015, hydrogen vehicles were introduced in Southern California.

2015 and Beyond

There is no doubt that the 21st century will witness increased use of alternative fuels and marketable hydrogen vehicles. Recently developed V2V (vehicle-to-vehicle communication) technology most likely will bring autonomous cars to the market by 2020. Plug-in hybrids, carbon-fiber composites, hydrogen fuel cells—a wide range of innovative technologies—could make our future vehicles more efficient and less reliant on oil. Smartphone ridesharing apps (e.g., Uber, Lyft) will continue to evolve for future solutions in the urban taxi industry worldwide. With the development of string power, flying cars will be a reality. Perhaps even the vacuum train will be in operation by the end of this century.

AERIAL TRAMWAYS

Aerial tramways stand out among tourist attractions because they reach into the skyline and provide one-of-a-kind views of the surrounding landscape. Tramways around the world have been providing experiences to tourists and residents for decades, and include the Schmittenhöhebahn tramway located in Zell am See, Austria, which has been transporting visitors during summers and winters for more than four decades. Other such iconic aerial tramways include the Roosevelt Island Tramway in New York City, which opened in 1976, and the Sandia Peak Tramway in Albuquerque, New Mexico, which began operations in 1966. These tramways serve thousands of tourists each year and have made significant impacts on local economies and businesses. In some cases tramways serve as the only means to access unique destinations.

Many systems that have had continuous service for decades still use the original power and cable service, although issues have occurred in recent years; the issues primarily stem from the environments in which the aerial tramways operate. Although many transport systems are capable of pushing on with minor repairs and replacements, many other tramways operate in saltwater environments and in locations where elevations change significantly, and must be completely replaced after approximately 40 years.

The Sandia Peak Tramway in New Mexico shines as a system that has served its area residents and tourists for almost 50 years. It was updated and modified with new tramcars in 1986 and new cables in 1996. Like many tramway systems, there is a restaurant at the top of the system. Additionally, it is one of the most frequently mentioned points of interest in the visitor information for Albuquerque.

In recent years, several new aerial tramways have been constructed and have begun operations, most notably in Tatev, Armenia, where the world's longest single-span passenger tramway crosses the Vorotan Gorge. The span is 3.54 miles long. The system's cost of construction was approximately $45 million, and it connects the highway near Armenia's capital with the ninth-century monastery. The tramway was inducted into the *Guinness Book of World Records* on October 16, 2010, as the longest nonstop cable car.

Some of the older and more iconic aerial tramways in recent years have suffered due to outdated and failing systems. Most recently, the Roosevelt Island Tramway was replaced by a new system developed by Poma (a French company that specializes in cable-driven lift systems), which aided in renewed service in November 2010. The renovation was due after one of the original cars stranded passengers

Aerial tramway moving up tropical jungle mountains, 2013. (06photo/Dreamstime.com)

mid-air for more than 11 hours, and subsequent tests failed to solve the problem. Other such stories of trams either being replaced or discontinued altogether include those of the Jackson Hole Ski Resort in Wyoming, and Pipestem Resorts State Park in West Virginia.

The Jackson Hole ski lift in Wyoming has operated since 1966. It finally started to show its age, and resort owners determined that at some point it would operate at a compromised efficiency. Stakeholders determined that without funding from outside sources a new lift would not be a reasonable investment. This underscores the long lifespan of aerial tram systems, but also shows the cost to private businesses when the systems are not publicly owned. After securing funding from states and investors, Jackson Hole Ski Resort decided to open a new lift after all, ensuring that all existing ski runs were available in 2008. Similarly, the Pipestem tram system in West Virginia, first opened in 1971, provides exclusive access through the Canyon Rim Center to the bottom of the Bluestone Canyon and into the Mountain Creek Lodge and Restaurant. When the tram needed renovation funding was secured. Clearly, outside funding sources are plausible means for tramways to continue to provide necessary services.

With rapid advancement in technologies relevant to aerial tramway systems, trams quickly are becoming more than just one-stop destinations. Companies such as Poma and Angelil/Graham/Pfenniger/Scholl specialize in tram system installations, construction, retrofit, and renovation, and have been working on large-scale projects around the world that are not simply tourist destinations. Portland, Oregon, has the world's only commuter tram and it travels 3,300 feet, at 22 miles per hour. The trip takes only about four minutes and the tram operates continuously, ensuring that commuters do not have to wait long for a car to arrive.

As a tourist attraction and ski resort necessity, aerial tramways have a solid footing and are sure to be around long into the future. As innovative uses of aerial tramways emerge, their place in the urban framework will continue to change, especially if the cost varies significantly from that of light-rail or other major transit improvements.

Dylan Coolbaugh

See also: Elevators; Funiculars; Rack Railways; Ski Lifts.

Further Reading

The Gordon Project. http://gondolaproject.com/2010/04/24/technologies-module-5-aerial
-trams/. Accessed August 8, 2015.
Neumann, E. S. 1999. "The Past, Present, and Future of Urban Cable Propelled People Movers." *Journal of Advance Transportation* 33(1): 51–82.

ALASKA HIGHWAY (CANADA AND UNITED STATES)

The Alaska or Alaska-Canada (ALCAN) Highway was built during World War II to connect the United States to the territory of Alaska. This was considered necessary in light of the possibility that Japanese forces could sink ships moving between Pacific seaports and Alaska. A number of routes were considered for the highway, with the easternmost of these selected. U.S. military forces began construction in March 1942—initially without the knowledge or permission of the Canadian government. The road to Alaska was completed in October 1942.

The initial road was a rough dirt road, passable only in summer. It has been under reconstruction ever since, and now is entirely paved and can be used all year. Reconstruction has involved rebuilding and widening the road, easing curves and reducing the steepness of hills, replacing temporary bridges with more substantial bridges, and bypassing curvy sections with straighter roads. These changes have reduced the length of the Alaska Highway. When built it was 1,700 miles long, but it now is less than 1,400 miles long.

The ALCAN was more than just a road; it provided access to a set of airfields along the Lend-Lease route from the United States to the Soviet Union. These airfields ultimately were of much greater strategic significance than was the highway. Pipelines also were built along the highway to supply the airfields. By the time the highway was open to traffic the threat to Alaskan shipping had subsided, but the highway continued to make an important contribution to the war by helping to keep the air route open. The road remained under U.S. military control and then was turned over to Canada after the war ended. It became a public highway and is now Yukon Highway 1 and British Columbia Highway 97. In Alaska it is part of State Highway 2.

The road officially begins in Dawson Creek, British Columbia, because that was the point on the Canadian railroad system that was nearest Alaska. It runs to Delta Junction, Alaska, although Fairbanks sometimes is considered the endpoint. Whitehorse, Yukon Territory, is the largest town along the highway and smaller towns having some services are scattered along the highway. Much of the road is located in the interior plains of Canada. It does pass through the northern Rockies in the Yukon Territory, but the highest elevation along the road is only 4,250 feet above sea level.

The length and location of the Alaska Highway can be surprising. Dawson Creek is more than 460 miles north of the U.S. border, and 744 miles by main roads to the closest major border crossing. It actually is farther north than the southernmost part of Alaska. People who want to drive to Alaska from the lower 48 states therefore first must drive a great distance to Dawson Creek before officially starting on the Alaska Highway. Only about 15 percent of the highway is in Alaska. At the northern end (Delta Junction) it still is 334 miles to Anchorage, the state's largest city. Although the most famous road in Alaska, it actually is one of the more lightly used state highways because it does not lie between the major centers of population. It is almost 600 road miles from Delta Junction to the Arctic coast of Alaska at Prudhoe Bay.

The road remains a popular and challenging recreational route to Alaska and the Yukon Territory. The best guide to the Alaska Highway is *The Milepost*, an annual travel guide. It is not the only way to drive to Alaska, however; the Alaska Marine Ferry provides frequent ship service between Seattle, Washington, and ports in Alaska. Travelers also can take the Klondike Highway, which enters Canada farther north. This route passes through the town of Dawson City, Yukon, Canada.

Joe Weber

See also: Alaska Railroad (United States); Blue Ridge Parkway (United States); Ghan Train, The (Australia); Trans-Amazon Highway (Brazil).

Further Reading

The Milepost. 2015. http://www.themilepost.com/. Accessed January 13, 2016.
Twitchell, Heath. 1992. *Northwest Epic: The Building of the Alaska Highway*. New York: St. Martin's Press.

ALASKA RAILROAD (UNITED STATES)

The Alaska Railroad is a 467-mile-long railroad in the state of Alaska. It was one of a number of railroads built in Alaska in the early 20th century to connect seaports with gold mining areas in the interior of the territory. It later served interior cities, tourists, and additional mining activities. The Alaska Railroad is one of only two railroads in the state that survived the end of the mining booms. The other is the narrow gauge White Pass and Yukon Route, a tourist line operating between Skagway and Whitehorse, Canada.

Construction of the Alaska Railroad began in 1903 at the port of Seward. A work camp farther north grew into the city of Anchorage, and now is the headquarters of the railroad. Because of the importance of the railroad to the territory of Alaska, in 1914 construction was taken over by the U.S. government. The railroad continued north and was completed in 1923 at the town of Nenana, Alaska. President Warren Harding (1865–1923) was present for the completion of the railroad, which took place several weeks before Harding died in office. In 1985, the State of Alaska purchased the railroad and continues to own it, although no public subsidies are provided. It remains an important transport link within the heavily populated south central region of the state known as the "railbelt."

The Alaska Railroad is unique in being the only full-service railroad in the United States, meaning it has both freight and passenger service. Passenger service remains important to the railroad. This includes regularly scheduled passenger service between Anchorage and Seward, from Anchorage to Whittier, and from

President Warren Harding driving the last spike on the Alaska Railroad at Tanana River, 1923. (Library of Congress)

Anchorage to Fairbanks. The Anchorage-to-Fairbanks train is the longest line, with a stop inside Denali National Park. A popular rail destination, 36 percent of visitors to Denali arrive by train. Before 1957, when the first highway reached the park, the railroad was the only way to get there.

The Alaska Railroad also carries many cruise ship passengers in railroad cars owned by different cruise lines. Many of these passengers board at Seward directly from cruise ships; these passengers actually comprise more than two-thirds of those arriving at Denali National Park by train.

An unusual service of the Alaska Railroad is a local train, the Hurricane Turn, which provides service to rural areas north of Anchorage. Local residents stand by the tracks to flag down a train to board or to receive freight, and many residents rely on the train as their only connection to the outside world, as there are no roads in the area.

Today, freight traffic on the line includes gravel, coal, and oil along with containers. Coal is shipped to Seward for export to Chile and several Asian countries. The Alaska railroad has railcar ferries enabling freight cars to be shipped between the state and ports in the lower 48 states. The railroad continues to grow and a new line to Port McKenzie on Cook Inlet currently is being built. An extension to the lower 48 states via a new rail line through Canada has been discussed but seems unlikely under present conditions.

Joe Weber

See also: Alaska Highway (Canada and United States); Ghan Train, The (Australia); Trans-Amazon Highway (Brazil).

Further Reading

Alaska Railroad Corporation. 2012. http://alaskarailroad.com/. Accessed January 13, 2016.

ALPINE TUNNELS (EUROPE)

Tunnels have been dug in the Alps—the enormous mountain range that stretches across south-central Europe—since the first years of the 18th century. Early travelers had to follow high-elevation passes through the mountains and progress was slow at the best of times. During the winter much of the region became inaccessible. Digging through the mountains provides another alternative.

The first Alpine tunnel on record was the 210-foot-long Urnerloch, which was constructed from 1707 to 1708. This tunnel allowed easier passage along the St. Gotthard Pass Road and linked the Swiss cantons (administrative districts) of Uri and Ticino. The Urnerloch was planned and supervised by Italian engineer Pietro Moretti, whose workers used black powder to blast their way through the rock.

It would be more than a century and a half before more sophisticated techniques made it possible to dig longer tunnels in the region. During that time engineers learned to support the enormous weight of rock and soil with temporary tunneling shields and to deploy rail-mounted drill carriages. Dynamite—a safer alternative to gunpowder—became available in the late 1860s, and workers began using pneumatic and hydraulic drills shortly afterward.

Work on the first major Alpine tunnel, the 8.1-mile Fréjus Rail Tunnel (Mont Cenis Tunnel), began in 1857 but was not completed until 1871. Lengthened to 8.5 miles a decade later, the Fréjus provided a vital link between France and Italy. The Gotthard Tunnel, which opened in 1882 after a decade of construction, lies in the same area as its 18th-century predecessor, the Urnerloch, and runs 9.3 miles through the Gotthard massif. The next major tunnel dug in the Alps was the Arlberg Railway Tunnel, which linked the provinces of Tyrol and Vorarlberg in western Austria and was opened in 1884.

The first of two Simplon tunnels (linking Switzerland and Italy) opened in 1906 after eight years of construction. Due to the tunnel's great length—more than 12 miles—authorities decided to run electrical trains rather than steam trains through it. Completed in 1913, the 9-mile Lötschberg Tunnel extended a rail connection to Brig near the northern terminus of the Simplon Tunnel and connected the Swiss cantons of Berne and Valais. Nine years later, a second Simplon Tunnel was completed, running parallel to the first. At 12.31 miles it remained the longest railway tunnel in the world until 1985.

Vehicular tunnels began to supplement railway tunnels in the mid-20th century, by which time far more sophisticated methods of drilling were available. The first such major tunnel in the Alps was the Great St. Bernard. Completed in 1964 after six years of labor, it enables traffic to pass between Switzerland and Italy. The following year, the Mont Blanc Tunnel—which runs beneath the famed mountain of the same name—opened to traffic between Italy and France.

The Arlberg Road Tunnel (located, like its railroad predecessor, in western Austria) became the country's longest vehicular tunnel when it opened in 1978. Two other vehicular tunnels—the St. Gotthard in Switzerland and the Fréjus between France and Italy—were completed two years later near their earlier rail counterparts.

Although vehicular tunnels became more common in the Alps in the 20th century, the age of railway tunnels in the region is not over. The Gotthard Base Tunnel has been under construction since 1996 and is scheduled for completion in 2016. Consisting of two single-track tunnels running at a low elevation through the Gotthard massif for 35.4 miles, it will be the world's longest railway tunnel. Along with the Ceneri Base Tunnel (projected to open in 2019), the Gotthard Base Tunnel will form part of one section of a larger system known as AlpTransit or New Rail Link through the Alps (NRLA). The Lötschberg Base Tunnel, which lies west of the Gotthard and opened in 2007, comprises part of another section. Together the sections will improve transportation between northern and southern Europe, with passenger trains on the Gotthard section expected to move at a maximum speed of 155 miles per hour.

Tunnels have made travel through the Alps much easier, but their construction often comes at a high human cost. Two lives were lost in the construction of the Urnerloch, but more than 200 workers were killed during the construction of the first Gotthard Tunnel, and dangers remain—even for travelers. In 1999, nearly 40 people died when a truck carrying margarine caught fire and exploded in the Mont Blanc Tunnel. The following year at least 150 skiers died when a cable car they were riding through an Austrian tunnel stalled and caught fire.

Grove Koger

See also: Darjeeling Himalayan Railroad (DHR) (India); Guoliang Tunnel (China); Orient Express (Europe); Railroads.

Further Reading

AlpTransit Gotthard AG. 2015. We're Building the Heart of the Rail Link through the Alps. http://www.alptransit.ch/en/home.html. Accessed October 25, 2014.
Johnson, Stephen, and Roberto T. Leon. 2002. *Encyclopedia of Bridges and Tunnels.* New York: Checkmark Books.
Sharp, Paul. 1979. "Tunnels through the Alps." *The Alpine Journal* 84: 122–131.

ALTERNATIVE FUELS

The world's dependence on fossil fuels as an energy source is detrimental to the environment. Unlike fossil fuels, alternative fuels do not emit greenhouse gases. Presently, there are several alternative fuel options, including electricity, biodiesel, ethanol, propane, natural gas, and hydrogen. Each option has different benefits and potential drawbacks, and there are different geographic areas that would greatly benefit from the use of specific alternative fuel sources. What is a sound alternative fuel for one location might not work as an alternative fuel source in another place. This entry focuses on the United States and how alternative fuel sources can have economic, political, cultural, and environmental effects.

The hybrid electric vehicle already exists and uses electricity as a power source instead of using only gasoline. This is a step in the right direction and it decreases the use of fossil fuels. The next step is the plug-in electric vehicle, which relies on electricity completely and requires no gasoline. This option would work for almost any area in the United States because electricity already is available almost every-where. The greatest hurdle to overcome is increasing the availability of electric charging stations, although much of the infrastructure already is in place.

Biodiesel is a renewable fuel source that is created from vegetable oil, animal fats, or recycled restaurant cooking oil. It can be used to power vehicles that typically use diesel fuel. Ethanol is a renewable fuel source made from corn and can be used in flex-fuel vehicles. When it comes to biodiesel and ethanol as alternative fuel sources, proximity to refineries plays an important part. The closer to refineries that fueling stations are located, the lesser the transportation costs of transporting that fuel and the lower the fuel prices. This factor is most important for biodiesel and ethanol because they are transported via trucks and railways instead of pipelines. Both of these fuels are produced domestically, however, which makes their use a good political choice because it reduces U.S. dependence on other countries for oil.

Propane—also known as liquefied petroleum gas—is a good option as an alternative fuel source because it is relatively inexpensive, it burns clean, and it can be produced domestically like biodiesel and ethanol. Propane vehicles currently cost several thousand dollars more than comparable gasoline-powered vehicles, but propane is a cheaper fuel source making it a better long-term economic decision. Natural gas can be used as an alternative fuel source for vehicles in two different forms: compressed natural gas and liquefied natural gas. Natural gas is a clean-burning fuel source that mostly is produced domestically. Renewable natural gas is a newer option in which fuel is created from decaying organic materials such as landfill contents and animal and plant waste. One benefit of natural-gas vehicles is that they are similar to gasoline vehicles in how much power they have and how they drive.

Hydrogen is an alternative fuel source that still is in the development phase. Hydrogen can be found in great amounts in water, hydrocarbons, and other organic matter. The problem is efficiently extracting the hydrogen for use as a fuel source. Another issue with hydrogen use is distributing it when it is ready to be used for fuel. Presently, hydrogen tends to be used in close proximity to where it is produced, but if more pipelines are built then hydrogen can be transported to wherever it is needed.

One of the main reasons to use alternative fuels instead of fossil fuels is the benefit to the environment. Fossil fuels produce greenhouse gases, thus destroying the atmosphere. Alternative fuel sources burn cleaner, which reduces harm to the environment. An economic factor that encourages the use of alternative fuels is the price of gasoline and diesel fuel. If alternative fuel sources are more affordable because of increased oil prices, it would encourage the use of more environmentally friendly options. A major political factor includes state incentives encouraging the use of vehicles that don't use fossil fuels. All vehicles must meet emissions requirements to pass inspection each year, but the alternative fuel sources listed above would eliminate much of the emissions currently produced by using fossil fuels.

There also are political and economic factors that work against alternative fuel uses. Large corporations that make money from oil and gasoline are not going to support an alternative fuel vehicle that does not use their products. These corporations can influence government through their donations to different political parties. A cultural factor affecting whether alternative fuels would be successful is the access to and ease of use of fueling/charging stations. No one considers it strange to go to a gas station and pump gas—it's something every driver is accustomed to doing. But in the early stages of alternative fuel use, would there be positive or negative connotations associated with getting fuel at an alternative fueling station instead of from a gas station?

Many alternative fuel sources are being developed and some are ready for transportation use. Reducing dependence on oil, decreasing greenhouse gases, and reducing fuel costs are all reasons for the United States to embrace alternative fuel sources instead of continuing to use gasoline and diesel.

Selima Sultana

See also: Battery Swapping; Biofuels; Biofuel Vehicles; Hydrail; Plug-In Hybrids.

Further Reading

Johnson, Caley, and Dylan Hettinger. 2014. "Geography of Existing and Potential Alternative Fuel Markets in the United States." National Renewable Energy Laboratory. http://www.afdc.energy.gov/uploads/publication/geography_alt_fuel_markets.pdf. Accessed June 5, 2015.

U.S. Department of Energy. Alternative Fuels Data Center. http://www.afdc.energy.gov/fuels/. Accessed June 5, 2015.

AMBULANCES

An ambulance is a specially equipped emergency transportation vehicle used for transporting sick and injured people to a treatment center. The evolution of the ambulance has had a strong, significant relationship with military science and the culture of military-medical administration. Although it is hard to determine exactly when such specifically designed vehicles were introduced, the job of lifting and carrying sick and injured people has existed throughout the history of humankind. The earliest known written report indicates that the first ambulance was devised in 900 CE by the Anglo-Saxons. It was a hammock-based cart used to transport people with psychiatric disorders. Military use of such a device for sick and wounded soldiers did not begin until the 15th century. By the 1700s, surgeons were using two-wheeled wagons on battlefields to take wounded soldiers to the hospital as well as to bring the surgeons to the battlefield. The "ambulance"—a term derived from the Latin word "ambulare," meaning to walk or move from one place to another—first was used by the French surgeon Dominique-Jean Larry to provide early medical care supported by ambulance service and first aid for wounded soldiers during the Napoleonic Wars (1803–1815). By mid-century, almost every European army used some sort of stretcher-bearers and ambulance wagons.

In the United States, General Jonathan Letterman—a Union military surgeon during the Civil War—developed the first completely successful system for treating and transporting injured soldiers. The Ambulance Corps Act was passed by Congress in 1864 to establish a uniform system of ambulances in the U.S. Army. This act also changed the way ambulances were operated. The head of the medical department within a jurisdiction became responsible for all aspects of the ambulance services—such as administration, personnel, and vehicles—instead of

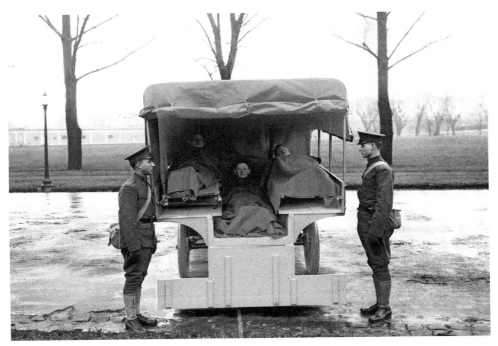

Red Cross ambulance, circa 1920. (Library of Congress)

serving as a part of general transportation system. Under this new administrative structure of ambulance operation, special uniforms were introduced to the ambulance corps to identify them to the general public. This uniform system in the ambulance corps later became a model for ambulance service throughout the world until World War I.

A similar ambulance system was not available to the general population until 1865 when the Commercial Hospital of Cincinnati (known today as Cincinnati General) created an emergency transport system to reach civilians. Unlike military ambulances, the civilian ambulances were smaller, lighter, and faster. Their interiors were designed to include a mattress, pillows, and blankets to offer better comfort to the civilian patients. This "wagon"-like ambulance, however, had a lack of storage space for medicine and other emergency equipment. As a result, better design of wagons for comprehensive ambulance services was emphasized. By 1883, Bellevue Hospital in New York City had ambulance wagons well equipped with stretchers, handcuffs, strait jackets, a box of brandy (for pain), two tourniquets, six bandages, sponges, splint material, and a small bottle of persulfate of iron. These ambulances also included the word "AMBULANCE" painted on them in black and dark green, and had direct telegraphic and telephonic communications services to police departments and hospitals. These types of ambulance designs were quickly adopted in other major urban environments in the United States. The New York state legislature also passed an act that gave the right-of-way to ambulance drivers. The first motorized ambulance was introduced in Chicago by Michael Reese Hospital in 1899, and traveled only 16 miles an hour.

The emergency transport system in the United States exists in both government-supported and privately supported systems. Most governmental emergency management systems usually are based out of local fire or police departments, but in large counties or cities can be a completely separate entity. Private ambulance services usually partner with municipal services that provide some or all of the care patients receive before reaching a hospital. Some local hospitals offer their own emergency services. It's important to note that ambulance services differ and that many local first responders are police or firefighters, depending upon service location. For the majority of the United States, the emergency service phone number is 911.

Selima Sultana

See also: Autonomous Vehicles; Complete Streets (United States); Wheelchairs.

Further Reading

Haller, John S. 1990. "The Beginning of Urban Ambulance Service in the United States and England." *The Journal of Emergency Medicine* 8: 743–755.
Haller, John S. 1992. *Battlefield Medicine: A History of the Military Ambulance from the Napoleonic Wars through World War I.* Carbondale: Southern Illinois University Press.
McCallum, Jack E. 2008. *Military Medicine: From Ancient Times to the 21st Century.* Santa Barbara: ABC-CLIO Inc.

AMTRAK (UNITED STATES)

Amtrak is the brand name for the National Railroad Passenger Corporation, which is the only national passenger rail operator in the United States. In 2012, Amtrak carried 31.2 million passengers to more than 500 destinations on 21,000 route miles in 46 states. Amtrak collected $2 billion in ticket revenue and covered 88 percent of its operating cost. Federal government appropriations for Amtrak totaled $1.4 billion. Amtrak's fleet consisted of 2,090 passenger cars and 485 locomotives.

Amtrak's operations are divided into three business lines. The Northeast Corridor includes the heavily trafficked routes between Boston, Massachusetts, and Washington, DC, where passenger rail is still an important part of daily life for many residents. Long-distance routes, as the name implies, are trains that connect widely dispersed cities across the country and can take as many as two days to complete the journey. Many of these routes, such as the Empire Builder (between Chicago and Portland/Seattle) and the Crescent (between New York and New Orleans), trace their names back to legendary trains of the heyday of passenger railroading in the early 20th century. State-supported and short-distance operations provide service within limited areas (such as Illinois or California), generally requiring less than eight hours to complete their routes, and often receive financial support from the governments in the states where they operate.

Passenger railroad use peaked in the United States around the time of World War I. As more Americans purchased cars more highways were paved, and air travel became more affordable, causing trains to gradually lose riders to these other travel modes. Following a final upward spike caused by the demands of World War II, rail passenger use plummeted. Regulations implemented earlier in the century to protect customers when railroads were the primary mode of transportation required the railroads to maintain money-losing passenger and freight routes. This caused a serious decline in the economic fortunes of the railroad companies and, ultimately, a crisis of railroad bankruptcies in the late 1960s and early 1970s. Popular support for passenger trains, however, prevented federal leaders from relaxing those regulations.

The political compromise that government leaders reached to resolve this situation was the formation of a new government-supported private corporation that would take over most passenger rail service in the United States, allowing the private railroads to focus on profitable freight service. This compromise also gave the new company a mandate to keep only the routes with the greatest levels of ridership while maintaining a national network. Congress passed the Passenger Rail Service Act of 1970 and Amtrak began service on May 1, 1971. As part of the compromise, the private railroads transferred much of their passenger equipment to Amtrak and were required to allow Amtrak to use their tracks. During the 1970s and 1980s, Amtrak gradually phased out much of the old equipment by purchasing new locomotives and passenger cars, although some legacy baggage cars were still in operation in 2010 and beyond.

This compromise has been viewed unfavorably by many political leaders as an inappropriate use of taxpayer money to support a service used only by a limited number of riders and that competes with private carriers, such as airlines and bus companies. Rail advocates counter that all modes of transportation require public investment in infrastructure, such as highways and airports. This conflict has resulted in a number of instances in which continued financial support for Amtrak has been threatened by Congress. Efforts to completely defund and shut down Amtrak have not been successful, but this conflict also has prevented Amtrak from making the investments needed to bring American passenger rail service up to the levels enjoyed by riders in many Asian and European countries.

Because Amtrak only owns track in a limited number of locations, Amtrak must share tracks owned and operated by the private freight railroads. This causes scheduling difficulties for both the host railroads and Amtrak, often resulting in Amtrak trains being delayed. Despite these challenges, Amtrak has experienced fairly consistent growth in the number of riders over the years, and has managed to make meaningful system improvements. Passenger traffic in 2013 was 6.8 billion passenger miles, compared with 3.9 billion passenger miles in 1975. In 2000, Amtrak's Acela Express debuted. The Acela Express is a premium service train that travels between Boston, Massachusetts, and Washington, DC, and attains speeds of 150 miles per hour. Although funding remains uncertain, Amtrak retains a focus on the future with plans to upgrade infrastructure in the high-traffic northeast corridor ultimately to support faster, more-frequent operations in parts of the country where such systems will help meet America's mobility demands in the 21st century.

Michael Minn

See also: High-Speed Rail; Indian Railway System (India); Qinghai-Tibet Railway (China); Railroads.

Further Reading

Amtrak. 2015. http://www.amtrak.com. Accessed June 5, 2015.
Hilton, George Woodman. 1980. *Amtrak: The National Railroad Passenger Corporation.* Washington, DC: American Enterprise Institute for Public Policy Research.
Wilner, Frank N. 2012. *Amtrak: Past, Present, Future.* Omaha, NE: Simmons-Boardman Books.

AMUSEMENT PARK TRAINS

A common activity that people can enjoy during a visit to a theme park is a train ride, typically in a vehicle that makes a large loop around the park. The ride can serve to transport guests to another part of the park and provides an opportunity for riders to sit down and take a welcome break from the frenzy of a large theme park. Compared to the other attractions and amusements of a theme park these trains are unlikely to receive much attention. Yet these trains are different from the other park attractions in that they often started off as part of actual transportation systems. Many of these engines once hauled freight or passengers on small railroads before being retired and ending up at amusement parks.

The railroad at the Cedar Point amusement park in Sandusky, Ohio, operates six narrow gauge steam engines. The oldest was built in 1910 and spent much of its career hauling sugar cane trains in Louisiana. Three of the engines were built in the 1920s and had careers with freight railroads before ending up at Cedar Point. With their bright paint and shiny brass they do not look like hard-working freight haulers, and visitors could easily mistake them for modern replicas of real railroad equipment. The Tweetsie Railroad is an amusement park in the mountains of North Carolina and was built around a former narrow gauge railroad. That railroad, the East Tennessee & Western North Carolina Railroad, was called the Tweetsie because of the sound made by the steam whistles. It operated as a freight and passenger railroad between 1881 and 1950. Rather than abandoning the site, an engine was restored and an amusement park was created around the narrow gauge line.

Joe Weber

See also: Aerial Tramways; Darjeeling Himalayan Railway (DHR) (India); Narrow Gauge Railroads; Ski Lifts.

Further Reading
Cedar Point and Lake Erie Railroad. 2015. http://www.cplerr.com/. Accessed August 2015.
Tweetsie Railroad. 2015. http://tweetsie.com/. Accessed August 2015.

APPALACHIAN TRAIL (UNITED STATES)

The Appalachian Trail (AT) is a long-distance hiking trail running along the crest of the Appalachian Mountains from Georgia to Maine. The southern end of the trail is Springer Mountain in Georgia, and the northern terminus is Maine's Mount Katahdin. It runs for 2,200 miles through 12 states between these endpoints, although the southern end requires many miles of hiking to reach. There have been attempts to lengthen the trail, most notably at the southern end. The trail has a varied geography. At the southern end it passes through Great Smoky Mountains and Shenandoah National parks, and the highest point of the trail is Clingman's Dome on the Tennessee/North Carolina line at 6,643 feet above sea level. Except in Virginia, the AT runs well to the west of the Blue Ridge Parkway, a parallel recreational highway that was developed at the same time as the AT. Farther north the trail runs through lower elevations and less spectacular scenery before encountering the Green and White Mountains in New England.

The idea of the trail was originated by Benton MacKaye (1879–1975) in 1922. Sections were built beginning in 1923 and for decades after that. MacKaye's role in the trail was recognized by the naming of the Benton MacKaye Trail, opened in 2008 to connect the southern terminus in Georgia to the Pinhoti Trail in Alabama. This latter trail runs to the southernmost Appalachian peak more than 1,000 feet in elevation.

The AT is an extremely popular hiking trail, whether for short hikes or longer excursions. Thru-hikers are those who attempt to hike the entire trail in one season. These usually start in the south in early spring and head north through the warmer months. A thru-hike might take five to seven months for an average hiker. Shelters are located at intervals representing a day's journey. The trail crosses many highways and passes near numerous small towns, enabling hikers to pick up supplies. Harpers Ferry, West Virginia, is one such trail town.

Many other long-distance trails have been constructed since the Appalachian Trail's creation, and since 1968 the National Park Service has been involved in planning these trails. The National Park Service now administers 11 National Scenic Trails, of which the AT was first. The Pacific Crest Trail is a western equivalent, running 2,663 miles from the Mexican border to Canada along the Sierra Nevada and Cascade mountains. The Continental Divide Trail also runs from Mexico to Canada but along the continental divide through the Rocky Mountains. These western trails, however, do not attract nearly as many long-distance hikers as does the AT. There

Hikers and their dog on the Appalachian Trail in southwestern Virginia, circa 1990. (Library of Congress)

also are 19 National Historic Trails that follow historic routes such as the Oregon Trail, and more than 1,140 additional National Recreation Trails for local use.

Joe Weber

See also: Bicycle Lanes; Blue Ridge Parkway (United States); Natchez Trace Parkway (United States); Run-Commuting; Sidewalks.

Further Reading

Appalachian National Scenic Trail. 2015. http://www.nps.gov/appa/index.htm. Accessed August 21, 2015.
Appalachian Trail Conservancy. 2015. http://www.appalachiantrail.org/. Accessed August 21, 2015.

AUTO TRAIN (UNITED STATES)

The Auto Train is an Amtrak passenger train that runs between Lorton, Virginia, and Sanford, Florida. The route is 855 miles long and takes about 18 hours to travel from end to end. A daily train departs each way in the afternoon and arrives the next morning. The train is unique among U.S. passenger trains for several reasons. One reason is that it carries both passengers and their motor vehicles. It therefore provides a way for people to travel between the cities of the Northeast Corridor and Florida without the need to rent a car or be dependent on public transportation once at their destination. Passengers ride in standard passenger cars with special freight cars at the rear of the train for vehicles. Railroad employees load the automobiles, which takes about four hours, and unload them at the destination, which takes about two hours. The train operates nonstop service with no intermediate stops. It does not operate between downtown train stations of large cities but between stations in suburban areas outside of Washington, DC, and Orlando, Florida. The Amtrak Silver Service and Palmetto trains are passenger-only trains that run on similar routes between the Northeast Corridor and Florida, with many intermediate stops. Similar services exist elsewhere. The Channel Tunnel between France and the United Kingdom has both automobile and passenger service. Around the world there are car-shuttle trains; cars and their drivers park on railroad cars to be carried through tunnels.

Joe Weber

See also: Amtrak (United States); Channel Tunnel (France and United Kingdom); Northeast Corridor (United States).

Further Reading

Amtrak. 2015. "Auto Train Transports You and Your Car." http://www.amtrak.com /auto-train. Accessed August 31, 2015.

Ely, Wally. 2009. *Images of Rail: Auto-Train*. Arcadia Publishing.

AUTOBAHN (GERMANY)

The Autobahn (German for "auto way") is the main system of highways in Germany. It was the world's first limited-access road system and has served as a model for other countries, including the Interstate Highway System in the United States. The Autobahn system has 7,936 miles of highway and boasts the claim that a route is within 6.2 miles of any point in the country. The system is a staple of transportation for both commuters and the transportation of goods around the country and beyond. The Autobahn has received attention for decades, as German highways have no speed limits but yet provide a safe and efficient driving experience.

Since its initial construction began in 1913, the Autobahn has garnered a rich history of political and cultural significance and remains important today. The first section of an experimental 11-mile intersection-free road network was constructed in Berlin exclusively for automobiles. World War I diverted attention and funding from the project for several years, but the Autobahn was revived when the Nazi Party came to power with the leadership of Adolf Hitler in the 1930s. There was debate over whether the idea of building roadways solely for automobiles was justified given the fact that the automobile use in Germany was low compared with countries at the same level of industrialization prior to 1930s. In fact, in 1932 there were 8 cars for every 1,000 people in Germany (compared to 183 per 1,000 in the United States). The post–World War I economic struggle led to the rise of Adolf Hitler and the Nazi Party. Both strongly embraced "the Autobahn" as part of the *Reichsarbeitsdienst* or "RAD initiative" to recover from economic depression.

To reinforce the sense of the Autobahn as "public works" or "the people's community," it was necessary to produce cars for all the people of Germany and "not just for the rich," hence another major project was launched by Hitler: the development of what became known as the Volkswagen or "people's car." The ideal people's car was economical at speeds of up to 60 miles per hour with a carrying capacity of five passengers, and could be purchased with small weekly payments. Although none of the newly designed Volkswagen cars were delivered to the 360,000 Germans who paid for them—production was halted to work on military vehicles for Word War II—many people were employed through the auto plants. The Autobahn was used to create so many economic opportunities during the years under Hitler that it ultimately led him to gain public support for World War II.

World War II brought a halt to the Autobahn project in 1941, because of strained economic and energy resources due to fighting the war. The effectiveness of the Autobahn during the World War II is controversial, with some people arguing that it helped Germany fight a war on two fronts but others stating that it had little impact during the war. Some sources state that, at the outset of World War II in Europe, the Autobahn proved to be a primary asset to Germany in defeating Poland in 1939; Belgium, Luxembourg, and the Netherlands in 1940; and the Soviet Army in 1941. Others argued that tanks did not travel well on the Autobahn because the weight of the tank and their caterpillar tracks destroyed the roads. As a result, most troops were sent via rail to their various locations, dismissing the Autobahn as a tactical advantage for Germany.

The design of the Autobahn itself is another story of technologically innovative architectural achievement. Unlike Germany's railway—which was blamed for destroying the organic landscape—the guiding vision of the Autobahn was to reconcile nature and technology so that particular characteristics of each individual landscape could be preserved with sensitivity as to where the Autobahn's routes were laid out. Whether these goals were genuinely achieved is questionable as, ultimately, engineers and not the environmentalists determined the routes of the Autobahn. More recently, the Autobahn has been the center of political debate as some environmental interest groups have proposed eliminating the unrestricted speed limit zones, sparking outrage from some citizens in the country. The unrestricted zones on the Autobahn are said to be a danger to the environment as they enable vehicles to consume more fuel and create more greenhouse gases. The Autobahn's effects on the environment will continue to be an issue in Germany in the 21st century.

Nonetheless, the Autobahn was intended to be an iconic transportation infrastructure blended with nature and up-to-date technology and it succeeded on these terms. Known worldwide, driving enthusiasts make great efforts to travel to Germany only to experience this amazing feeling where safe speed limits are determined largely by skill and not by regulations.

Selima Sultana

See also: Autostrade (Italy); Freeways; Interstate Highway System (United States); National Highways Development Project (India).

Further Reading

Autobahn website. "Autobahn History." http://www.german-autobahn.eu/index.asp?page =history. Accessed March 2015.

AUTOMOBILES

An automobile, commonly referred to as a "car," is a self-propelled vehicle that typically has four wheels and two axles and is used for personal transportation. Automobiles primarily travel on roads, are steerable, and can seat up to nine people. They house an engine to give the vehicle power; a chassis to support the body, engine, and wheels and their components; and a power train to transmit the power from the engine to the wheels. Today's engines are fueled by gasoline, diesel fuel, or electricity. Karl Benz of Germany is credited with having invented the first modern automobile, the Benz Patent-Motorwagen, in 1886. Today, there are more than 1 billion automobiles in use worldwide. The United States leads in number of vehicles (approximately 240 million) as well as number of drivers (more than 210 million).

Prior to Benz's invention of the first gasoline-powered automobile, a steam-powered vehicle for human transportation was invented in France in 1768, followed by the first vehicle with an internal combustion engine fueled by hydrogen in 1807. Electric-powered cars also were introduced in this era. Immediately following the release of Benz's vehicle, manufacturers in both Germany and France began perfecting the automobile. One such invention, the Mercedes, designed by Wilhelm Maybach for Daimler Motoren Gesellschaft in Germany and considered the first modern motorcar, was produced in 1901. It ran on a 35-horsepower engine and achieved a top speed of 53 miles per hour. Mercedes automobiles were expensive to make, however, and required 700 workers to manufacture fewer than 1,000 vehicles each year.

Commercial production of automobiles began in 1890 in France and in about 1900 in the United States. At the time, 2,500 vehicles were made in 1899 by 30 American manufacturers, a number that catapulted to 485 companies just 10 years later. Although much more affordable than the typical German vehicle of the time, U.S. auto quality was lacking. Whereas German manufacturers tended to make

Karl Benz (right) test-drives the first automobile powered by a gas combustion engine in Mannheim, Germany, October 1885. The three-wheeled vehicle was named "Velociped." (AP Photo)

entire cars themselves, U.S. companies bought parts from individual suppliers, and quality varied.

Considered the first affordable vehicle, the Model T was created by American Henry Ford and the Ford Motor Company in Michigan in 1908. The Model T, nick-named the "Tin Lizzie," was the first vehicle produced by Ford's efficient assembly-line process. Ford borrowed the assembly-line techniques in auto making from Ransom Olds, inventor of the Oldsmobile, but rolled cars out on an unheard-of scale. In an assembly line, standard interchangeable parts are added to the vehicle one piece at a time as it travels from worker to worker. This process is faster and requires less labor than building a stationary vehicle, it can use an unskilled labor force, and it contributes to mass production. The process was adopted by other U.S. manufacturers—including General Motors (GM), also founded in 1908—and became the standard assembly method. Healthy production made vehicles accessible and affordable to Americans. Installment sales, introduced in 1916, further increased affordability.

To make cars more affordable in Great Britain—where Ford had begun importing his Model A as early as 1903—Englishman William R. Morris of Morris Motors Ltd. wished to undertake Ford's production methods around 1912, but British

Google Prototype Self-Driving Car

People have dreamt of having automated cars since the 1930s, if not earlier. Self-driving cars were displayed in the famous Futurama attraction at the 1939 World's Fair in New York City and since then have made countless appearances on TV, in movies, and in comic books. They finally are becoming a reality.

Google has developed equipment that can be installed in a variety of cars that enables the car to drive itself. The capability is based on stored street map data and on-board sensors that indicate the car's position both on the road and relative to other vehicles. Since 2012, self-driving cars have been legal in Nevada, and they are undergoing testing in several other states. Vehicles still include the option to drive the car manually, and so far manual driving seems to be the cause of most accidents involving these vehicles. It remains to be seen how widespread this technology will become, and how it impacts society.

Joe Weber

engineering firms were reluctant to commit to mass-producing parts. Morris ended up buying parts from America, but the advent of World War I put a halt to auto-manufacturing operations altogether. After the war, Morris, as well as British automaker Herbert Austin and Frenchmen Louis Renault and André-Gustave Citroën, resumed making low-priced, reliable cars. Great Britain's tax on horsepower encouraged automakers to develop small engines, which was atypical of in-demand styles around the world. This led to lost sales overseas and low production levels. To keep up, Morris and other British carmakers were compelled to adopt the assembly-line process to be able to compete in the marketplace.

By the 1920s, most auto innovations already had occurred, including creation of all-steel bodies, high-compression engines, hydraulic brakes, gear-changing transmissions, and low-pressure air-filled tires. By 1929, 80 percent of the U.S. auto industry was controlled by Michigan's Ford, GM, and Chrysler (created by Walter P. Chrysler in 1925). Many of the smaller independent concerns collapsed because they could not keep up with the "big three" automakers. A similar pattern in Europe emerged, where the three firms of Morris, Austin, and Singer took a 75 percent share of the British market. In France, Peugeot, Renault, and Citroën led the pack before market-dominating Citroën was bought by Michelin Tire. Germany's industry suffered greatly after the nation's defeat in World War I. Benz and Daimler merged and GM acquired Opel. Italy became the leader in highly engineered sports cars but never reached mass scale.

World War I and World War II brought myriad changes to the automotive industry. During World War I, auto factories switched their focus to building military airplanes, equipment, tanks, and other motor vehicles. World War II saw an escalation in such contributions. At the onset of war, British factories immediately ramped up military production, but the United States held back until after the 1941 bombing of Pearl Harbor. Ultimately, however, the U.S. contribution was immense.

Both the United States and Germany produced a simple but rugged vehicle for the war effort that became popular after the war—the German Volkswagen and the U.S. Jeep (or "GP," meaning "general purpose"). A tenfold increase in production was experienced by U.S. automakers after the war, but the European industry was faced with recovering from the devastating losses resulting from the war. Great Britain, for one, concentrated on exporting its cars to help the country's suffering economy.

Planned obsolescence was introduced by Alfred Sloan of General Motors in the 1920s and 1930s to combat market saturation. This plan—which included changing styles and gadgets to motivate consumers to trade up their vehicles—continued into the postwar era. By the middle of the 1960s and into the 1970s, as nonfunctioning style took a backseat to engineering, quality diminished. At about the same time, the typically large U.S. vehicles put a strain on dwindling oil reserves. Federal standards for emissions, energy consumption, and auto safety—including mandatory seatbelts (1968), prompted by Ralph Nader's *Unsafe at Any Speed*—contributed to decreased sales of U.S. vehicles. This paved the way for foreign manufacturers such as Volkswagen, Honda, and Toyota—and even autos made by previously nonexistent players such as South Korea—to take hold of the U.S. market with their smaller fuel-efficient cars. To compete, U.S. manufacturers underwent massive reorganizations and began modernizing plants and making cars smaller, safer, and more economical. By the end of the 20th century, however, the U.S. Big Three automakers lost considerable market share and even faced bankruptcy. Meanwhile, foreign automakers began building vehicles in North America to keep up with demand. In 1982, nearly 28 percent of U.S. auto sales were Japanese cars versus 7 percent of U.S. cars. Japan, whose auto industry barely existed prior to World War II, still is the world's leader and Toyota is the largest manufacturer. By 1980, motor vehicles were owned by more than 87 percent of U.S. households.

An important motivator of change in the modern world, the automobile is credited with eliminating social isolation of rural Americans and homemakers and allowing city dwellers to move beyond urban boundaries, and thus creating a need for suburbs. Auto manufacturing also increased the massive growth of other industries, specifically petroleum, rubber, and steel. More autos on the road created a need for gas stations and repair shops, and for infrastructure such as roads and highways. Ancillary industries including hotels and motels, restaurants, insurance, and tourism blossomed. The modern world's dependence on the automobile, however, also is blamed for social ills that include urban sprawl, the rise in obesity, and an increase in accidental deaths.

Automobiles come in all shapes and sizes—from Smart minicars to giant sport utility vehicles and vans—and are available in all price ranges. With each model year, cars become evermore technologically advanced. Newly produced vehicles commonly are equipped with Wi-Fi, GPS tracking devices, backup cameras, and self-parking mechanisms. Robots even help build cars. Vehicles now run on petroleum, electricity, or a combination of electricity and gasoline (for cars known as "hybrids"). Automakers are experimenting with alternative fuels to further reduce emissions and reliance on petroleum. Prototypes of fully autonomous, or driverless, vehicles are in the works, with the Google driverless car being tested for possible commercial release in coming years.

Rosemarie Boucher Leenerts

See also: Electric Vehicles; Minicars; Plug-In Hybrids.

Further Reading

Eckerman, Erik. *World History of the Automobile*. Warrendale, PA: SAE International, 2001.
Ford, Henry. 1922. *My Life & Work*. Garden City, NY: Garden City Publishing.
Newton, Tom. 1999. *How Cars Work*. Vallejo, CA: Black Apple Press.

AUTONOMOUS VEHICLES

Autonomous vehicles employ electronic systems to perform some or all of the tasks of driving a car. Automakers use a variety of technologies to enable autonomous cars to detect their environment, including cameras and radar sensors. Other frequently encountered names for autonomous cars include self-driving cars, driverless cars, and automated cars.

Autonomous vehicle technology is beginning to appear in car dealers' showrooms. For instance, Mercedes currently (2015) sells an add-on package that will keep a car within its lane and maintain a safe following distance from a vehicle in front of it, as long as the driver keeps his or her hands on the steering wheel. This is representative of the current wave *of partially autonomous* technology—a human driver is still required, but is able to "hand over" control in some circumstances, such as freeway driving or a slow-speed traffic jam.

Road safety will be improved by systems such as autonomous emergency braking, in which a car begins to brake when a crash is imminent even if the human driver has yet to press the brake pedal. Broadly speaking, autonomous cars will be safer because human drivers suffer from a variety of shortcomings that electronics do not—all of us take time to perceive and react to driving situations, and some drivers send and receive text messages while driving, drive after drinking alcohol, and glance away from the road to change the radio station, to name a few dangerous behaviors. Just as computers now can defeat the best human players at the game of chess, autonomous cars now are better at navigating race tracks than are expert human drivers.

When and if fully autonomous cars become available, it will be possible for people that cannot drive (such as children, blind people, or the elderly) to travel by car without a human chauffeur. Even people that can drive will be able to go

Vehicle-to-Infrastructure (V2I) Communications Technology

Vehicle-to-infrastructure (V2I) communications technology is a component of what often are called "intelligent transportation systems" (ITS). This technology—still under development—would allow vehicles such as passenger cars or trucks to communicate with sensors along the road. This works in two ways. Traffic control signals can make use of real-time information from vehicles in traffic to better synchronize lights and improve the flow of traffic. This information also can be used to detect accidents or congestion and to alert authorities. The roadside equipment also will communicate with vehicles about congestion on their route, alternate routes, and the best speed to drive to ensure getting green lights at intersections. By ensuring a free flow of traffic the accident rate can be reduced, as can the environmental impacts of travel.

Joe Weber

online to do work or relax while they are in the car, if they wish. This vision of a fully autonomous future will take many years to be realized, however. Though it cannot be known for certain, it is unlikely to happen before 2030.

The impacts of autonomous cars will be far-reaching, and experts in this area have very different views of how society will adapt. One vision is that we will not even *own* a private car, and instead simply will summon an autonomous taxi (dubbed an "aTaxi" by Princeton University's Alain Kornhauser) exactly when and where we need it via an app on a smartphone. Private cars typically sit idle for 23 hours each day, and therefore "sharing" aTaxis could lessen the need to dedicate valuable urban real estate for parking.

Autonomous cars might reduce traffic congestion if cars can follow each other more closely. Real-world testing has shown that very large increases in the capacity of freeways are possible, particularly if cars "talk to each other" by exchanging electronic messages. If more people travel by car, however, and if we each travel longer distances because we do not have to actually do the driving, then overall traffic congestion might not be reduced at all. Autonomous vehicles might even lead to greater suburban sprawl if people are willing to commute longer distances because they can work productively while in their car.

Many hurdles must be overcome as cars become increasingly autonomous. The legal framework for motor vehicles is built around the assumption that a human driver is "in control" of his or her car at all times. As this becomes less true, it is likely that automakers will need to accept a greater share of monetary responsibility when crashes occur. A side effect of safer roads will be structural shifts in the car insurance industry. Attempts at cyberattacks against autonomous cars are inevitable, and robust protections must be engineered. A number of U.S. states already have passed laws that regulate autonomous cars for testing purposes. As is frequently the case in the field of transportation, California is a leader in developing the new laws and rules to govern the rollout of this emerging technology.

Autonomous cars (and autonomous trucks) clearly will play a major role in the transportation system of the future, although precisely how this will happen is still unclear. Many experts think that the dawn of the autonomous car will prove to be as consequential as the advent of the automobile itself in the early 20th century. A complex set of interacting factors—technological, legal, social, and consumer desires—will combine in innovative and unpredictable ways to determine what the future will hold.

Scott E. Le Vine

See also: Automobiles; Bicycle Libraries; Intelligent Transportation Systems; Minicars.

Further Reading

Anderson, James M., Nidhi Kalra, Karlyn D. Stanley, Paul Sorenson, Constantine Samaras, and Oluwatobi A. Oluwatola. 2014. "Autonomous Vehicles: A Guide for Policymakers." Santa Monica: RAND Corporation. http://www.rand.org/content/dam/rand/pubs/research_reports/RR400/RR443-1/RAND_RR443-1.pdf. Accessed April 1, 2015.

Fagnant, Daniel J., and Kara M. Kockelman. October 2013. *Preparing a Nation for Autonomous Vehicles: Opportunities, Barriers and Policy Recommendations.* Washington DC: Eno Center for Transportation. http://www.enotrans.org/wp-content/uploads/wpsc/downloadables/AV-paper.pdf. Accessed April 1, 2015.

Townsend, Anthony. 2014. *Re-Programming Mobility: The Digital Transformation of Transportation in the United States.* New York University's Rudin Center for Transportation Policy and Management. http://reprogrammingmobility.org/wp-content/uploads/2014/09/Re-Programming-Mobility-Report.pdf. Accessed April 1, 2015.

AUTOSTRADE (ITALY)

The autostrade (plural for "autostrada") is an Italian major highway/motorway system that was built by the government, public institutions, and private companies and runs 1,842 miles throughout Italy. The first of these roads was constructed north of Milan in the early 1920s with a length of 23 miles, and the system eventually linked all regions of Italy with more than 4,163 miles of highways. The term "autostrada" literally means "automobile road" and first was coined in 1922 by engineer Piero Puricelli in his project "Road Network for Motor Vehicles," which he described as "Autostrada dei Laghi" or motorway of the lakes. With a current maximum speed of 80 miles per hour, this network has transformed the way of life for millions of Italians and will continue to shape Italy's future in the global arena.

By 1945, approximately 932 miles of motorways extended throughout Italy, linking cities and rural towns at a time when Italy was also beginning a rapid economic boom. Italian infrastructure could not support the number of vehicles

Autostrade Liguria landscape view, Italian Riviera, 2015. (Anna Hristova/Dreamstime.com)

needed to transport goods produced in northern factories, nor enable tourists to travel to desired locations. Traffic congestion became such a major issue that goods could not be shipped to buyers in a timely fashion, and the competitive edge of Italian companies dwindled. Additionally, southern Italy, largely due to its remote location and rugged terrain, remained isolated and did not experience the economic growth that the north enjoyed. It became evident that a national highway system was needed in Italy to address the inadequacy of the nation's road infrastructure.

As the 1960s approached, the autostrade network was designed to meet the needs of the new decade. The government-owned Institute for Industrial Reconstruction (IRI) joined with private companies to build a network of highways that could expand and support a growing economy and rapid growth of motorization. By the end of 1977, 3,100 miles of the highway network were completed that connected throughout Italy. Businesses such as hotels, restaurants, shops, and service stations were established along the major highways to cater to the increased number of travelers. From 1968 to 1975, more than 600 industries were established along the Autostrada del Sole alone and more facilities were needed to keep up with the demand. Many sectors of the Italian economy prospered—including petroleum companies—due to the highways that were being built and used by an increasing number of travelers.

The autostrada had an economic impact on the entire country; it also changed Italy's social-cultural makeup. Highway construction also gave southern impoverished farmers easy access to northern markets. Many poor people in southern Italy moved to the northern region to find employment and became a much needed source of cheap labor for northern businesses, although this also resulted in over-crowded urban areas and organized crime. Criminals, who had exploited southern farmers at harvest time, organized themselves into what was known as the "Calabrian Mafia." In fact, the construction of the autostrade has been called the "catalytic event" that enabled the spread of organized crime throughout Italy. Many of the businesses that were designed to serve the needs of motorists were victimized by organized crime, as was the construction industry. The mafia had so much control over this sector that many companies, when bidding on a job, added 15 percent to the bid and labeled the additional cost as the "Calabrian risk."

Most importantly, the ease of transportation due to the autostrade impacted Italian politics. Prior to the autostrade, the people in the south were so isolated from the politics in the north that the southern population was unwilling to cooperate with a remote body of legislators. As the autostrade connected remote locations in the southern region, communication and transportation were established and more people in the south were able to participate in and interact with the Italian government.

Today there is political tension due to the tolls on the autostrade. Political parties in the north want laws to be passed that eliminate free roads in the south so that the costs are distributed equally throughout the nation. The southern Italian parties disagree, however, and point to the fact that southern Italy is more impoverished and lags behind the north in development.

Presently, there is considerable discussion on the autostrada's detrimental effect on the environment. Tolls are based on the number of axles per vehicle without consideration to the emissions that each vehicle creates, which raises concerns. There is a demand for environmental policies including varying tolls based on the amount of pollution a vehicle produces, adding green spaces along the network, and promoting the recycling of materials. The recent plan to expand the autostrade across the Apennine Mountains—which would bring more than 50,000 automobiles to this area daily—has faced strong opposition from environmentalist groups.

Selima Sultana

See also: Autobahn (Germany); Interstate Highway System (United States); National Highways Development Project (India); National Trunk Highway System (China).

Further Reading

Atlantia S.p.A. 2005. "Autostrade Group Social and Environmental Report 2004." United Nations Global Compact. http://www.globalcompactnetwork.org/files/partecipanti/cop/2004_autostrade.pdf. Accessed June 8, 2015.

Ginsborg, Paul. 2003. *A History of Contemporary Italy: Society and Politics, 1943–1988*. New York: Palgrave Macmillan.

Moliterno, Gino. 2003. *Encyclopedia of Contemporary Italian Culture*. London: Routledge.

Pacione, Michael. 1974. "Italian Motorways." *Geography* 59(1): 35–41.

BAMBOO TRAINS (CAMBODIA)

In the heart of the Cambodian Village of Battambang is a very unique mode of transportation known as a "bamboo train." These trains are cheap to build and can be made quickly using local resources. After the fall of the Khmer Rouge in 1979, many local residents needed a means of transportation within the country. The existence of unused rail lines inspired a transport solution in the form of home-made trains, consisting of two axles, a small motorcycle motor, and a flatbed. Local resources—such as bamboo—are used to create the flatbed that supports the people and goods being carried. The engine turns belts on the back axle to propel the train. These bamboo trains (also known as "Nori") can carry up to 20 people and can travel as fast as 25 miles per hour. Trains are easily assembled and disassembled, which is an important feature in instances when two bamboo trains meet. The common rule for such circumstances is that the train with the fewest passengers is disassembled to let the other train pass.

Bamboo trains are an important means of transport for both people and goods in areas of Cambodia that otherwise are unserved. The trains also provide a form of income for the region because they attract many tourists that want to experience this unusual mode of transportation—even though the tracks do not go to a special destination and sometimes are overgrown with weeds. For as little as $2 per day, passengers experience not only the unique bamboo train ride, they also see the beautiful Cambodian countryside—including rice paddies. People in the villages that the railway passes through generate income from tourists, thus these trains help to support many people and their way of life. Because of the rehabilitation of the Cambodian Railway Network, however, people believe that the bamboo trains will disappear as tracks are replaced.

Selima Sultana

See also: Camel Trains (Mali); Chicken Buses (Central America); Chiva Express (Ecuador); Jeepneys (Philippines).

Further Reading

Bergner, Audrey. 2013. "The Bamboo Train in Battambang: The Good and the Ugly." *That Backpacker* (blog). http://thatbackpacker.com/2013/04/18/bamboo-train-battambang-cambodia-the-good-and-the-ugly/. Accessed June 9, 2014.

Kaushik. "Bamboo Trains of Cambodia." *Amusing Planet*, June 12, 2011. http://www.amusingplanet.com/2011/06/bamboo-trains-of-cambodia.html. Accessed June 8, 2015.

Villagers near Battambang riding on a bamboo train, 2011. (Lauren Cameo/Dreamstime.com)

BANGLADESH RAILWAY

Bangladesh inherited its railway system—commonly known as Bangladesh Railway (BR)—from the British when Bangladesh was part of British India. Rail journey is not only relatively comfortable but also is affordable for a majority of the passengers in this densely populated country of approximately 160 million people. The BR is a state-owned rail transport agency, under the jurisdiction of the ministry of railway.

Although the history of the BR goes back to the British colonial period, the government of Bangladesh continued the expansion of the BR—albeit slowly—after the country became independent in 1971. The Dhaka-Mymensingh section was constructed in 1985 and the East-West Railway connectivity was established over the river Jamuna in 1998. The expansion of service continued as direct communications between Dhaka and Rajshahi, and Dhaka and Kolkata, India, were established in 2003 and 2008, respectively. Dhaka-Kolkata train service is known as "Maitree Express."

The BR covers a length of 1,788 miles and has a total of 25,083 regular employees. In Bangladesh, 44 of 64 districts are directly connected by railway system. The BR offers three different gauges: meter gauge (MG), broad gauge (BG), and dual gauge (DG). The railway divides the country into two zones—"East Zone" and "West Zone." The East Zone has 787 miles of MG track and 21 miles of DG track. The West Zone has 332 miles of MG track, 409 miles of BG track, and 233 miles of DG track. There are 440 stations in the country.

Semaphore and color-lights are two different types of signals used by BR with a domination of color-light signals. Of other types, both fixed and moving block signaling are used by BR. Despite its expansion of services, BR crossing gates still are operated manually. The gatekeepers operate the level crossing manually after the departure and arrival of a train. Such manual crossing systems are both inefficient and inaccurate. Optical fiber–based digital telecommunication network has been implemented since 1984, but it covers only 1,118 miles of rail network connecting about 250 railway stations.

Currently, the BR has been suffering from many problems. Although the BR expanded its service in certain locations and bridged the connection between East and West zones in the postindependence era, its coverage is miniscule as compared to the expansion of the road network system in Bangladesh. In some cases, branch lines were closed instead of constructing new rail lines. A lack of qualified manpower, lack of modernization and development, poor maintenance of schedules, maintenance of trains, misuse of tickets, and cleanliness are few of the major problems that the BR has faced for years. These problems have been coupled with two more imminent threats: the declining financial support by the government, and the tremendous competition posed by road transport. As a result, the service coverage and efficiency of the railway system in Bangladesh gradually have been declining.

Taslima Akter and Bhuiyan Monwar Alam

See also: Indian Railway System (India); Narrow Gauge Railroads; Qinghai-Tibet Railway (China).

Further Reading

Government of the People's Republic of Bangladesh. 2010. "Bangladesh Railway." http://railway.portal.gov.bd/sites/default/files/files/railway.portal.gov.bd/page/1ccd6ead_159b_4ee6_8a49_db0072629403/4.%20Opportunites%20for%20%20BR.pdf. Accessed March 15, 2015.

Government of the People's Republic of Bangladesh. 2014. "BR in Short." http://railway.portal.gov.bd/site/page/ce7dd6af-c7c8-4811-86b3-ba871e2e406e. Accessed March 15, 2015.

Hasan, M. D. Raqibul. 2009. "Problems and Prospects of a Railway: A Case Study of Bangladesh Railway." *Journal of Service Marketing* 4: 124–136. http://www.slideshare.net/sohagmal/a-case-study-of-bangladesh-railway-33311962. Accessed March 15, 2015.

BATTERY ELECTRIC VEHICLES

See Plug-In Battery Electric Vehicles (BEVs).

BATTERY SWAPPING

Battery swapping is an integrated system encompassing electric vehicles with removable batteries, battery-swap stations (BSS), and extra batteries charged and waiting at stations. Battery-swapping electric vehicles (BSEV) can be plugged in and charged, but the batteries also can be replaced quickly at special stations resembling automated car washes. The time required for a swapping operation is comparable to that of filling a gasoline vehicle. Batteries are located underneath the car and are attached by bolts or sometimes by latches. To swap a battery, a driver enters a BSS, swipes an electronic membership card, and remains in the vehicle while a track moves the car into position and robotic equipment removes the used battery and replaces it with a fresh one.

Battery swapping was conceived to address several concerns that were expected to suppress consumer demand for pure (nonhybrid) battery electric vehicles (BEVs). First, it extends the short driving range of electric vehicles (EVs) without the delay of charging the battery. Current BEVs charge far more slowly than the driving public is used to. Even Level 3, 480-volt DC "fast-chargers" take more than 20 minutes to achieve an 80 percent charge of an 80-mile battery. For longer trips, BEV drivers not only must stop more frequently but also must wait much longer at each charging stop. As a result, the first-generation BEVs are most attractive to multicar households that have a second gasoline- or diesel-powered car to use for longer trips. By slashing the time needed to leave a station with a fully charged battery, BSS technology makes it possible for a single-vehicle household to buy an EV. Additionally, drivers still can charge their BSEVs while doing other activities, such as housework, shopping, dining, or watching a movie. Thus, BSEV drivers potentially enjoy the ultimate in refueling convenience: They wake up each morning with a full battery; most days they don't drive enough miles to need to swap or recharge away from home at all; and when they do make longer trips they can swap out their battery in five minutes or less.

Equally important is how BSEVs can reduce the upfront cost of an EV and address other barriers to consumer adoption of EVs. Because the battery is the most expensive part of an EV, the purchase price can be reduced by separating the ownership of the car and ownership of the battery by charging drivers for a monthly subscription service based on miles driven. Many prospective buyers also worry about how long a battery will last, how much it will cost to replace, and whether it's better to wait until battery technology with superior performance and lower cost is available. By using switchable batteries, companies would guarantee a minimum battery capacity and—as they degrade—could replace older batteries with newer, better batteries. This diminishes prospective buyers' concerns about the future resale value of their BSEV.

Another benefit of battery swapping is to stabilize the effect of EV charging on the electric grid. With a single company coordinating the charging of thousands of batteries, the electric load can be managed optimally. Electric utilities can pass the savings on to the BSS company, further reducing the cost to consumers.

Battery swapping first was introduced in 1910 by Hartford Electric Light Company for electric trucks made by GeVeCo (a subsidiary of General Electric), and continued until 1924. Since the 1940s, the batteries used in indoor electric forklifts have been swapped manually. The start-up company Better Place, led by Israeli/Silicon Valley entrepreneur Shai Agassi, revived the battery-swapping concept in 2007. The Renault Fluenze ZE sedan was the first (and as it turned out, only) BSEV designed to work with the Better Place BSSs. Drivers paid a monthly subscription service based on miles driven for unlimited battery charging and swapping. Using more than $800 million in funding from investors and venture-capital firms, and grants from the European Union, Better Place launched its first stations and vehicle showrooms in Israel and Denmark in 2011. The company eventually opened 37 BSSs in Israel and 17 in Denmark, but sold fewer than 3,000 cars. Better Place was developing similar plans around the world when, in May 2013, it declared bankruptcy. The company was done in by delays caused by switching from a latch to a bolt design for securing the batteries, higher than ex-pected costs, low market penetration of the Fluenze sedan, and failure to attract other car manufacturers. Some experts expressed concern over compatibility of switching technology across manufacturers. Renault was working on a cheaper hatchback with latches and a flat battery when Better Place ran out of cash. Indus-try experts questioned the fundamental financial feasibility of the business model, due to the great number of extra batteries—the most expensive part of a BSEV—required.

The Silicon Valley EV company Tesla has picked up on the battery-swapping idea. Unlike in the Better Place business model, Tesla drivers are expected to own their battery, which might make it necessary for drivers to return for their original battery or pay additional fees along with adjustments to the battery's warranty. Researchers at University of California at San Diego have proposed the "Modular Battery Exchange and Active Management" (M-BEAM) system based on the idea that it is simpler and more practical to swap out lightweight battery modules. China currently is developing and testing a system of electric buses with switchable batteries.

Michael Kuby

See also: Biofuel Vehicles; Electric Vehicles; Hydrogen Vehicles; Plug-In Battery Electric Vehicles (BEVs).

Further Reading

Adner, Ron. 2013 (June 7). "Don't Draw the Wrong Lessons from Better Place's Bust." *Harvard Business Review*. https://hbr.org/2013/06/dont-draw-the-wrong-lessons-fr. Accessed November 10, 2014.

Avci, Buket, Karan Girotra, and Serguei Netessine. 2012. "Electric Vehicles with a Battery Switching Station: Adoption and Environmental Impact." Faculty and Research Working Paper, INSEAD, Fontainebleau, France. http://www.insead.edu/facultyresearch /research/doc.cfm?did=49338. Accessed November 10, 2014.

Bullis, Kevin. 2013 (June 19). "Why Tesla Thinks It Can Make Battery Swapping Work." *MIT Technology Review*. http://www.technologyreview.com/news/516276/why-tesla -thinks-it-can-make-battery-swapping-work/. Accessed November 10, 2014.

BAY AREA RAPID TRANSIT (BART) (UNITED STATES)

Bay Area Rapid Transit (BART) is a heavy-rail public transportation system serving the San Francisco Bay Area in Northern California. Its system connects the city and county of San Francisco with East Bay communities in Alameda and Contra Costa counties, as well as northern San Mateo County, on the peninsula to San Francisco's immediate south. The five BART routes encompass 44 stations along 104 miles of rail line. It serves nearly 400,000 daily weekday riders and between 120,000 and 180,000 daily weekend riders, making BART the fifth busiest heavy-rail rapid transit system in the United States.

The idea of BART was conceived by San Francisco's business and civic leaders in 1946, when the Bay Area prepared for a post–World War II population boom and the associated surge in vehicles on area roads and bridges. In 1951, the San Francisco Bay Area Rapid Transit Commission—composed of 26 representatives from each of the nine communities surrounding the bay—was created. In its 1957 report, the commission recommended forming a five-county rapid transit district to oversee a high-speed rapid transit line traversing San Francisco and neighboring communities that would become the solution to reducing residents' dependence on automobiles. The result was the San Francisco Bay Area Rapid Transit District, originally made up of the counties of Alameda, Contra Costa, Marin, San Francisco, and San Mateo. To pay for construction of the new rail system, including an

Bay Area Rapid Transit (BART) train speeds into the station in San Francisco, California, 2011. (Eric Broder Van Dyke/Dreamstime.com)

underwater tube to the East Bay, an initial tax of $0.05 on every $100 of property values was assessed.

When a final engineering plan was proposed to leaders of the five counties, San Mateo County supervisors were put off by the high cost of the system. They also thought that the existing Southern Pacific commuter trains served their county adequately and therefore withdrew from the transit district. Marin County, across the bay north of San Francisco, also withdrew its support in early 1962, citing that its tax base could not adequately support the county's share of BART's projected cost. Also discouraging Marin County leaders was a controversy brewing over how to get trains across the Golden Gate Bridge, the main link between San Francisco and Marin County. The plan was placed before voters in the remaining three counties. Voters approved a $792 million bond issue to pay for a high-speed transit system to travel 71.5 miles and consisting of 33 stations serving 17 communities. The proposal also called for rebuilding 3.5 miles of the San Francisco Municipal Railway (MUNI) and linking the MUNI streetcar lines with BART. Tolls along Bay Area bridges would help pay for additional costs of the Transbay Tube to the East Bay.

Following a lawsuit by taxpayers that resulted in a six-month delay, construction finally commenced on June 19, 1964, with U.S. president Lyndon B. Johnson presiding over the groundbreaking ceremonies of the Diablo Test Track between Concord and Walnut Creek in Contra Costa County. A joint venture among three well-known engineering firms, a collaboration known as Parsons-Brinkerhoff–Tudor–Bechtel, managed all technical and construction aspects of the BART project. The equipment contract for the project was won in a lowball bid by Westinghouse Electric Corporation.

Construction of the 3.8-mile Oakland subway began in January 1966. Ten months later, 57 giant steel and concrete sections of the Transbay Tube were lowered to the bottom of the bay by boats and barges. The tube was completed in August 1969 for $180 million. The tube became the fourth longest vehicular tunnel in the United States and won dozens of engineering awards. Steep inflation at the time—as much as 7 percent a year, or double that which was projected—increased the cost to complete BART and the transit district had to request $150 million in additional funds. The funds were granted by the state legislature and paid for by a half-cent sales tax in the five counties, so the district turned back to construction after three years of financial uncertainty. Other delays included retrofitting stations to include elevators to accommodate the disabled in compliance with new federal laws and alterations to original plans, such as the Berkeley subway system, a more expensive design than the originally slated subway/aerial plan.

When the system finally was built, the first 250 cars out of a total 450 were ordered from Rohr Industries in Chula Vista, California. Federal funding helped improve the project and sped up its completion. The first prototype cars were run around the clock on the Fremont line and IBM prepared to install the first set of fare-collection machines. Passenger service began on September 11, 1972, between the MacArthur station in Oakland and the city of Fremont, the system's southernmost end. BART reported carrying 100,000 passengers in its first five days.

BART trains run on electricity and are powered by a third rail consisting of 1,000 volts of DC electricity. The trains can travel up to 80 miles per hour but typically travel at speeds of 33 miles per hour. The track is 37 miles long within the subway and tunnel system, 23 miles long above ground, and 44 miles long on surface track. Cars operate on seven separate lines that are labeled "A," "M," "W," "Y," "R," "C," and "L." Tickets are plastic cards with a magnetically encoded value. Fares are charged according to distance traveled.

To ease congestion along the heavily traveled north-south corridor on the San Francisco Peninsula, the Santa Clara Valley Transportation Authority in 2013 began a four-year, 16-mile extension project that would bring BART to the city of San Jose and the rest of the Silicon Valley. BART's environmental impact in reducing air pollution in the Bay Area is estimated to be enormous. According to figures produced by the Bay Area Council during a series of labor union strikes in 2013, traffic congestion brought on by the additional automobiles on Bay Area roads due to the shutdown of BART was equivalent to 16 million pounds of carbon and 800,000 gallons of gasoline daily.

Rosemarie Boucher Leenerts

See also: Light Rail Transit Systems; San Diego Trolleys (United States); San Francisco Cable Cars (United States); TriMet MAX, Portland (United States).

Further Reading

Bay Area Council. 2015. "BART Strike Having Costly Environmental Impact on Bay Area." http://www.bayareacouncil.org/economy/bart-strike-having-costly-environmental-impact-on-bay-area/. Accessed May 20, 2015.
Bay Area Rapid Transit. "Projects." http://www.bart.gov/about/projects. Accessed May 20, 2015.
Bay Area Rapid Transit. 2015. "System Facts." http://www.bart.gov/about/history/facts. Accessed May 20, 2015.
Richards, Gary. 2014 (March 9). "BART Extension to San Jose Is Moving Right Along." *San Jose Mercury News.* www.mercurynews.com/bay-area-news/ci_25302661/bart-extension-san-jose-is-moving-right-along. Accessed May 20, 2015.

BEIJING SUBWAY (CHINA)

The Beijing Subway, the second largest subway network in the world (327.46 miles)—after the Shanghai Metro—first opened in 1969, more than 100 years after the world's first subway route opened in London in 1862. As early as September 1953, Beijing's planning committee submitted a proposal for building subway routes in the city. In 1965, the first subway route began construction. Since then, the Beijing Subway experienced four development stages—the slow beginning, a period of stability with two lines in service for two decades, important changes for the Olympics, and continuing expansion since then.

The first subway route was built in 1965 and put into use in 1969. It ran from Fushouling to Beijing Railway Station with a length of 13 miles and 16 stations, and then was expanded from Beijing railway station to Gongzhufen in January 1971, to Yuquanlu in November 1971, and then to Pingguoyuan in April 1973. In 1969, the second subway route with a U-shape began to be constructed from Fuxingmen to Jianguomen along the old wall of Beijing inner city. During this period, the subway was controlled by the People's Liberation Army and mainly served the military. In January 1971, the subway first opened to civilians who had appropriate documentation from their work units and charged a fare of $0.02 (¥0.10) for a single trip.

In April 1981, the Beijing Subway company took over subway operations, and the subway began to serve the public instead of the military. The lines opened to the public one by one, with the subway route from Fushouling to the Beijing Railway Station opening in September 1981, the second subway from Fuxingmen to Jianguomen opening in December 1987, and the whole route of Line 2 opening in September 1984. The Batong line opened to the public in September 1999 and then connected to the Line 1 in June 2000. By 2000, the mileage traveled by the Beijing Subway increased to 34 miles and the annual ridership increased to 434 million people. The fare increased from $0.02 (¥0.10) for a single trip in 1981, to $0.03 (¥0.20) in 1984, $0.08 (¥0.5) in 1991, $0.31 (¥2) in 1996, and then to between $0.47 (¥3) and $1.09 (¥7) per trip in 2000, depending on the line.

The 2008 Summer Olympics in Beijing accelerated the expansion of the subway in Beijing, with $10 billion investment increase in subway projects from 2002 to 2008. In October 2007, Line 5—the first subway route with a platform screen door, and the first north-south line—came into operation. The fare of subway travel decreased to a flat fare of $0.31 (¥2). Three new lines were built for the Olympics, at which time the system reached 124 miles, and all the existing subways were connected forming a whole network. The annual ridership reached 1.2 billion.

Under the impacts of the financial crisis of 2007–2008, the central government of China submitted strategies concerned with "stimulating economic development by investment" with $525 billion investment in 2009–2010 to promote the development of the Beijing Subway. During this period, Lines 4, 15, 9, 14, 7, 8, and the 10 loop opened to the public. This provided service not only in the urban area but also in the suburban region of Beijing. In 2014, Beijing became the city with the second-largest subway network in the world, at 327.46 miles in length. In December 2014, the subway fare changed to a variable-rate fare from the $0.31 (¥2) flat-rate fare, with $0.47 (¥3) for trips within 4 miles, $0.63 (¥4) for trips of 4 to 7.5 miles, $0.78 (¥5) for trips of 7.5 to 13.5 miles, $0.94 (¥6) for trips of

13.5 to 20 miles, $1.09 (¥7) for trips of 20 to 32 miles, and $1.25 (¥8) for trips of 32 to 45 miles.

Jingjuan Jiao

See also: Busan Subway (South Korea); London Underground/The "Tube" (United Kingdom); Seoul Metropolitan Subway (South Korea).

Further Reading

Beijing Subway Company. 2012. *History of Beijing Subway* [in Chinese]. Beijing: Beijing Publishing House.
Beijing Subway website [in Chinese]. http://www.bjsubway.com/corporate/dtdsj/. Accessed December 31, 2014.

BELTWAYS

A beltway is a major highway that encircles an urban area. Although even a two-lane, undivided road with unrestricted access could serve a city as an encircling "beltway," the common association with the term indicates a higher-speed, multi-lane, divided highway, usually with restricted access (as evidenced by entrance/exit ramps.) In the United States, beltways typically are part of the Interstate Highway System and enclose one particular city. The existence of beltways greatly aided the rapid growth of American suburbanization after World War II.

Beltways enable vehicular traffic to avoid passing through a city center, preventing traffic overcrowding in a central business district. Before the interstate highway era, long-distance travelers and large trucks drove through a city's downtown, increasing congestion. Since the passage of the National Interstate and Defense Highways Act in 1956, construction of beltways has been widespread in the United States. Suburban growth was encouraged due to efficiencies in transportation—residents could live in new suburbs with lower-cost housing and drive on a high-speed beltway to jobs in the central city. Additionally, although an individual interstate highway would connect different cities and encourage suburban growth along a single linear path, beltways allowed for radial suburban growth around the rounded beltway, allowing for a greater number of people to relocate to suburbs. One effect of this radial suburbanization has been traffic congestion on many beltways during morning and evening "rush hours."

Beltway suburbanization was aided by complex processes of pressures and reactions: returning soldiers from World War II affected a postwar "baby boom," which resulted in demand for new housing. This housing demand was met through favorable home loans for low-cost homes—homes that were available just beyond many urban areas due to the mass-produced nature of suburbanization. William Levitt's multiple "Levittown" suburban developments in the northeastern United States enabled rapid home construction in assembly-line fashion. New residents filled new suburbs that were connected to wider urban systems by new beltway

The ring road beltway, or loop, encircling Canberra in Australia. (Inavanhateren/Dreamstime
.com)

highways. Today, as jobs increasingly have relocated from city centers to beltway
suburbs, "edge cities" have developed around beltways, enabling residents to live
and work near the once-suburban beltway.

Several morphological types of beltways exist. A beltway can loop around entire
urban areas in a very large circular pattern. Interstate 275, for example, is more than
80 miles in length and surrounds the Cincinnati area, traversing parts of Ohio, Indiana, and Kentucky. Conversely, a beltway can be very small, encircling only a
downtown core area, such as Interstate 440 in Raleigh, North Carolina. Beltways
can be considered "beltways" even if major portions of a beltway circle have not yet
been planned. The Interstate 840 beltway to the south of Nashville, Tennessee, is
missing a northern half and there is no immediate construction timeline.

A beltway also does not necessarily require a closed circle. Interstate 495 surrounding the Boston, Massachusetts, urban area forms a large semicircle, with the
Atlantic Ocean serving as an eastern border. Many U.S. beltways were constructed
in the early years of the interstate highway era (Interstate 465 surrounding Indian-
apolis, for example, was constructed throughout the 1960s). A few urban beltways
are very recent in inception and construction, however, such as Interstate 485,
which encircles Charlotte, North Carolina (with major construction during the
1990s and 2000s). Multiple beltways can exist within a single city: Houston, Texas,

features an inner beltway loop (Interstate 610) and an outer loop (The Sam Houston Parkway), with city locations referred to as "inside" or "outside" a particular loop. In many U.S. metropolitan areas, beltways allow for highway access to airports, as airports require large tracts of land away from tall buildings in a city center. The southern edge of Interstate 285 in Atlanta, Georgia, for example runs along the edge of Hartsfield International Airport.

A beltway is also known as a "loop" or "ring road" in Europe. "Beltway" sometimes is used synonymously with "bypass," although a bypass is defined as a temporary split away from a main road to avoid a downtown in contrast to an encircling beltway. The capitalized term "Beltway" also can be associated with the specific circular Interstate Highway 495 in Maryland and Virginia that encircles Washington, DC. Interstate 495's enclosure of Washington, DC, is so identified with the U.S. capital city that Washington, DC–centric issues and politics often are described as "inside the Beltway." Other beltways are so strongly associated with a particular culture that the beltway name carries another meaning. The name of Massachusetts's "Route 128"—referring to an inner loop beltway to the west and north of Boston—has become synonymous with the high-tech industry that is located near the highway.

Patrick D. Hagge

See also: Autobahn (Germany); Autostrade (Italy); Freeways; Interstate Highway System (United States).

Further Reading

Kaplan, David H., Steven R. Holloway, and James O. Wheeler. 2014. *Urban Geography*. 3rd ed. Hoboken, NJ: Wiley.

Lewis, Tom. 1997. *Divided Highways: Building the Interstate Highways, Transforming American Life*. New York: Penguin.

BERLIN U-BAHN/S-BAHN (GERMANY)

During more than a century of operations, the U-Bahn and S-Bahn mass transit systems have served both to connect and divide the German capital of Berlin. Today the U-Bahn—short for "*Untergrundbahn*," meaning "underground railway"—carries city residents across more than 90 miles of track on 10 separate lines. The S-Bahn—short for "*Stadtbahn*," meaning "cross-city railway"—runs across more than 200 miles of mostly elevated tracks on 15 separate lines. Although the two systems currently connect to form the backbone of Berlin's public transportation system, this interconnectedness greatly hindered development during the Cold War.

The S-Bahn transit system evolved out of a much older mass transit network called the "Ringbahn" ("circular railway"), which was completed in 1877. The Ringbahn consisted of steam-powered trains that traveled around the city on a mostly street-level rail line punctuated by bridges and embankments. At first passengers were reluctant to use the Ringbahn, as fares were significantly more expensive than those of horse-drawn carriages for hire throughout the city. As of 1881, the Ringbahn transported just 2 million passengers, and 63 million passengers travelled by horse-drawn carriage.

The first true S-Bahn line opened in 1882 and cut across the Ringbahn on an east-west axis, passing through the heart of the city on 731 viaduct arches. Although it was only seven miles long, the S-Bahn provided greater connectivity for transport across the city and quickly became a prominent architectural feature of Berlin, influencing urban development around it. As Berlin continued to expand outside the Ringbahn through the turn of the 20th century, additional radial mass-transit lines were built. When long-distance traffic increased during World War I, two of these lines were converted exclusively to long-distance use, and the remaining Ringbahn and the Stadtbahn lines became a dedicated local rail system. This network was electrified between 1926 and 1929, and in 1930 was consolidated under the single name of S-Bahn.

The U-Bahn entered into the transport fray in 1897. Beginning in the late 1890s, urban growth had accelerated to such a degree that the Ringbahn and Stadtbahn transit systems were no longer adequate—particularly as increasingly congested roadways hindered further construction of aboveground rail lines. City planners initially dismissed the idea of the underground rail system proposed by private contractor Hochbahngesellschaft, forcing the company to begin its project as an aboveground network instead. But need soon trumped reservations, and not long after the U-Bahn's official opening in 1902, underground rail lines also were being built. Today 80 percent of the U-Bahn runs underground.

The S-Bahn and U-Bahn systems gradually became integrated through the 1920s and 1930s, with city residents switching between them depending on their destination. By the start of World War II, the dual system formed the backbone of public transportation in Berlin, which by then had exceeded 4 million people.

It was World War II, however, that nearly destroyed Berlin's mass transit system. By 1943, power outages and infrastructure damage from Allied air raids began to cripple both the U-Bahn and, especially, the S-Bahn. The destruction of bridges and embankments crippled the S-Bahn, and both aboveground and below ground U-Bahn stations also eventually saw major damage. By the end of the war, both were in disarray and the impending division of East and West Berlin only served to worsen the situation.

After the fighting ended, Allied forces gave responsibility for S-Bahn network operations to railroad authorities in the east—even though property rights for western portions of the system remained under West Berlin's control. West Berlin ran the U-Bahn network, even though much of it passed under what was now Soviet-occupied East Berlin. The resulting hodge podge of partial transport connections and cut off lines severely hindered public transport in both parts of the city. East and West Berlin came to use different railway gauges, equipment, and scheduling, rendering the two transit systems less and less compatible.

Finally, when the Berlin Wall was constructed in 1961, both the U-Bahn and the S-Bahn were divided along with the rest of the city. Services were suspended on several lines that ran through both parts of the city. Elsewhere, rail lines ended at the Wall. U-Bahn trains that zigzagged through both West and East Berlin were allowed to continue through, but could not stop at the East Berlin stations, leading to the creation of so-called ghost stations where passengers could neither enter nor exit. East Berlin authorities destroyed many of the U-Bahn's aboveground stations in an effort to discourage East Berliners from thinking about the West Berlin residents travelling beneath their feet.

Eventually the S-Bahn became an important part of transportation infrastructure in East Berlin, as did the U-Bahn in West Berlin. West Berlin's S-Bahn system fell into disarray. By the time the Berlin Wall came down in 1989, two very different transport systems had emerged on either side. Reuniting them proved a daunting task, but by the late 1990s they slowly were merging once again into a cohesive public transportation network.

Today the U-Bahn's 143 stations and the S-Bahn's nearly 170 stations accept the same passenger tickets but are run by separate companies. Both are part of the Transport Association Berlin-Brandenburg (VBB) and are interlinked with city buses and trams. This hodge podge of public transport—cobbled together not only since the Cold War but from its very beginnings—can be confusing, but it also is seen as a symbol of a city and a nation reunited.

Terri Nichols

See also: Mass Rapid Transit (Singapore); Mass Rapid Transit, Bangkok (Thailand); Mass Transit Railway, Hong Kong (China); Railroads; Rapid Transit.

Further Reading

Fabian, Thomas. 2000 (October). "The Evolution of the Berlin Urban Railway Network." *Japan Railway & Transport Review* 25: 18–24.

Kunst, Friedemann. 2013 (June 19). "Challenges and Answers: The Berlin Transport Strategy. Berlin High-Level Dialogue on Implementing Rio+20 Decisions on Sustainable Cities and Transport." https://sustainabledevelopment.un.org/content/documents /3708kunst.pdf. Accessed September 2, 2015.

BEVs
See Plug-In Battery Electric Vehicles (BEVs).

BICYCLE LANES

A bicycle lane is a designated and demarcated portion of a roadway reserved for cycling. Line markers, colored pavement, or standard symbols distinguish a bike lane from the rest of the road and indicate the direction of travel. Bike lanes improve

Dedicated bike lanes are increasingly commonplace, as seen in this photo from 2011. (Ying Feng Johansson/Dreamstime.com)

safety by reducing the interactions of cyclists and automobiles, and enhancing opportunities for bike commuting and recreational riding. There are several designs for bike lanes, standardized by transportation engineers for both urban and highway settings. Bicycle lanes separate part of an existing roadway, usually 3 to 7 feet on the right of the automobile lane. Parking for trucks and autos typically is to the right of the bikeway. Additional markers are required for intersections, crosswalks, and transit stops.

A paved shoulder is a section of roadway outside of the motor-vehicle lane, sometimes separated by a rumble strip. It commonly is installed on highways that are part of a state or regional bike trail system. In places where a dedicated lane is unavailable, transportation engineers recommend shared lanes to alert motorists that there might be cyclists on the road. A "sharrow" sign—showing a bicycle symbol with arrows— often is painted on the pavement, and "share the road" signage is placed to inform drivers that bikes will be on the road. Sidepaths are specially designated off-road cycling paths, connecting neighborhoods, running alongside motorways, or providing access to parks and recreation areas. A rail-to-trail sidepath is one created using an abandoned railway route. Sidepaths are very popular with walkers, skaters, and disabled users because motorized vehicles (in most cases) are not allowed. In some places, all-terrain vehicles (ATVs) and snowmobiles are allowed on sidepaths. In some urban areas, bicycle boulevards have been designated to give cyclists priority over motor vehicles. They are common in neighborhoods adjacent to central business districts, so that bike commuters can move throughout the city with fewer restrictions.

During the cycling craze of the 1890s, bike clubs around the world promoted better roadways for bikers. Paved roads were safer, smoother, and more conducive to riding, and worked to reduce the interactions of cycles with pedestrians and horse-drawn carriages. Albert Pope, a leading bicycle manufacturer from Boston, Massachusetts, led the way for the creation of state and federal roadways, and funded the road engineering department at the Massachusetts Institute of Technology. Bike clubs promoted the "Good Roads Movement"—an innovative model for private roadways that became the standard for public and toll roads.

As automobiles developed into viable means of transportation in the early 20th century, there was a repeat battle for the contested space on roadways, this time between motorized vehicles and slower horse-drawn vehicles and bicycles. Eventually, through encouragement and subsidization, motor vehicles in the United States gained control of the roadways, and cycles were relegated to become mainly children's toys and recreational hobbies. In Europe, conditions were much more favorable for the mainstream acceptance of cycles and cycle paths as being affordable and efficient transportation. Dutch nationalism during World War I helped secure a bicycle tax to pay for a nationwide sidepath system that remains the paragon of cycle pathways today.

The modern American bike lane system was developed in Davis, California, in the early 1960s. The Davis situation is unique because it is a small, confined city on flat ground and has a large university; however, it often is cited as beginning a nationwide movement. Cyclists represent 25 percent of the commuters in Davis, and the number is growing across the county, especially in similar towns such as Boulder, Colorado (11 percent), Palo Alto, California (8 percent), and Somerville, Massachusetts (8 percent). Large cities with aggressive biking policies have seen tremendous growth in bike commuting; for example Portland, Oregon. New York City has more than 46,000 regular bike commuters, perhaps as a response to more than 245 miles of new bike lanes across the city. Advocacy actions such as the Green Lane Project have helped install nearly 150 protected bike lanes in more than 40 U.S. cities, and there are now 8,700 miles of bike lanes in the United States that serve 786,000 riders.

Bikeways are given credit for creating healthier environments, reducing teenage obesity, easing traffic congestion, and generating substantial economic impacts. Advocates for bike lanes and side paths often cite statistical evidence for a reduction in bike-related fatalities, fewer incidents involving cars and pedestrians, and an increase in business along the routes. Bike lanes form a key component of the National Safe Routes to Schools program and part of the assessment of the Bicycle Friendly America assessment.

The major criticisms of bike lanes have to do with their expense and safety records. They can be expensive—up to $110,000 per mile for lanes and $355,000 per mile for side paths. There also is evidence that bike lanes actually increase accidents at road crossings and places where cars turn or park across the lane. The allocation of public funds toward bike lanes is intended to encourage more cyclists, yet that leaves open the argument that bike lanes serve a privileged sector of society—those who can afford to bike for recreation. In the United States, less than 1 percent of commuters travel by bike, yet 2.1 percent of the annual federal transportation funding goes toward bicycling and pedestrian projects. A response to this criticism is that building bike lanes is 70 to 100 times less expensive than building new roads for motorized traffic. The contested space of publically funded transportation appears to be shifting in favor of promoting continued growth of cycling lanes, especially as anthropogenic climate change concerns and sustainability initiatives drive the political realities in highly congested urban areas.

Doug R. Oetter

See also: Bicycle Libraries; Bicycles and Tricycles; Cykelsuperstier (Denmark); Velomobiles.

Further Reading

Alliance for Biking and Walking. http://www.bikewalkalliance.org/. Accessed January 13, 2016.

American Association of State Highway and Transportation Officials. 2012. *Guide for the Development of Bicycle Facilities,* 4th ed. Washington, DC: AASHTO.

Forester, John. 2012. *Effective Cycling,* 7th ed. Cambridge, MA: MIT Press.

BICYCLE LIBRARIES

A bicycle library is a bike-share model that enables individuals to borrow bicycles in a manner similar to how books are borrowed from a book library. Like book libraries, memberships often are free or low cost, providing an alternative to programs unlocked through credit card payments or costlier paid memberships. Further distinction from common bike-share programs can be attributed to the period of access. Although most bike-share program bicycles require hourly or half-hourly payments, bike libraries offer extended lending periods, typically anywhere from several hours to several months of borrowing time. Lastly, to differentiate from common bike-share programs, bicycle libraries can be used by all ages of cyclists and other programs are restricted to adults.

Critics argue the without an extensive and substantial payment system bicycle library bikes are more likely to be lost or stolen. This criticism extends from one of the first bike-share programs instituted in Amsterdam, Netherlands, where free-to-access bicycles resulted in bikes that were not returned and that ended up damaged, lost, stolen, or polluting local canals. Bike libraries attempt to overcome this issue by closing the system of the library and bringing bikes into the program that were previously lost, stolen, or abandoned, thus limiting the impact of loss if the loss of a bicycle does occur.

Several bicycle libraries currently are in operation across North America including in Hamilton, Ontario; Edmonton, Alberta; Iowa City, Iowa; Portland, Oregon; Fort Collins, Colorado; and Santa Cruz and Arcata, California. The first and largest bicycle library is located in Arcata, California, and has been in operation since 1997. The only bicycle library to operate out of and on the same borrowing system as a book library is located at Mills library in McMaster University, Hamilton, Ontario. The first bicycle library targeted specifically toward youth who are less than 16 years old is located at Dr. Davey Public School in Hamilton, Ontario, Canada. One bicycle library, located in London, United Kingdom, is mobile, utilizing a converted double-decker bus to service patrons.

Charles M. Burke

See also: Bicycle Lanes; Bicycles and Tricycles; Car Sharing; Cykelsuperstier (Denmark); Sidewalks.

Further Reading

Arcata Library Bike. 2015. http://www.arcata.com/greenbikes/. Accessed June 5, 2015.
The Bicycle Library. 2015. http://www.thebicyclelibrary.com/. Accessed June 2015.
Start the Cycle. 2014. http://www.startthecycle.ca. Accessed June 5, 2015.

BICYCLES AND TRICYCLES

A bicycle is a two-wheeled human-powered vehicle that dates from the early 19th century. It is closely related to the unicycle (one wheel), tricycle (three wheels), and quadricycle (four wheels). The development and growth of cycling is one of the most important advances in human transportation, as these vehicles enable people to move themselves freely and quickly over great distances. Cyclists also are free from the burden of feeding and caring for animals that provide transportation. Additionally, cycling technology and popularity led the way for inventions and improvements in automobiles, airplanes, motorcycles, and roadways.

There are several varieties of cycles used for recreation, racing, touring, and commuting. For the last 20 years, the most popular style of cycle has been the mountain bike and its hybrids. This style of bicycle features a durable steel or aluminum alloy frame, wide rubber tires, and elaborate power systems enabling a rotating pedal crank to use upward of 20 gears on all types of surfaces and inclines. The rider sits upright using straight handlebars that provide agility in tight turns over bumpy terrain. Bicycle motocross (BMX) bikes offer even more maneuverability and durability, and are used for racing over extremely rugged dirt tracks, extreme downhill courses, and for acrobatics. Touring and racing bikes are designed for smooth roadways, featuring narrow tires and dropdown handlebars to reduce air resistance. Professional riders on elite racing bicycles can achieve speeds greater than 31 miles per hour (mph) for great distances, such as Bradley Wiggin's world record of 34 miles in one hour, set in 2015. Recumbent bicycles reduce wind drag even more by positioning the rider low to the ground, seated with legs in front, and body reclining in a chair. The world bicycle speed record of 83 mph was established by Sebastiaan Bowier in 2013 using a recumbent bike encased in aerodynamic composite shielding. A recent development is the e-bike—electric motor–powered bicycles—which can be recharged by pedaling or plugging into an outlet.

It might be surprising that the revolutionary concept of linking two wheels in a straight line to form a sleek vehicle took so long

Paris policeman with his bicycle, circa 1900. (Library of Congress)

to develop. After all, carts and chariots with two wheels on a single axle had been available for thousands of years. Spurred on by a critical shortage of horses in 1817, German inventor Karl Drais delivered the world's first cycle. His device essentially was a wooden horse connecting two wheels such that a rider could push off on the ground and coast forward. The "draisine," as it later was named, was terribly bumpy and difficult to manage. Early editions were favored by aristocratic gentlemen and so they also were called "dandy horses."

The first of several early cycling crazes came and went quickly, as the reputation for the machines failed to gain acceptance. By the middle of the 19th century, however, carriage makers and tinkerers had developed foot-propelled tricycles and quadricycles, usually for the benefit of children or the disabled. An important improvement allowed a rider to move the vehicle forward by pushing on foot pedals attached to rods that propelled one or more of the wheels. In the mid-1860s, a second great explosion in biking took Paris by storm and quickly spread across Europe and the United States. An important contribution, generally attributed to Pierre Michaux, was the addition of foot pedals to a rotary crank attached directly to the front wheel of the velocipede. This simple improvement enabled riders to achieve greater speed and distance, launching a worldwide cycling craze. The "high-wheelers" were

Human-Powered Mass Transport

Human-powered mass transit (HPMT) is a concept that mixes monorail and recumbent cycle technologies. In this transit system, a power pedal pod is suspended under the monorail line. This results in a cost-effective, sustainable transport solution for getting around in congested cities at speeds of up to 50 miles per hour (mph). A "pod" is a capsule covered with transparent plastic sheets that have ventilation holes. Because this vehicle is human powered, it provides the health benefits of bicycle riding, and at the same time provides protection from weather conditions and safety from traffic on congested streets. Most importantly, HPMT offers on-demand service as opposed to the preset scheduled of conventional transit, and offers the speed, flexibility, and comfort of driving a car in the morning commute. That said, HPMT has great potential to offer safer and energy-efficient alternative solutions to cars—and any other mass transit—hence it removes fossil fuel–burning vehicles from the roadways. There are number concerns about the system, however, such as the cost of building infrastructure and the system's safety.

The concept of HPMT originally came from Shweeb, which is why this type also is known as the "Shweeb" monorail technology, or simply, "Shweeb." This technology already has received attention, especially in adventure and theme parks for transit and fun. The HPMT or "Shweeb" presently is fully operational in Agroventures Park in Rotorua, New Zealand, and is capable speeds up to 28 mph. A North American company also is interested in using the HPMT concept and technology in Niagara Falls.

Selima Sultana

notoriously dangerous, however, and in the mid-1880s, John Starley introduced the "safety" bicycle, which included a chain-driven rear wheel, tension spoke wheels, caliper brakes, and suspension systems to soften the ride. The 1886 Rover—which essentially has the same design as modern bicycles—helped launch a global wave of bicycling for common people. Biking became accepted and promoted for health, recreation, and general transportation. Tricycles and quadricycles also were produced en masse, popular with both Victorian ladies and deliverymen. Throughout the 20th century, cycling fads have come and gone, yet the utilitarian benefits of cycles continue to develop as newer technologies—such as gearing improvements and suspension designs—lead to ever-greater applications for cycling.

New bicycle sales in the United States totaled 27.5 million in 2013, with revenues at $5.8 billion. Both these figures have remained fairly consistent for the last 10 years, but U.S. production has dropped dramatically in that same period. Imports from China and other producers have come to dominate the domestic market, primarily through the sale of moderately priced mountain bikes targeted for recreational users and sold through major department and outlet stores. There are about 40 million cyclists in the United States, down from more than 50 million in the mid-1990s, as there has been a decline in younger riders. The number of adults using bicycles has increased dramatically, however, as the League of American Bicyclists estimates that more than 4 billion trips were taken by bicycle in the United States in 2009.

Globally, cycling production is increasing dramatically, as India and Brazil experience biking booms among growing middle-class consumers. China produced more than 80 percent of the 130 million bikes made in the world in 2007, yet ridership there is dropping off partly due to efforts to restrict bike use in Shanghai and other globally prominent urban centers. As more middle-class Chinese select automobiles, the fleet of bicycles in that country has dropped below 500 million and the percentage of trips taken on a bicycle has dropped to 40 percent. Aggressive policies in other countries have spurred bike use among commuters and day laborers, for example in Latin America where the development of bike lanes and availability of low-interest loans for bicycles have increased daily-commute ridership to 10 percent. Europe continues to lead the world in bicycling promotion, especially in the Netherlands, where the average person cycles two miles per day, and 30 percent of all trips are made on a bike.

Cycling is growing vigorously worldwide, especially in rapidly developing countries. Bicycling provides several important benefits, including gender-neutral access to jobs, reduced urban congestion, healthy lifestyles, and better environmental quality. Compared to automobiles, bicycles reduce the cost of urban infrastructure by as much as one-seventh and are much easier to afford for low-income families. Owning a bike can increase worker productivity more than 35 percent by reducing travel time and expanding the daily range of the laborer. Several developing nations have found that low-cost loans for bikes can reduce poverty. Cycling also decreases greenhouse gas emissions. Bike tourism, commuting, and recreational rides reduce stress, obesity, and the risk of cancer, and riding improves mental health. The economic benefits alone support reasons to increase cycling development, as the industry supports more than 1 million jobs in the United States and contributes $133 billion annually to the U.S. economy. Ridership in the United States remains among the lowest in the world; however, the popularity for bicycle and tricycle have been growing steadily.

Doug R. Oetter

See also: Bicycle Lanes; Bicycle Libraries; Cykelsuperstier (Denmark); Motorcycles.

Further Reading

Epperson, Bruce D. 2010. *Peddling Bicycles to America: The Rise of an Industry*. Jefferson, NC: McFarland.

Gardner, Gary. 2008 (November 12). "Bicycle Production Reaches 130 Million Units." Worldwatch Institute. *Vital Signs*. http://vitalsigns.worldwatch.org/vs-trend/bicycle -production-reaches-130-million-units. Accessed August 3, 2015.

Herlihy, David V. 2004. *Bicycle: The History*. New Haven, CT: Yale University Press.

BIOFUEL VEHICLES

A biofuel vehicle is a vehicle such as a bus, car, motorcycle, truck, or train that is propelled by energy derived from living organisms (i.e., plant- or animal-derived material). Nonterrestrial transport modes—such as aircraft, spacecraft, and water-craft—are capable of using biofuels in their propulsion systems. Economic and positive energy output, however, has not yet made biofuels practical for these transport modes. Biofuel vehicles use renewable energy sources derived from bio-mass converted into a convenient liquid fuel that suits most transportation needs.

Most biofuel vehicles use the two most common types of biomass-based fuels— ethanol and biodiesel. Ethanol is a form of alcohol derived from the fermentation of starches and sugars. High-carbohydrate biomass derived from corn and sugar-cane are the most common feedstocks used in the production of ethanol. The United States and Brazil are the world's primary users of fuel ethanol in the trans-portation sector. As of 2013, the United States and Brazil were the first and second leading producers of fuel ethanol (Renewable Fuels Association 2015). Together, the United States and Brazil were responsible for 84 percent of world ethanol pro-duction in 2013.

Vehicles with engines that operate on pure ethanol or biodiesel exist throughout the world. Ethanol-only engines use a fuel designated E100, which denotes that the content is 100 percent ethanol. Automobiles designed to run on either gasoline or E85—fuel with an 85 percent ethanol content—are called "Flexible-Fuel Vehi-cles." Biodiesel-only engines use a designation of B100, which indicates that the fuel content is 100 percent biodiesel. Biodiesel-propelled vehicles use a fuel that contains methanol, food-grade vegetable oil, animal fat, or recycled used cooking oil (i.e., "yellow grease").

A great portion of internal combustion engine vehicles operate with ethanol or biodiesel as a blending agent with hydrocarbon-based petroleum (i.e., gasoline/ petrol). The designation "E15" often is used to designate gasoline blended with 15 percent ethanol, and "B5" or "B10" are used to denote petrol-diesel blended with biodiesel. Displacing gasoline with a 15 percent ethanol blend has been mandated by the Renewable Fuel Standard program as required by the Energy Independence and Security Act of 2007.

The aim of using biofuel vehicles is to be carbon neutral. Biofuels have the po-tential to reduce greenhouse gas emission because they use renewable sources such as plant and animal materials for their feedstocks. In an ideal life-cycle system, the renewable material used to create biofuel will be replenished under carefully con-trolled production. Controversy exists, however, as to whether the energy ex-pended in production of biofuels negates their carbon neutrality. The energy needed to grow, collect, and operate the fermentation machinery required to create ethanol, for example, must be considered in what is known as the energy balance of ethanol fuel. Some research has shown that ethanol and most biofuels in general are energy negative; that is, more energy is required to create them than is con-tained in these fuels. Conversely, high yields have been achieved in Brazil where sugarcane is the main feedstock. Additionally, recent developments in cellulosic ethanol can increase yields and mitigate any negative energy contributing factors.

Aside from carbon neutrality, biofuels ideally must behave like gasoline and provide at least enough energy output to power vehicles. Of the two most common biofuels, ethanol receives the bulk of industry focus but is much less dense in energy as compared with regular gasoline. A liter of corn ethanol, for example, can propel a passenger car only 70 percent as far as the equivalent of gasoline. This illustrates why ethanol lacks sufficient power for commercial trucks or aircraft. Ethanol also becomes easily diluted with water from the environment and has been found to be corrosive in today's engines. Researchers are attempting to overcome these constraints by creating more complex alcohols from biomass that more closely resemble the hydrocarbons in petroleum.

No biofuel vehicle is completely free of negative environmental impacts. When combined with other sustainable practices such as conservation, however, resource use and greenhouse gas emissions can be reduced. The choice is ultimately the user's point of view as to which system is better for the environment. Corn ethanol uses large amounts of valuable land that is required to plant crops that are then further processed. Using plant waste to make cellulosic ethanol will improve the energy balance of this fuel. In biodiesel production, it is necessary to consider the process of making oil, there is less concern, however, if recycled oil (yellow grease) is used.

Joe Di Gianni

See also: Biofuels; Hydrogen Vehicles; Plug-In Battery Electric Vehicles; Plug-In Hybrids.

Further Reading

Canackci, Mustafa. 2007. "The Potential of Restaurant Waste Lipids As Biodiesel Feedstocks." *Bioresource Technology* 98(1): 183–190.

Kinver, Mark. 2006 (September 18). "Biofuels Look to the Next Generation." *BBC News.* http://news.bbc.co.uk/2/hi/science/nature/5353118.stm. Accessed January 5, 2015.

National Renewable Energy Laboratory. 2014. "Learning About Renewable Energy. Biomass, Biofuel Basics." http://www.nrel.gov/learning/re_biofuels.html. Accessed January 4, 2015.

Renewable Fuels Association. 2015. "World Fuel Ethanol Production." http://ethanolrfa.org/pages/World-Fuel-Ethanol-Production. Accessed January 5, 2015.

BIOFUELS

Biofuels are alternative fuels that use material derived from living or recently living organisms. Biofuels primarily are converted into a liquid for use as transportation fuels but also can be used in stationary engines such as those used to generate electricity. There are a variety of biofuels that can be produced from biomass, the basic feedstock material of biofuels. As of 2012, biofuels accounted for approximately 7.1 percent or 13.8 billion gallons of total transport fuel consumption in the United States (USDA 2015).

The two most common types of biofuels are ethanol and biodiesel. A third kind of biofuel—bio-oil—produces a fuel similar to biodiesel by using a process called "flash pyrolysis." Bio-oil is suitable primarily for use in stationary engines used to generate electricity. Ethanol is the most common biofuel. In 2009, more than 7.3 billion gallons were used as an additive to gasoline to meet United States biofuel requirement to reduce air pollution and dependency on foreign oil supplies. Ethanol is a form of alcohol derived from the fermentation of starches and sugars, similar to the process of making beer. High-carbohydrate biomass derived from corn and sugarcane are the most common feedstocks used in producing ethanol. The United States and Brazil are the world's primary consumers of fuel ethanol in the transportation sector. As of 2013, the United States and Brazil also were the first and second leading producers of fuel ethanol. Together, the United States and Brazil were responsible for 84 percent of world ethanol production in 2013 (Renewable Fuels Association 2015).

There are two processes for creating ethanol: fermentation and the synthetic method. Fermentation uses corn or other biomass (e.g., sugarcane in Brazil) as a feedstock. Fermentation is the most typical process for creating ethanol and accounts for more than 90 percent of all ethanol produced. The synthetic method uses ethylene—a byproduct of petroleum—and primarily is used for industrial purposes. Current research on an alternative method of producing ethanol—called cellulosic ethanol—uses twigs, grasses, and stover (i.e., the parts of the corn plant left in the field after harvest). It is anticipated to produce biofuel that has greater greenhouse gas–reduction potential than that of corn ethanol. Cellulosic ethanol solves issues such as using food and cropland for fuel.

The second most common biofuel, biodiesel, is created from food-grade vegetable oil, animal fat, or recycled cooking oil. The process used to make biodiesel involves combining organically derived oils with an alcohol—such as ethanol or methanol—and a suitable catalyst to form ethyl or methyl ester. Like ethanol, biodiesel is a clean-burning biofuel made from renewable sources. Biodiesel often is used as an additive in petroleum-based diesel to reduce particulate levels. Most biodiesel produced in the United States is made from soybean oil or "yellow grease," recycled oil that had been used in restaurants to prepare food. The primary agricultural feedstock in Europe is rapeseed. Biodiesel fuel is created using a chemical process called transesterification, which separates glycerin from the fat or vegetable oil. The transesterification process produces two products: methyl esters (i.e., biodiesel fuel), and glycerin (used in soaps and other cleaning agents).

Biobutanol is butyl alcohol produced from the same feedstocks as ethanol (e.g., corn, sugar beets). Biobutanol also can be made from algae to make algal biofuels. Like other biomass used in biofuel production, algae releases CO_2 removed from the atmosphere via photosynthesis. The benefit of algal biofuels is that by using terrestrial feedstocks (such as corn) they can be produced with minimal impact on arable land.

The fermentation process, which is part of corn ethanol production, focuses on one of its chief criticisms because of the intensive use of water. Pimentel and Patzek (2005) state that 15 liters of water are used during the fermentation process for each 2.69 kg of corn. Another criticism of corn ethanol production is the amount of prime arable land devoted to transportation fuel instead of food. Shapouri et al. (2002) calculated the amount of U.S. land required to replace only one-third of the gasoline used with corn ethanol. The result of the study concluded that fueling one automobile for one-third of the year with corn ethanol requires more cropland than is needed to feed one person. Similar debates have centered on biodiesel that also is subject to the "food versus fuel" dilemma. Farmers in poor countries reportedly have diverted farmland to more lucrative fuel crops, causing uncertainty in world food markets. These same farmers, however, have seen increases in their overall incomes.

The aim of using biofuels is to achieve carbon neutrality. Unlike hydrocarbon-based petroleum that uses finite resources, biofuels are made from materials replenished with careful production methods. Like fossil fuels, biofuels also can contribute to greenhouse gas emissions when they are combusted. There is research supporting the idea that, because biofuels are derived from recent biomass, their net carbon dioxide emissions will decrease as they replace fossil fuels.

Joe Di Gianni

See also: Biofuel Vehicles; Electric Vehicles; Hydrogen Vehicles; Plug-In Hybrids.

Further Reading

Pimentel, D., and T. W. Patzek. 2005. "Ethanol Production Using Corn, Switchgrass, and Wood; Biodiesel Production Using Soybean and Sunflower." *Natural Resources Research* 14(1): 65–76.

Renewable Fuels Association. 2015. "World Fuel Ethanol Production." http://ethanolrfa .org/pages/World-Fuel-Ethanol-Production. Accessed January 5, 2015.

Shapouri, H., J. Duffield, and M. Wang. 2002. *The Energy Balance of Corn-Ethanol: An Update.* Washington, DC: U.S. Department of Agriculture.

U.S. Department of Agriculture (USDA), Economic Research Service. 2015 (June 5). US Bioenergy Statistics. http://www.ers.usda.gov/data-products/us-bioenergy-statistics.aspx. Accessed August 3, 2015.

U.S. Department of Energy (DOE). *Biomass Energy Data Book.* 2011. Oak Ridge, TN: National Renewable Energy Laboratory.

BLUE RIDGE PARKWAY (UNITED STATES)

The Blue Ridge Parkway is one of the most famous and popular roads in America. It runs 469 miles from the Newfound Gap Road in the Great Smoky Mountains to

Shenandoah National Park in Virginia. The entire length of the road goes along the ridges of the Blue Ridge Mountains, often alternating between the east and the west sides of mountains, with many parking areas for motorists to stop and enjoy the scenic views. The road's elevation varies from 649 feet above sea level at the James River crossing to 6,053 feet on Richland Balsam Mountain. The southern end generally is higher than the portion in Virginia. The terrain posed many construction challenges and required 26 tunnels and many bridges.

The Blue Ridge Parkway was built by the National Park Service, and construction began in 1935. The Parkway was a new type of unit in the national park system. Rather than being a large park with scenic roads, it was designed to be a scenic road buffered by nice views. Land was purchased not just to build the road but to include scenic views of hillsides and pastures. The road was designed so motorists maintain a constant driving speed—there are no stoplights or stop signs, and no commercial vehicles are permitted. Vehicles enter and exit the Parkway only at selected locations along its length. It has proven a tremendously popular road and regularly ranks as among the most visited national park units.

The busiest portion of the highway is the section that passes the city of Asheville, North Carolina, which carries about 7,000 vehicles per day. This number is miniscule compared to urban highway traffic (which can be more than 100,000 vehicles per day). Sections near the cities of Roanoke and Blowing Rock also carry more traffic and, in general, the northern half of the Parkway has more traffic than does the southern half. Some rural stretches might have only 500 vehicles per day or fewer. Relatively few visitors drive the highway end to end. June through October are the most popular months for travel on the road. Traffic can be significant on October weekends, when people come to the higher elevations along the road in search of vistas of fall colors. During the winter months traffic falls off, and parts of the Parkway often are closed due to snow and ice. A check of Parkway road conditions is advised before traveling it at any time of the year.

There are many interesting historic and recreational sites along the road. At Mabry Mill in Virginia, visitors can see an old gristmill, blacksmith shop, and other pioneer structures. Yankee Horse Ridge has a replica of an old narrow gauge logging railroad. The Linn Cove Viaduct in North Carolina was both the last section to be completed (in 1987) and one of the most famous landmarks along the highway. Mount Mitchell, the highest peak east of the Mississippi River, is in a North Carolina state park, a short distance off the Parkway. Several campgrounds and many hiking trailheads are found along the road.

The endpoints of the Blue Ridge Parkway are not the end of the road. Skyline Drive in Shenandoah National Park continues onward from the end of the Parkway for another 105 miles. Construction started on this road a few years before the Parkway build was initiated, and it is built to the same basic standards. From the driver's point of view, it is the same road and simply has a different name. At the southern end the Newfound Gap Road gives access to the highest elevations in Great Smoky Mountains National Park. There also are several other noteworthy parkways in the southern Appalachian Mountains including the Cherohala Skyway, which passes over the mountains between North Carolina and Tennessee.

Joe Weber

See also: Natchez Trace Parkway (United States); National Scenic Byways (United States); Parkways.

Further Reading

National Park Service. U.S. Department of Interior. 2015. Blue Ridge Parkway. http://www .nps.gov/blri/index.htm. Accessed December 19, 2015.
Whisnant, Anne Mitchell. 2006. *Super-Scenic Motorway: A Blue Ridge Parkway History.* Chapel Hill: University of North Carolina Press.

BORDER TUNNELS (MEXICO AND UNITED STATES)

More than 150 illicit cross-border tunnels have been discovered running between the United States and Mexico. These tunnels are constructed as a new transport route for narcotic trafficking operations. Besides discovering ample loads of narcotics destined for illegal U.S. markets, humans also have been caught coming through the tunnels. Whether they are migrants or involved in the drug industry is not always clear. Further speculation associates the tunnels with potential human and weapons trafficking, and concern about their presence has been conflated with potential access points for terrorists and weapons of mass destruction. This conflation has driven the first national legislations of subterranean border space (U.S. Congress 2007, 2008, 2012) to target a specific transport route that is less visible and thus more problematic to control than is surface crossing.

The very first tunnel discovered in 1990 ran between a luxury home in Agua Prieta, Sonora, Mexico, and a small warehouse in Douglas, Arizona. Two years later, a second tunnel was uncovered by Mexican authorities, who found documents detailing tunnel logistics during investigation of a safe house in Tijuana. The third tunnel then was discovered in 1995 where it surfaced in the worship area of an abandoned church in downtown Nogales, Arizona. These first tunnels seemed random, but as the decade ended six more tunnels were discovered—a second one between San Diego, California, and Tijuana, Mexico; one in Naco, Arizona; and then four more in the Nogales area—and patterns began to emerge. More than 80 tunnels are clustered along a 4 mile stretch that divides Nogales, Arizona, and Nogales, Sonora, and more than 40 tunnels have been discovered along the 12-mile border between San Diego County, California, and Tijuana, Baja California. A small scattering of discoveries more recently have been uncovered in other communities—12 across the Calexico, California, and Mexicali, Baja California, border; 2 in the Tecate, California, and Tecate, Baja California, areas; 3 between San Luis, Arizona, and San Luis Rio Colorado, Sonora; and 2 under

U.S. Immigration and Customs Enforcement agent guards an entrance to a tunnel found along the U.S.-Mexico border at a warehouse in Otay Mesa, California, 2006. The tunnel runs 2,400 feet and was used to funnel drugs into the United States from Mexico. (AP Photo/Denis Poroy)

the Rio Grande River at El Paso, Texas, and Ciudad Juarez, Chihuahua. The cumulative number of tunnel discoveries per year also increased dramatically with peak years in 2006 (15 tunnel discoveries in California), in 2007 (17 tunnel discoveries in Arizona), and in 2011 (14 tunnel discoveries in Arizona).

National security agencies commonly associate the increase in tunnel discoveries to heightened border security at crossing points on the surface. Changes in border policy initiated in 1994 with Operation Gatekeeper in San Diego, California, and Operation Hold the Line in El Paso, Texas, were responsible for this initial shift to heightened security. These policies infused border security efforts with additional resources, personnel, and funds to build more formidable border barriers. Dilapidated chain-link fences were replaced with steel landing-mat walls, and more recently with cement-rebar bollard-style walls. Increasing difficulty in crossing at the standard surface locations explains the increased efforts to dig underground pathways.

The correlation between border-policy shifts, fortified barriers, and increased tunnel discoveries is very clear. Every tunnel discovery in the 1990s except the very first was at a location where chain-link fences had been replaced or were in the process of being replaced with steel landing-mat walls. By the year 2000, when stronger barriers lined the boundaries of most Southwestern transborder communities, the number of discoveries increased greatly.

Tunnels fall into three categories—sophisticated, interconnecting, and rudimentary. Sophisticated tunnels largely are discovered in the San Diego area. They are longer (up to a half mile long), deeper, and usually contain added infrastructure such as electricity, railways, drainage equipment, pulley systems, and air conditioning. Tunnels discovered in the Otay Mesa industrial zone in eastern San Diego County typically fit this category. The U.S. side of the industrial zone contains manufacturers, wrecking/recycling yards, nondescript warehouses, and storage facilities to hold and manage distribution of goods from Mexico. If these buildings stand empty or are underutilized, they become vulnerable to drug-front businesses that use them to effectively hide digging operations underneath. These warehouses also serve as storage spaces for contraband awaiting distribution. On the Mexican side, industrial parks have been developed directly south of the border, alongside the municipal airport, and also to the southeast. Unmarked trucks or trailers used to distribute cocaine, marijuana, or methamphetamine blend in with legitimate semitruck or trailer traffic that passes through the whole transnational industrial complex.

In contrast, interconnecting tunnels connect with underground infrastructure and are found more commonly in the Nogales, Arizona, area, where more than three miles of underground drainage tunnels run under the downtown of the community and its southern neighbor Nogales, Sonora. Numerous storm drains are located within the vicinity to channel runoff into the underground drainage tunnels. Traffickers use the underground drainage infrastructure to their advantage, guiding their tunnel digging alongside the drainage tunnels, and often tunneling into or out of inlets and outlets situated alongside roads. Lastly, rudimentary tunnels are found in both areas. They are shorter in length, run closer to the surface, and are small in size. They typically do not have additional infrastructure.

Thus far, tunnels have been discovered in urbanized sections of the border where it is easier to hide the displacement of dirt resulting from tunnel construction and to camouflage the larger trafficking operations that include drug storage, pick up, and transfer. Depending on the type of tunnel, they also can require energy sources—which are difficult to access in remote portions of the border. Traffickers rely on the same highway system for distribution that legal industries depend upon, thus tunnels are likely to appear near major ports of entry. For these reasons, no tunnels have been discovered on the New Mexico border with Chihuahua, which has no major transborder urban formations. The Texas border seems problematic for tunnel construction as well, given that it is bounded by the Rio Grande River and potential water seepage into the alluvial plain makes tunnel construction underneath the river difficult. Only at El Paso, where the river is channelized with concrete and the ports are well linked to major highways, have tunnels been discovered.

Cynthia Sorrensen

See also: Channel Tunnel (France and United Kingdom); El Camino del Diablo (Mexico and United States); Tünel (Turkey).

Further Reading

Serna, Joseph. 2015 (May 1). "The Ins and Outs and U.S.-Mexico Border Tunnels." *Los Angeles Times*. http://www.latimes.com/local/lanow/la-me-ln-border-tunnels-201505 01-htmlstory.html. Accessed August 3, 2015.

BOULEVARDS

Boulevards are tree-lined streets containing separate realms for faster-speed vehicle traffic in the center and slower-speed local vehicle and pedestrian travel at their outer edge. Grand boulevards such as the Avenue des Champs Élysées in Paris are among the most famous streets in the world. Boulevards are central to the spatial organization of many European cities. Lined with retail shops, sidewalk cafés, apartments, and public buildings, and furnished with grand monuments and formal landscaping, boulevards are quintessential places to enjoy urban life. From Europe, the concept of the boulevard diffused around the world so that boulevards now also can be found in North America, South America, Asia, Africa, and Oceania. Boulevards are constructed less frequently today, but some transportation planners argue for their rediscovery because they serve multiple functions so well.

The term "boulevard" comes from the Dutch word "*bulwark,*" because the first boulevards were constructed along obsolete city wall fortifications. Prior to the late 16th century, trees were rare in cities. Urban tree planting began in the 1580s in Dutch cities, when trees were planted on tops of wide earth-filled city walls, creating shady elevated promenades. In the 1660s, the French followed the Dutch practice when the Tuileries promenade in Paris was extended through a field called the Champs Élysées, creating the first tree-lined boulevard. Baron Haussmann implemented a street modernization program in Second Empire Paris (1853–1870) that built a network of broad boulevards to replace the city's cramped medieval streets. In Paris, those who frequently strolled or lingered on the city's boulevards were called "boulevardiers." Unter den Linten, a broad, tree-shaded boulevard, is Berlin's most important street. Unter den Linten, literally "under the Linden trees," is home to important buildings such as the Berlin State Opera, the Berlin Cathedral, Humboldt University, and the Brandenburg Gate.

Boulevards inspired the design of both parkways and freeways. In contrast with the more pastoral feel of a parkway, boulevards offer a distinctly urban experience. In contrast with the freeway's emphasis on high-speed vehicular traffic, the boulevard is both a thoroughfare and a destination. In the design of streets and roadways, transportation planners recognize the trade-off between access and mobility. A freeway provides maximum mobility by reducing access. A local street maximizes access to origins and destinations but only can handle modest traffic volumes and at relatively slow speeds. A well-designed boulevard provides both accessibility and mobility, combining features of local streets and freeways.

To combine mobility and accessibility in an attractive environment, a boulevard requires a wide right-of-way. Paris's Champs Élysées is 220 feet wide, accommodating 10 lanes of fast-moving traffic in the center, landscaped medians at both sides, local access roads at both edges, and 36-foot-wide sidewalks along the building frontages on both sides. An example of a boulevard that gracefully balances the needs of through traffic and local residents is the Paseo de Gracia in Barcelona, Spain. This boulevard is 200 feet in cross section, allowing generous 36-foot-wide sidewalks on both sides. The sidewalks of the Paseo de Gracia are lined with stores and are outfitted with benches and decorative light fixtures. The street is filled with people enjoying a stroll, especially at night. The pedestrian realm on both sides of the boulevard is protected from the central roadway by a wide park median and a local access road with diagonal vehicle parking. Four parallel lines of soaring trees give the boulevard a protective canopy. The New Urbanist designers Andres Duany and Elizabeth Plater-Zyberk's Avalon Codes set forth 14 different street patterns including two boulevard designs, a simpler version with a central landscaped median and a wider one resembling the Paseo de Gracia.

Prominent boulevards around the world include Keshavarz Boulevard in Tehran, Iran; C. G. Road in Ahmedabad, India; the Rajpath in Delhi, India; Roxas Boulevard in Manila, Philippines; Broadway in Galveston, Texas; Sunset Boulevard in San Francisco, California; and Wilshire Boulevard in Los Angeles, California. Wilshire Boulevard is perhaps the Los Angeles metropolitan area's greatest street, stretching 15.5 miles from downtown through Beverly Hills, past many of the city's film-industry landmarks and premier luxury retail venues, and ending at the ocean cliffs at Santa Monica.

A boulevard is more than a wide street, it is a destination as well as a thoroughfare—characterized by its size, landscaping, and multiple functions. Boulevards are comfortable places for walking and shopping as well as moving heavy vehicle traffic. Designed to serve the needs of both through traffic and neighborhoods, boulevards suggest the possibilities for harmonizing local environmental quality and regional needs.

Mark Bjelland

See also: Freeways; Greenways; Las Vegas Strip (United States); Parkways.

Further Reading

Jacobs, Allan B., Elizabeth MacDonald, and Yodan Rofé. 1995. *Great Streets*. Cambridge, MA: MIT Press.
Jacobs, Allan B., Elizabeth MacDonald, and Yodan Rofé. 2003. *The Boulevard Book: History, Evolution, Design of Multiway Boulevards*. Cambridge, MA: MIT Press.

BUSAN SUBWAY (SOUTH KOREA)

The Busan subway system is the second-largest subway network in South Korea and serves the Busan metropolitan city and the surrounding cities of Gimhae and Yangsan. The first Busan subway train began operation in 1985 and runs between Beomeosa and Beomnaegol. Since then, the subway network has expanded to 128 stations on 5 lines and 83 miles of routes. The network consists of 4 Busan subway lines operated by the Busan Transportation Corporation and owned by the city of Busan, and the Busan-Gimhae Light Rail Transit (BGL), which was constructed by private investors. The modal share of the Busan subway increased from about

14 percent of all trips in the city in 2010 to nearly 17 percent in 2013, with 1.1 million daily riders in 2013.

Busan is the second-largest city in South Korea, with a 2014 population of 3.6 million. The most urbanized districts are densely concentrated along a number of valleys due to several mountains. Thus, most traffic uses the linear trunk roads and only limited alternative routes are available. This causes severe congestion. Additionally, the seaport of Busan—the fifth-busiest seaport in the world—creates freight-related traffic on the already congested roads. Considering the limitations of widening roads to handle this traffic, Busan has expanded the subway network to provide for heavy traffic across the linear urban area.

The Busan Metro lines contribute to reducing traffic congestion as well as providing fast connections between urban centers. Line 1 is the most crowded metro line in South Korea (except for Seoul's subway lines), carrying 840,000 daily passengers. This line connects central business districts in Nampo-dong and Seomyeon to subcenters such as Dongnae, Yeonsan, and Hadan, running north and south. It also connects major transportation terminals such as the Busan station (for Korean high-speed rail), the Bujeon station, and the Busan Central Bus Terminal.

Line 2 connects emerging subcenters of Busan—such as Deokcheon, Sasang, Daeyeon, and Haeundae—to Seomyeon, running east to west. It was extended to Yangsan, a city adjacent to Busan, which improved the accessibility of Yangsan.

Line 3 is a relatively short line (11 miles) but improves network efficiency by crossing Line 1 and Line 2. It provides an express route for the wide U-shaped Line 2, between the Deokcheon station and the Suyeong station.

Line 4 is the first public Light Rail Transit (LRT) project for commercial operation. The LRT train for Line 4 runs on a concrete track and uses rubber tires instead of steel wheels. The line is called Korean Automated Guideway Transit (K-AGT). It serves the northeastern residential districts such as Anpyeong, Bansong, and Geumsa by connecting with Line 1 at the Dongnae station and Line 3 at the Minam station. The Busan–Gimhae Light Rail Transit (BGL) is another type of LRT system but it uses steel wheels. It acts as a commuter rail system between Busan and Gimhae and as an airport rail system for the Gimhae International Airport (9.67 million annual passengers in 2013).

The expansion of the Busan subway network is focused on both improving connectivity and connecting neighboring cities, such as Ulsan, Changwon, Gimhae, and Yangsan. The south terminus of Line 1 is being extended to Dadae in 2016. Additionally, the Sasang-Hadan LRT will provide a fast connection between Line 1 at the Hadan station and Line 2 at the Sasang station, which will improve the current lengthy detour within the subway network. The Yangsan LRT Line will connect between the Nopo station (the north terminus of Line 1) and Bukjeong in northern Yangsan via the Line 2 Yangsan Sports Complex station. Two conventional railways also will serve as commuter rail service to neighboring cities composing sections of the subway network. The Donghanambu Line will serve commuter trains between

Busan and Ulsan—which is expected to promote interaction between the two metropolitan cities. Additionally, the Busan-Masan Line will provide commuter service between Busan to Changwon via Jangyu in southern Gimhae.

The Busan subway has various convenient services for passengers. The color-LCD screens in the platform areas and at the gates inform passengers of the arrival schedule and the location of the next trains. All of the stations have elevators and escalators for passengers with disabilities and others. For safety and air quality in the station, platform screen doors (PSDs) are installed at all of the stations on Line 3, Line 4, and the BGL, and have been expanding to Line 1 and Line 2. The transfer stations try to minimize transfer times by scheduling similar arrival and departure times between trains of the different lines serving the station. Additionally, passengers can enjoy the developed IT infrastructure of South Korea using the Wi-Fi, Wibro, and 4G LTE networks available in the trains and on the platforms.

The Busan subway system also enables travelers to reach the international convention facilities and various attractions in Busan. The multilanguage voice announcements—made in English, Chinese, and Japanese—used at all train stops assist foreign passengers. Passengers can buy a one-day pass for the price of 4,500 KRW (Korean won) (approximately US $4.00) to enjoy various attractions.

Hyojin Kim

See also: Light Rail Transit Systems; New York City Subway (United States); Seoul Metropolitan Subway (South Korea).

Further Reading

Busan Transportation Corporation. 2015. http://www.humetro.busan.kr/english/main/. Accessed June 5, 2015.
Jang, Seok Yong, Hun Young Jung, and Sang Keun Baik. 2013. "A Study on the Improving the Services for Users of LRT (Light Rail Transit) by Structural Equation Model-Focus on Busan Gimhae Light Rail Transit and Busan Subway Line 4 (Bansong Route)" [in Korean]. *Journal of the Korean Society of Civil Engineers* 33(1): 261–272. *ResearchGate.* http://www.researchgate.net/publication/264134438. Accessed June 5, 2015.
Koo, Donghoe. 2013. "A Study on the Changes of Commuting Area in the Busan Metropolitan Area" [in Korean]. *The Journal of the Korean Geography Society* 48(4): 533–544.
Park, Junsik, and Seong-Cheol Kang. 2011. "Network Connectivity of Subway Stations in South Korea." Presented at The 9th International Conference of Eastern Asia Society for Transportation Studies: 204.

BUSES

The term "bus" is derived from the Latin word *omnibus*, meaning "for all." A bus is a land vehicle that is designed to transport many passengers at once. Buses are vital to transportation systems and carry more people daily than does any other mode of public transit. The size of buses varies, but most are large, single-deck, rigid vehicles 35 to 40 feet long and 8 to 9 feet wide. They are tall enough that people can easily walk inside. Other bus types include double-deckers, with two levels; articulated buses, comprising two rigid sections joined by a pivoting middle section; mini- and midi-buses, more economical and with fewer seats than traditional buses; and coaches, or motor coaches, often used for comfortable long-distance travel.

Buses can be publicly or privately owned and available for hire. Traditional buses in metropolitan areas are run by public transportation agencies and require riders to pay with coins or tokens upon entry or by displaying a prepaid fare card. Transit buses run on a schedule, allowing for pickup and drop-off of passengers at marked stops. School districts also employ bus companies or have their own fleet to transport children to and from schools in the bright-yellow vehicles. Buses within the tourism industry tend to be intercity or interstate coaches that carry passengers a significant distance between cities, towns, states, and provinces. Stops often are at bus depots in both in metropolitan or rural areas. Many coaches are equipped with restrooms and high-backed seats.

Bus transportation dates back to the mid-17th century, when the first horse-drawn public bus was introduced in Paris, France, by Blaise Pascal in 1662. Though at first popular, service was discontinued after just 15 years because of poor ridership due to an increase in ticket prices, making the service accessible only to the wealthy. There is no record of any revived bus service until the early 19th century, when the "omnibus" appeared in Bordeaux, France, in 1812, followed by omnibus services in Paris, France; New York, New York; and London, England. In this age, buses looked like a combination carriage and stagecoach and riders often would sit inside, outside, and even on top of the vehicles.

Tourist bus visits the tomb of Ulysses S. Grant, circa 1910. (Library of Congress)

Straddling Buses

The straddling bus is a fascinating public transport idea started in China to reduce urban congestion and utilize the existing infrastructure. It uses giant buses that run on small tracks between the lanes of traffic and also roll above other congested areas. This reduces congestion without the need to build completely independent track systems. It is faster and cheaper to build than a subway or a monorail system. It requires only a simple modification of existing roads with elevated bus stops. At 20 feet wide and 15 feet tall, straddling buses can straddle two lanes of traffic; they also can fit under most existing overpasses. A pilot project for constructing straddling buses was to be started in Beijing but appears unlikely.

Selima Sultana

The first recorded bus station was in Nantes, France, located in front of a hatter's shop. It was called "Omnès Omnibus," meaning "all for all." The name "omnibus" stuck. The shortened term "bus" first was used in London in 1829. Buses powered by steam and electricity were developed in the 1830s, and buses with internal combustion engines were developed in the late 1800s, at the time of the first motorcars.

The first true public bus service is believed to have been started in 1824 by a Pendleton, England, toll agent named John Greenwood, whose service started at Market Street in Manchester and ran to the suburbs of Pendleton, Ardwick, and Cheetham Hill. Rides on the horse-drawn carriage cost six pence for riders to be inside the bus or four pence if they rode on the outside of the bus. Only 11 passengers at a time could be transported on the bus, which ran three times a day and stopped at any point along the route to load and unload passengers. Unlike stagecoaches, riding Greenwood's bus did not require reservations. As is the case for modern-day bus stops, passengers simply waited for the vehicle to pull up for boarding. Many others followed Greenwood's lead, but the fierce competition ended in 1865, when Greenwood and the other operators created the Manchester Carriage Company.

The first recorded intercity bus travel in the United States was in 1914, when Swedish immigrant Carl Eric Wickman transported miners between Hibbing, Minnesota, and Alice, Minnesota, for $0.15 per ride. Wickman went on to found Greyhound Lines, which currently is owned by British transit corporation FirstGroup. Modern interstate fleets offer free Wi-Fi, power outlets, and increased legroom.

The first school bus was developed by George Shillibeer in London in 1827. The bus was a horse-drawn carriage that carted 25 children. The first American school buses were horse-drawn carriages manufactured by Wayne Works in 1886. These buses featured rear-entry doors so that the children exiting or entering the bus would not scare the horses. The rear door remained part of school bus designs even after internal combustion engines were manufactured. In the United States,

44 safety measures were introduced in 1939 by a conference convening at Columbia University in New York. One measure called for painting school buses a bright yellow. Today, all 50 U.S. states adhere to the law. Besides Wayne Works, other bus companies include Crown Coach, Gillig Bros., and Bluebird.

The Routemaster—the iconic red double-decker buses associated with England—were, in fact, specifically designed for use on London streets. Developed in the late 1940s and early 1950s by London Transport, and built to take on double the passengers of a standard bus, the Routemasters were manufactured by the Associated Equipment Company and Park Royal to replace the failing electric trolley buses. The Routemaster not only could seat more passengers than standard buses, but also were made of a lighter-weight material and were more fuel efficient, making them to be easy to drive and maintain. Nearly 2,900 Routemasters were built between 1957 and 1968. Because they were required to be overhauled every five years, the fleet maintained its integrity and there are still 1,000 Routemasters in existence today, although the vehicles were withdrawn from regular service in 2005. In an attempt to revive the Routemaster, in 2007 London's Mayor Boris Johnson announced a competition to update the Routemaster. The result was more than 600 redesigned Routemasters, which will enter service in 2016. Other double-decker buses are in common use as mass transit vehicles in the United Kingdom, Europe, and Asia, and in former British colonies and protectorates.

It is estimated that a bus can take 40 cars off the road, creating less pollution and reducing traffic congestion. Some bus companies even are using hybrid vehicles that run on gasoline and electricity. The 2006 CRA International report in Australia reported that if 10 percent of drivers would switch to bus transportation for a year, 400,000 tons of greenhouse gas emissions would be reduced.

Rosemarie Boucher Leenerts

See also: Chicken Buses (Central America); Double-Decker Buses; TransMileno, Bogotá (Colombia).

Further Reading

Bus Stuff. "A Brief History of the Bus." http://www.bus-stuff.co.uk/2011/01/brief-history
-of-bus.html. Accessed June 5, 2015.

Greyhound. "Historical Timeline." http://www.greyhoundhistory.com. Accessed June 5,
2015.

Hutchinson, Sean. 2013 (January 30). "Why Are School Buses Yellow?" http://mentalfloss
.com/article/33535/why-are-school-buses-yellow. Accessed June 5, 2015.

Hylton, Stuart. 2013. *Little Book of Manchester.* Gloucestershire, UK: The History Press.

Museum of Bus Transportation. http://www.busmuseum.org. Accessed June 5, 2015.

Transport for London. "New Routemaster." https://www.tfl.gov.uk/modes/buses/new
-routemaster. Accessed June 5, 2015.

C

CABLE CARS (UNITED STATES)

A cable car is an urban rail passenger vehicle that is similar to a streetcar but is powered by an underground cable rather than an electric motor and overhead wires. Cable cars do not carry their own engine or propulsion. Its cable is located in a trench between the rails and actually is a very long loop of cable moving at a constant speed and powered from a central location. To move forward the car's conductor closes a mechanical grip on the cable, pulling the cable car at the same 9 miles-per-hour speed (uphill or downhill) that the cable moves. To stop, the conductor releases the cable grip and applies wheel brakes. A car can move in only one direction on a track, therefore separate tracks and cableways on each side of the street are used. A turntable can be used to turn cars around at the end of their lines.

The San Francisco cable cars are the most famous example and are a popular tourist attraction. It often is assumed that cable cars were built in San Francisco, and are still found today, because of the steep hills of the city; this is not true. Rather, cable cars previously were found in many cities across the United States. Chicago—a city with very level terrain, once had the largest mileage of cable car lines. Cable cars were an attempt to provide a better form of propulsion than using horse or mule cars. Cable cars were built only during a short period, as the cables were replaced quickly with electricity after electric streetcars were invented.

It is perhaps fitting that San Francisco's premier transport attraction runs on wire ropes, as similar ropes are also used in the city's famous suspension bridges, including the Golden Gate Bridge. The city's cable car system was initiated in 1871 by Andrew Hallidie, an engineer with experience building suspension bridges. The construction of cable car lines up the city's steep hills opened these areas to residential development. As elsewhere, the proliferation of cable car lines in the city was cut short by the arrival of electric streetcars, which were much easier to install and were capable of varying their speeds and directions. The 1906 earthquake and fire did extensive damage to many cable car lines, giving the electric streetcars an opening to take over most of the city's transit needs. Only the steepest hills retained cable car lines, and many of these eventually were replaced by buses. In 1947, the city planned to retire the last cable cars but many residents resisted. This led to a 1971 city law guaranteeing that the three surviving lines would be allowed to continue service. These lines are popular with tourists, but serve largely as a tourist attraction rather than a way for locals to get around.

Joe Weber

Cable car in historic Chinatown in San Francisco, California, 2012. (Library of Congress)

See also: Funiculars; Stairways; Streetcars (United States); Suspension Bridges.

Further Reading

Thomas, Lynn. 1992. *San Francisco's Cable Cars.* San Francisco: H. S. Crocker Co.

CABLE-STAYED BRIDGES

Cable-stayed bridges are a rapidly growing but relatively recent bridge design. They are similar to suspension bridges in that a roadway is suspended from cables, but differ in that the cables run directly from the towers to the deck rather than being hung from larger cables. The cables from the towers support the weight of the bridge deck and are called "stays." These bridges can have one or more towers from which the cables are hung. A two-tower bridge most closely resembles a suspension bridge.

The first cable-stayed bridge in the United States opened in Sitka, Alaska, in 1972. Because of its relative newness the bridge design was not available when the Interstate Highway System was being built. The design, however, has been widely used for newer bridges and for replacing older highway bridges in the United States. The Sunshine Skyway across Tampa Bay is one of the best known of these bridges. Opened in 1987, it replaced a steel cantilever bridge that collapsed. The dramatic bridge has a main span that is 1,200 feet long with towers that are 430 feet tall. Since 1980, many cable-stayed bridges have been built over the Ohio River, all replacing steel cantilever or suspension bridges. As more bridges from the Interstate Highway System require replacement due to age, it is likely that cable-stayed bridges will become more and more familiar to Americans.

Much of the development of cable-stayed bridges has taken place in Europe and East Asia, and it is there that most of the longest bridges are found. Only 4 of the 100 longest bridges are located in the United States. The current record holder as the longest bridge is Russky Bridge in Petropavlovsk, Russia. This has a main span of 3,622 feet and was opened in 2012. Seven of the 10 longest bridges are located in China, as are many of the top 100; and bridges are increasingly common on the China National Trunk Highway System and new expressways elsewhere in Asia and Europe.

Not all cable-stayed bridges are for highways. The Sundial Bridge across the Sacramento River in Redding, California, is for pedestrians and bicycles. The bridge takes its name from its single tower, which serves as the world's largest sundial. The bridge, which opened in 2004, was designed by Santiago Calatrava (1951–), a world-renowned architect and designer of many spectacular bridges around the world. The Reedy River pedestrian bridge in downtown Greeneville, South Carolina, is another well-known example that is both a form of transportation as well as a local attraction.

Joe Weber

See also: Darjeeling Himalayan Railway (DHR) (India); National Trunk Highway System (China); Suspension Bridges.

Further Reading

Whitney, Charles S. 2003. *Bridges of the World: Their Design and Construction*. New York: Dover.

CAMEL TRAINS (MALI)

One of the most interesting aspects of Mali—with its mix of cultures and people—is its use of camel trains as a mode of transportation. A camel train basically is a group of camels that carries goods between points. It is an important mode of transportation for the salt trade in Mali. A solid slab of salt can be sold for $3 at the northern Taudenni salt mines, but the same amount can be sold in southern markets—such as Timbuktu—for $12.

Camels are used to move the salt from the Taudenni salt mines of the northern regions to southern markets because the camels are well matched to the desert conditions. The environment in which the camel train operates is a harsh one. Much of Mali is located in the Sahara Desert, which is the largest and one of the driest deserts on earth. Summer temperatures can reach up to 120 degrees Fahrenheit. Camel

Nomadic herders look on as a southbound camel caravan, heavily laden with blocks of salt, passes en route to Timbuktu, several days' journey to the south, and about 10 days' journey from the salt mines of Taoudenni to the north, in the Sahara Desert, Mali, 2000. (AP Photo/ Brennan Linsley)

trains carrying salt have to cross this sandy terrain of dunes where there is little access to water. On average, many of the camel trains must travel more than 450 miles carrying salt blocks that weight nearly 40 pounds each. It would be nearly impossible for humans to complete this task on their own. Camels can withstand extreme temperatures such as the hot days and cold nights of the desert. They can last up to 30 days without water and they actually have double rows of long eyelashes that help keep the sand out of their eyes. Camels primarily are used as pack animals, which makes them perfect for carrying heavy loads of salt.

It takes anywhere from 14 days to a month to get from salt mines such as Taoudenni to Timbuktu. Although typically a large group of men is involved in a camel train, but there is only one leader. The leader must have excellent tracking abilities and a special ability to read the desert—including the winds, the sands, the sun, and the stars—as making a navigational mistake of even a few miles can lead to certain death for the whole group. Another valuable skill is the ability to know where there could be small areas of water or shade—an ability learned from experience with the desert. Almost all of these men find their way much like the people did thousands of years ago, with no technology and only their instincts and experience to guide them.

The availability of access to motorized vehicles such as lorries and 4x4 trucks have impacted the camel trade. Although camel caravans take close to a month to cross the desert, these trucks need only a few days to make the same crossing. The trucks also can carry more loads to the market than the camels can. Although trucks help the economic side of the salt trade by reducing prices and enabling a quicker turnaround, they have had a negative effect on the culture of the people. Many people believe a camel train to be a kind of pilgrimage. Others believe that the trucks will drive out what remains of the camel trade and take with it a large part of the history and culture of the land. The decline has been apparent. As recently as 100 years ago it was not uncommon to see a caravan of 100 camels. Now, many camel trains are reduced to closer to 20 camels.

Selima Sultana

See also: Camels; Chicken Buses (Central America); Chiva Express (Ecuador); Jeepneys (Philippines).

Further Reading

Pilkington, John. 2006 (October 21). "Dying Trade of the Sahara Camel Train." *BBC NEWS, Mali.* http://news.bbc.co.uk/2/hi/programmes/from_our_own_correspondent/6070 400.stm. Accessed June 30, 2015.

Rainier, Chris. 2003 (May 28). "In Sahara, Salt-Hauling Camel Trains Struggle On." *National Geographic News.* http://news.nationalgeographic.com/news/2003/05/0528 _030528_saltcaravan.html. Accessed June 30, 2015.

CAMELS

The camel (genus *Camelus*) has been used as a form of transportation for centuries. Their ability to survive for extended periods without food and water and their strength for carrying loads make them the perfect animal for their desert habitats. The long-necked animals can carry 200 pounds across 20 miles of harsh desert terrain and have been nicknamed the "ships of the desert." Camels are swift and can travel as fast as horses—reaching speeds of 40 miles per hour.

There are two types of camels: one-hump dromedary camels and two-hump Bactrian camels. About 90 percent of camels worldwide are dromedaries (also known as Arabian camels). Nearly all dromedaries are domesticated animals, whereas the Bactrian (or Asian) camels can be either wild or domesticated. Dromedaries generally are 6.5 feet tall at the shoulder and weigh between 880 and 1,325 pounds. Bactrians range between about 5.6 feet and 6 feet at the shoulder and weigh between 992 pounds to 2,200 pounds. Wild Bactrians are smaller than domesticated Bactrians.

Several features make camels perfect for transportation in hot, arid climates. The camel stores its fat in its humps, which keeps the fat from being displaced throughout the body, and minimizes the insulating effect of body fat. The stored fat also is an energy source; when the fat tissue metabolizes, it yields 1 gram of water for every gram of fat processed. A camel can go without water for about a week and can survive without food for several months. A camel rarely sweats, but can lose 25 percent of its body weight through perspiration before experiencing dehydration as compared to the average mammal, which can withstand only a 12 to 14 percent loss. Camels also produce concentrated urine and dry dung, which conserves moisture. The camel's red blood cells are oval in shape instead of the typical round shape, facilitating the flow of red blood during dehydration and making the blood cells less able to rupture when camels drink significant quantities of water all at once. A camel can drink 26 to 40 gallons of water in one sitting, and the water is processed so quickly that the camel's stomach can be empty again within minutes.

Camels' external features also aid in their ability to travel in the desert. Camels can see in desert sandstorms, for example, because their eyes have a thin nictitating membrane, a clear inner eyelid that protects their eyes from sand. Also protecting their eyes are double rows of very long eyelashes that provide shade and keep out sand. Fur inside their ears shields them from sand. They can close their nostrils to the swirling sand, and the nose traps moisture from the breath and absorbs it in the nasal membranes. Camels have two-toed feet, not hooves, and the feet are large, broad, and heavily callused. The feet expand when they touch the ground and contract when lifted, making the camel able to move nimbly through sand and snow. Camels move both legs on one side at the same time, and this, plus the broadness of the feet, prevents the camel from sinking into sand or snow. Camels' long legs keep their bodies away from the hot ground too.

Dromedaries are native to North Africa and the Middle East. They also can be found in Australia, but these animals are feral and were brought to Australia in the mid-to-late 19th century to be used as pack animals in the desert regions of that continent. About 10,000 to 12,000 camels were imported to Australia between 1860 and 1907. Bactrian camels are native to the Gobi Desert in China and the Bactrian steppes of Mongolia. They grow a shaggy coat in winter to protect themselves from freezing temperatures and then shed the coat in the warmer months. They can survive in temperatures ranging from as low as 20 degrees Fahrenheit to as high as 120 degrees Fahrenheit.

Camels are clever at finding the food they need to survive in sparse regions. Their upper lip is split, which helps them move close to the ground to eat the shortest grasses. Their lips also are tough and their teeth are extremely sharp, making them able to break off and eat vegetation that other animals might avoid because of the plants' thorns or salty flavor. Camels also are ruminants and regurgitate the food back up from the stomach to chew it again, which aids in digestion.

Based on ancient carvings in what today is the country of Jordan, scholars have traced the origins of the camel back to the Stone Age. In these carved representations, the animal appears to have a single hump. The camel apparently was domesticated in North Africa. In the ruins of Tall Halaf in Iraq, however, there is a picture of a camel being used as transportation that dates from approximately 3000 BCE.

Camels have been used for transportation for thousands of years, but before that the camel was a source of milk and meat. Evidence of dromedary camels having been domesticated dates back 3,500 years. The lives of the Bedouin of Arabia revolve around the camel, much as did the lives of their predecessors, the Nabataeans. For centuries, these nomadic cultures have moved in search of water and land to grow their food. They use camels for everything from transportation to shelter to food. They also earn their income by transporting goods and people across the desert in camel caravans.

The first camel saddle likely was made of blankets or mats. A saddle on a Bactrian camel is placed between the two humps and saddles on dromedaries can go either in front of the hump, on the hump, or behind the hump. Saddles come in various shapes and sizes depending on where they are to be placed, and are used both for riding and for carrying loads. It is estimated that a camel can carry three times the weight that a horse can carry.

Camels have been used by armies dating back to Arab king Gindibu, who employed camels at the Battle of Qarqar in 853 BCE. King Cyrus of Persia also used camels in 547 BCE in his battle against the Lydian cavalry; and the camel soon became commonplace in wars in Arabia and North Africa. Camel corps were used throughout history, including by the U.S. Army prior to the Civil War. The army imported camels from Algeria, Tunisia, Egypt, and present-day Turkey at the request of Secretary of War Jefferson Davis.

Rosemarie Boucher Leenerts

See also: Elephants; Llamas.

Further Reading

Baum, Doug. 2013 (April 30). "The Camel Saddle: A Study." Paper presented at The Camel Conference, SOAS University of London. http://www.soas.ac.uk/camelconference 2013/file88887.pdf. Accessed June 5, 2015.
San Diego Zoo. "Mammals: Camel." http://animals.sandiegozoo.org/animals/camel. Accessed June 5, 2015.

CAR SHARING

Car sharing is a form of short-term car rental allowing users access to cars on short notice and during off-hours. Users pay for a car share on an hourly or per-minute basis as opposed to a 24-hour period or a weekly basis, as is the case with traditional rental companies. Car sharing reduces the need for private car ownership. Hence, car sharing is most attractive to users living in densely populated areas where parking regulations and the increased cost of insurance make ownership prohibitive. A car-sharing customer typically only uses a car occasionally or has a need to use different types of vehicles on short notice. Car-sharing organizational structure in North America can take three forms.

1. For-profit: Operated by large for-profit companies; for example, Flexcar, Zipcar, Hertz on Demand, Enterprise CarShare.

2. Non-profit: Programs incorporated as tax-exempt organizations; for example, City Car Share (San Francisco), and PhillyCarShare (Philadelphia).
3. Cooperative: Car share operators run by its members who own a "share" in the organization.

Car-sharing services are accessible worldwide and available in many cities. As with many transportation topics, the international nature of car sharing causes some confusion regarding terminology. In North America, the common terms for short-term auto rental are "car sharing," "car-sharing," and "carsharing." In the United Kingdom, the service often is referred to as a "car club." The North American term "car sharing," however, has been gaining popularity in the United Kingdom, perhaps due to the entry of large American car rental companies such as Avis and Hertz into the market. Different terms exist in other regions of the Anglosphere. In India, for example, the term "self drive" refers to "car sharing."

The car-sharing organization cares for the vehicle and manages issues such as registration, inspection, insurance, and parking. The advantage of members of a car-sharing organization is that they dispense with these routine aspects of vehicle ownership and only pay an hourly rate for use of a car. Typically, members pay only a one-time registration fee of about $35 and—depending on the organization or plan selected—pay an hourly or per-minute rate. The per-minute rate for a using a Car2Go mini compact car in Washington, DC, for example, in 2013 was $0.38 per minute (Steinberg and Vlasic 2013).

A member begins the process by making a reservation for car. Reservations can be made in advance or on short notice. Depending on the company, this could be done by telephone, smartphone application, or online via the company's website. The majority of North American organizations use advanced or partially automated systems with supporting software packages. Some smaller operators use manual operations such as key boxes and log books. Members are prompted to answer questions such as time the car is required, the length of rental time, the location of pickup, and the type of car needed.

When the member arrives at the location of the reserved vehicle, he or she passes a membership card over an electronic reader (usually located on the windshield inside the car), and the card unlocks the car (but only if it has been reserved for the specific member). The ignition key usually is in the vehicle's glove box or elsewhere inside the vehicle. As with traditional car rental, the member should check for any vehicle damage and report it to the car-share organization before using the vehicle. Once the member finishes using the car, he or she returns it to the same location where it was picked up. Some car-sharing organizations allow drop-off at different locations or anywhere within a specified zone of a city. Gasoline is reimbursed to the member or a gas card is provided with the car. Members are responsible for any traffic or parking fine incurred while they are using a vehicle.

The concept of short-term car rentals began in 1948 in Switzerland where car ownership was prohibitive during the postwar recovery period. Early programs in

Europe started in France with "Procotip" (1971); "Witkar" in the Netherlands (1973); "Green Cars" in the United Kingdom (late 1970s); and "Vivalla Bil" in Sweden (1983). In the United States, Purdue University conducted research on car sharing from 1983 to 1986. The Short Term Auto Rental (STAR) was established in San Francisco by a private company and ran from 1983 to 1985. A common thread among these early programs is their quick failure rate—usually attributed to poor planning. Of the previously mentioned programs, only San Francisco–based STAR had positive results from the consumer's perspective.

Car sharing in its current form has roots in Switzerland and Germany, with programs dating to the late 1980s. Mobility Switzerland (1987) is the first sizeable car-sharing company based on the present model. Mobility Switzerland still is a leading car-share program in Europe. Statt Auto Berlin followed one year later (1988). Shortly thereafter, car-sharing programs began to proliferate throughout Europe. By 2004, there were 70,000 car-sharing members in Germany and 60,000 in Switzerland (Transportation Cooperative Research Program 2005). By 1994, the concept of car sharing arrived in North America in Quebec City with Auto-Com, which later became Communauto. In the United States, the first large-scale program—CarSharing Portland—grew rapidly from its beginning in 1998.

As of today, there are 20 active car-sharing programs in Canada and 23 in the United States with more than 1.5 million car-sharing members and 22,000 vehicles available to them. The four largest providers serve more than 95 percent of the membership (Shaheen and Cohen 2015).

Joe Di Gianni

See also: Bicycle Libraries; Hitching; Mopeds; Slugging (United States).

Further Reading

Self Drive. Self Drive Home Page. https://www.selfdrive.in/. Accessed January 9, 2015.

Shaheen, Susan A., and Adam P. Cohen. 2015. "Innovative Mobility Carsharing Outlook: Carsharing Market, Overview, Analysis, and Trends." http://tsrc.berkeley.edu/sites/default/files/Summer%202015%20Carsharing%20Outlook_Final%20(1)_0.pdf. Accessed January 4, 2016.

Steinberg, Stephanie, and Bill Vlasic. 2013 (January 25). "Car-Sharing Services Grow, and Expand Options." *New York Times*. http://www.nytimes.com/2013/01/26/business/car-sharing-services-grow-and-expand-options.html. Accessed January 9, 2015.

Transportation Cooperative Research Program. (2005). *Car-Sharing: Where and How It Succeeds.* TCRP Report 108, Transportation Research Board. Washington, DC.

CARMELIT (ISRAEL)

The Carmelit is an underground funicular railway in Haifa, Israel, that carries passengers on a 1.1-mile line from the lower city up Mount Carmel. The system services 6 stations and rises 886 feet. Total travel time from the lowest station (Paris Square) to the highest (Gan Ha'em) is usually 8 to 10 minutes.

Although commonly referred to as a "subway," the system is more of an underground elevator than a subway system like those of New York or Tokyo. The Carmelit consists of two sets of two cars connected by a cable-and-pulley system. The upward and downward cars run simultaneously in opposite directions on a single track with a short section of double track in the middle of the line that allows the upward and downward cars to safely pass each other. Engines that move the cables are located at both ends of the line and are controlled from a center in the Gan Ha'em station. The stations and the cars are located on a steep incline. Like some lines of the Paris Metro, the original Carmelit ran on rubber tires. A renovation in the 1990s converted the line to steel wheels on steel rails.

The Carmelit has six stations: Gan Ha'em (at the top of the mountain), Bnai Zion, Massada, HaNevl'im, Solel Boneh, and Paris Square (renamed in honor of the French company that built the original system). The underground stations also occasionally have served as shelters from rocket attacks.

The Carmelit was built by a French company under the leadership of Haifa's mayor, Abba Hushi, and opened to the public in October of 1959 at a ceremony that included Prime Minister David Ben-Gurion. The original system ran safely until it was shut down in 1986 due to concerns about the old, worn equipment. The system underwent a complete renovation in the early 1990s by the Swiss company Von Roll, with US$27 million spent to replace the pulley mechanism and cars, renovate and remodel the tunnels and stations, and install safety and surveillance systems. The upgraded system reopened to the public in August 1992.

When the Carmelit opened, it connected what was then the city's main centers—the industrial lower city at the foot of the mountain, the historic commercial center of Hadar in the middle, and the residential areas of Central Carmel at the top of the line. As industrial activity in Haifa declined over the years of the system's operation, however, fewer people had a reason to use the system. Ridership steadily declined from a peak of as many as 21,000 riders per day in the early 1960s to 11,000 riders per day when the system was closed in 1986. At the time of the renovation, it was estimated that 15,000 passengers per day would be needed for the system to pay for its own operations using fares. By 1995, ridership was down to around 5,000 riders per day compared to maximum capacity of approximately 3,000 passengers per hour. Ridership in 2010 was around 2,000 passengers per day, with many of those passengers being tourists riding the system primarily for enjoyment. Operations have been able to continue with a subsidy from the city.

Expansion of the system has been considered numerous times over the years, but the significant cost of such an expansion has prevented plans from going forward. More promising efforts are being made redevelop the downtown and western port area into a center of tourism and night life, along with parks, a beach promenade, and renovated commercial buildings. This would make the lower city a more attractive destination for both residents and tourists and would increase passenger demand for the Carmelit.

Michael Minn

See also: Darjeeling Himalayan Railway (DHR) (India); Elevators; Funiculars; Ski Lifts.

Further Reading

Buzi, Miriam, and Daniel Shefer. 1978. "The Carmelit: Transit Characteristics and Mode Choice Determinants." *Traffic Quarterly* 32(1): 145–167.

"Hacarmelit: The Subway of Haifa." http://www.carmelithaifa.com. Accessed June 5, 2015.

Lazaroff, Tovah. 2006 (July 23). "Haifa Residents Struggle through Rocket Barrages." *Jerusalem Post.*

Rudge, David.1990 (November 21). "Carmelit Subway Restoration Begins." *Jerusalem Post.*

Schrag, Carl. 1995 (February 10). "An Underground Movement Fights Traffic Jams in Haifa." *Jerusalem Post.*

Shefler, Gil. 2010 (April 9). "The Line That Time Forgot." *Jerusalem Post.*

CAR2GO

Car2Go bills itself as an "innovative mobility service for urban areas." A subsidiary of German automaker Daimler AG—manufacturer of the Mercedes Benz and Smart Fortwo ultra-compact automobiles used as Car2Go vehicles—the Car2Go format is a shared-use, on-demand system of transportation. Available in large urban areas

Car2Go is an eco-friendly car-sharing company that has small parking stations throughout service areas, where drivers can rent vehicles from its Smart Fortwo fleet. (Howesjwe/Dreamstime.com)

around the globe since 2008, Car2Go's concept features one-way point-to-point rentals and, for spontaneous urban dwellers, is an alternative to prearranged rental agencies.

Interested Car2Go users register online and pay a one-time fee in advance. If the prospective member has a clean driving record, an electronic card the size of a credit card and encoded with information about the driver is issued. Instead of the centralized office utilized by most rental car agencies, Car2Go customers can search for available vehicles nearby via a downloadable smartphone application. Cars can be reserved up to 30 minutes in advance but reservations are not required. Upon finding a nearby Car2Go vehicle available for rent, the user waves his or her electronic card in front of a card reader located behind the windshield. The driver-side door unlocks and the driver can access the vehicle's key once inside.

The renter may drive anywhere but the vehicle must be returned somewhere inside of the designated urban zone—or "home area"—in which the vehicle was picked up. When reaching his or her destination, the driver must park in any legal parking space or designated Car2Go lot. The car can be left for the next Car2Go driver to rent, or the driver may keep the key and return to the vehicle to continue his or her trip. A vehicle that is in use does not appear on the app as an available vehicle. In addition to a low fixed annual fee paid by the driver, there is a per-minute rate, with discounts for hourly or daily usage. Rates cover all costs including the rental, fuel, insurance, and maintenance.

The Smart Fortwo vehicles used by Car2Go are painted white with blue accents and have the Car2Go logo emblazoned on the hood. Models are either gasoline or electric powered. Fuel cards are left inside the cars in the event drivers need to refuel the vehicle while in use. Before renting a Car2Go, the driver can check the fuel level of the car online. If drivers anticipate a long drive or rent a vehicle that is low on fuel or has a low electric charge, then if they refuel or power up the cars they are awarded free usage minutes. Drivers are required to power up electric cars that have less than a 20 percent charge. A touch-screen navigation system inside the vehicle displays charging station locations. The 3-cylinder, 1.0-liter gasoline-powered cars have a combined 36 miles per gallon (mpg), and the electric cars reach a combined 107 mpg. Electric cars can drive 84 miles on a single charge. Cities offer both electric- and gasoline-powered fleets, gasoline-only fleets, or (as in San Diego, California, where fuel tends to be expensive) all-electric fleets.

The first U.S. Car2Go program began in 2011 in Washington, DC. As of late 2015, more than 1 million members participated in the Car2Go service offered in the following cities: Austin, Texas; San Diego, California; Portland, Oregon; Washington, DC; Miami, Florida; Seattle, Washington; Los Angeles, California; Minneapolis and Saint Paul, Minnesota; Columbus, Ohio; Denver, Colorado; and Brooklyn, New York. Outside of the United States, the service is offered in Canada in Vancouver, British Columbia; Toronto, Ontario; Calgary, Alberta; and Montreal,

Quebec; and in Germany, it's available in the cities of Ulm (where the service originated), as well as Düsseldorf, Hamburg, Berlin, Cologne, Stuttgart, Munich, and Frankfurt. Additional European markets include Amsterdam, Netherlands; Vienna, Austria; Lyon, France; London and Birmingham, England; Milan, Rome, and Florence, Italy; and Copenhagen, Denmark. The size of the fleet varies by locale. Saint Paul rents just 185 vehicles, for example, but Berlin rents 1,200—the largest fleet worldwide.

Rosemarie Boucher Leenerts

See also: Alternative Fuels; Bicycle Libraries; Electric Vehicles; Personal Rapid Transit, West Virginia (United States); Plug-In Hybrids.

Further Reading

Auto Rental News. 2014 (June 30). "Seattle Is Fastest-Growing U.S. Car2Go City with 500 Vehicles." http://www.autorentalnews.com/channel/rental-operations/news/story/2014/01/seattle-is-fastest-growing-u-s-car2go-city-with-500-vehicles.aspx. Accessed June 5, 2015.
Car2Go. 2015. https://www.car2go.com. Accessed June 5, 2015.

CHANNEL TUNNEL (FRANCE AND UNITED KINGDOM)

The 31.4-mile-long Channel Tunnel beneath the English Channel connects England and France and is the world's longest undersea railway. This daring project is a 19th-century feat completed just before the start of the 21st century. The idea for a connector between England and France first was proposed in 1802 by Albert Mathieu. He believed this would bring the two nations together after years of war and hatred. The idea was buried after only one year until the French engineer Thomé de Gamond proposed an underground train system that seemed far more practical than Mathieu's underground carriage tunnel lit by candles. Serious construction on the tunnel did not begin until 1881, but was yet again put on hold by the British two years later due to paranoia about invasion by the French. Conversation about the construction of Channel Tunnel resumed 100 years later, in 1980.

Although it was apparent that England and France were having difficulties compromising on how to fund the construction of the Channel Tunnel, Margaret Thatcher (then prime minister of England) and François Mitterand (then president of France) were able to collect private funds from companies and via donations to make the tunnel a reality. Thatcher and Mitterand then invited the public to create and submit designs for the tunnel. The winning design was by a group called the Channel Tunnel Group-France Manche. The design included three tunnels: two outer tunnels each containing a single, one-way railroad track with a 99-mile-per-hour speed limit for transportation of people and goods, and a central tunnel joined at intervals with the two outer tunnels for maintenance and safety. All three tunnels are built 44 yards below the seabed. The plan was to dig through the chalk stratum layer of sediment under the channel. This layer had a high clay content making it water resistant. Construction moved much more quickly on the English side due to easy access to the chalk stratum layer. Construction on the French side moved much slower because of the lack of clear sediment layers. The excavation finally was completed in June 28, 1991, with the help of 5,900 underground workers. The tunnel did not officially open for use and transportation until 1994, after eight years of construction work with $21 billion cost and involving 13,000 technicians, engineers, and workers.

The Channel Tunnel is a modern marvel. The American Society for Civil Engineers named the tunnel one of the Seven Wonders of the Modern World, along with the Empire State Building, the Panama Canal, and the Golden Gate Bridge. The tunnel is the only physical connection Great Britain has to the mainland of Europe. Each day an average of 50,000 passengers, 54,000 tons of freight, and 400 trains use the tunnel. Without the Channel Tunnel, transportation between Great Britain and the mainland would have to rely on ferries and the choppy, unforgiving waters of the English Channel. The tunnel not only has eased transportation for millions of people, it has also created an emotional bridge and a true relationship between Great Britain and France.

Selima Sultana

See also: Border Tunnels (Mexico and United States); Seiken Tunnel (Japan); Stormwater Management and Road Tunnel (SMART Tunnel) (Malaysia); Tünel (Turkey).

Further Reading

Chou, Michael. 2015. "A Railway under the Ocean: The Channel Tunnel Linking Britain and France." *Illumin* 16(1). http://illumin.usc.edu/printer/172/a-railway-under-the-ocean-the-channel-tunnel-linking-britain-and-france/. Accessed July 2, 2015.

Darian-Smith, Eve. 1999. *Bridging Divides: The Channel Tunnel and English Legal Identity in the New Europe*. Berkeley: University of California Press.

CHICAGO "L" (UNITED STATES)

The rapid transit system serving the city of Chicago, Illinois, and surrounding communities is affectionately known as the Chicago "L," for elevated train and referring to the majority of track in the system that is raised above ground level. With average weekday ridership of more than 725,000 people, the "L" is the third busiest mass transit system in the United States, surpassed only by the New York City Subway System and the Washington Metro in the District of Columbia. The "L" is operated by the Chicago Transit Authority (CTA) and has 145 stations situated along eight lines of track. It is one of just four U.S. heavy-rail rapid transit systems that run select cars 24 hours per day. The "L" is credited with fostering the growth of the inner core of Chicago.

Chicago is a sprawling city of 234 square miles. Throughout its history, various areas of the city have been serviced by streetcars, diesel buses, electric trolleys, subways, and elevated railways. Even within this one mode of transportation there have existed several operators. The "L," for instance, was not started as one system but rather as part of 70 companies that eventually merged. The first rapid transit line was produced by the Chicago and South Side Rapid Transit Railroad Company in 1888. Opening in 1892, the South Side Rapid Transit began at a terminal at Congress Street and ended at 39th Street—a distance of 3.6 miles. It traveled in a straight path along city-owned alleys and became known as the "Alley 'L.'" Traveling through alleys circumvented the need for the transit system to secure consent from property owners along the route, a requirement put forth in the Cities and Villages Act of 1872. Also quieting the opposition was the impressive fact that the "Alley L" could travel 34 blocks in less than 10 minutes. Originally a steam locomotive, the original "L" cars later ran on an electrical "third rail."

The first expansion of the "Alley L" occurred in 1890, when the trains were run to the fairgrounds at the World's Columbian Exposition at Chicago's Jackson Park. To enable trains to make turns, curved rail lines were added for the first time. Competing privately owned rail lines soon were built throughout the city by several companies, including the Lake Street Elevated Railroad Company, which began operations in 1893; the Metropolitan West Side Elevated Railroad Company and the Union Elevated Railroad Company, both operating by 1895; and the Northwestern Elevated Railroad Company, completed in 1899. These private concerns built the infrastructure of today's CTA lines that are known as the Red, Blue, Green, Brown, and Pink lines. These separate lines reached most Chicago neighborhoods and the outlying undeveloped areas by 1909, and in 1924 they merged to form the Chicago Rapid Transit Company (CRT). Financial distress—including bankruptcy—affected the CRT, however, and the U.S. District Court was appointed to manage operations, offering subsidies to bankroll the agency. Despite financial troubles, at the time the CRT still was able to carry more than 600,000 passengers on 227 routes daily.

Private ownership proved unprofitable for years to come, and in 1947 the transit systems were incorporated into a public municipal organization, the Chicago Transit Authority (CTA). The CTA immediately began improving the system, which was a tangled web of multiple lines, some of which ran on the same tracks and crossed each other's paths. To clean up the system, seven rail lines were removed and physical infrastructures were demolished. At the time, the CTA also assumed control of other modes of public transit including highways, which were being built to move urban dwellers to outlying suburbs. Improvements continued throughout the years, with air-conditioning being introduced in 1964, and the last of the pre–World War II cars was retired in 1973, the same year that new lines were built in expressway medians.

Today's "L" is composed of eight rapid transit lines laid out in a spoke-hub pattern, with transit focused mainly in the central business district of Chicago, known as "the Loop." Many of the 224 track miles are elevated, but trains also travel through subways, embankments, and at grade level. Average "L" ridership was nearly 230 million in 2013, with 726,000 rail miles traveled daily along 145 stations.

The "L" had a stable ridership for 40 years, until the Chicago Loop Flood in April 1992, which released 250 million gallons of water through a damaged utility tunnel beneath the Chicago River. The flood suspended "L" operations, with the most heavily traveled lines affected for several weeks, dropping ridership to 418,000 that year. Since that time, weekday ridership has increased about 25 percent.

The newest rail line is the 11.2-mile Pink Line, which began operating in June 2006. Improvements on the other lines continue. In his Sustainable Chicago 2015 plan addressing residents, Mayor Rahm Emanuel set out ways that the city could save on energy and improve its infrastructure, including transportation. Transit-related goals included increasing daily transit ridership by rebuilding Red Line service, modernizing the Red and Purple Lines, and completing reconstruction of Wilson station; accelerating transit-oriented development, and accelerating high-speed passenger rail projects. Other "L" projects include expansion of the Brown Line and Slow Zone Elimination, and replacing aging rail ties, plates, and tracks.

Rosemarie Boucher Leenerts

See also: Light Rail Transit Systems; Rapid Transit; Stairways.

Further Reading

Chicago Transit Authority. http://www.transitchicago.com. Accessed June 5, 2015.
City of Chicago. 2015. "Sustainable Chicago 2015." http://www.cityofchicago.org/city/en/progs/env/sustainable_chicago2015.html. Accessed June 5, 2015.
Eanich, Terence. 2011 (July 2). "Remnants of the 'L.'" *Forgotten Chicago.* http://forgottenchicago.com/features/remnants-of-the-l/. Accessed June 5, 2015.

CHICKEN BUSES (CENTRAL AMERICA)

Chicken buses are a mechanical innovation found in Central America. They are old American school buses sold at auction after 10 years of service in the United States. These vehicles then are converted into distinctively decorated colorful buses that serve as a major transportation system in Central American countries such as Honduras, Guatemala, El Salvador, Nicaragua, and Panama. The "chicken bus" name originated from the common practice of transporting live chickens, and also from passengers being crammed in the bus like the chickens. These buses carry people across the countryside, deliver chickens to the city, and bring cash crops to market—essentially connecting the regions and rural communities with cities. In

A multicolored chicken bus in Guatemala. (Brian Eberly/Dreamstime.com)

addition to the driver, each chicken bus might have one or two helpers or assistants, who usually do not have training in driving and are in charge of collecting bus fare and handling luggage, livestock, produce, and anything else that needs to be organized. Announcing arrival areas and destinations is also the responsibility of the helpers. There is no rush to keep the buses on schedule, as they do not run at any regular times.

The chicken buses are decorated in vibrant colors, which makes them stand out at any bus stop or terminal in Central America. The colors of the buses reflect the personal flare of the owners or drivers—sometimes reflecting national pride, religion, or a personal love for something. With their bright colors and the murals and names of the route painted on the buses, they definitely do not go unnoticed. They sometimes are a bit risqué, and you might see a religious reference alongside a sticker or photo of a "hot/sexy" woman. Chicken buses also have railings added to the roof, specifically for holding luggage or anything else that must be strapped to the top of the bus.

In addition to transporting goods, people, and luggage, chicken buses commonly carry live animals. This is not unusual in the local context, but might seem unusual to tourists from developed nations. When riding a chicken bus, tourists should not be surprised at sharing the ride with goats and poultry—the latter often being carried in a basket on a lap or on a rack above a seat.

Chicken bus drivers usually drive fast and do not wait for passengers to find seats before the bus starts moving. The drivers (or helpers) typically insist that passengers make room for additional passengers, they pass other buses dangerously, and they slam on the brakes when making complete stops. Although the fast-paced driving style might seem inconsiderate to most guests, many bus drivers still are very considerate and will stop for anyone on the side of the road even if the bus is packed full. Although visitors from developed countries might be horrified by this fast-paced driving style at first—especially when people literally are run over in streets by reckless driving or a wheel falls off of a bus—but, at the same time, visitors might be surprised to see skillful driving in an inadequate street network.

The lack of safety regulations are certainly huge concerns of chicken bus riders. The buses do not go through any type of safety inspections, and as a result many buses are in poor mechanical condition. Many of these buses' emergency doors are painted shut, for example, or they might not have doors, and helpers might be standing in an open doorway. These circumstances are nothing special to locals, but they certainly do violate the safety standards of the developed world. Despite this, the physical attractiveness of the buses is appealing to tourists, and the buses provide another interesting experience for visitors to Central America.

Selima Sultana

See also: Buses; Jeepneys (Philippines); Mammy Wagons (West Africa).

Further Reading

Gilliam, Thane. 2014. "Travelling in Guatemala by Chicken Bus." SaltySailors.com: Living and Cruising Aboard a Boat. http://www.saltysailors.com/articles-cruising/chicken-bus.html#.U5uBXUS53rM. Accessed June 13, 2014.

Kaushik. 2012. "Chicken Buses of Guatemala." Amusing Planet. http://www.amusingplanet.com/2012/10/chicken-buses-of-guatemala.html. Accessed December 2, 2015.

Stephanie. 2011 (April 3). "Chicken Bus Controversy in Panama." *The Travel Chica* (blog). http://www.thetravelchica.com/2011/04/chicken-bus-controversy-in-panama/. Accessed December 2, 2015.

CHIVA EXPRESS (ECUADOR)

The Chiva Express is an extremely popular mode of transportation in Ecuador, and draws people in from all over the world. The Chiva Express is a multicolored bus or single-coach train with plenty of comfortable seats on the roof that provide open-air sightseeing experiences. It runs on tracks that once were used by steam engine trains. Its route is surrounded by the mystical beauty of the Andean region and the coastal lowlands of Ecuador. Visitors on the Chiva Express quickly realize this is not your average sightseeing trip. The bus itself offers toilets and a bar. The open roof on the bus is for guests to enjoy breathtaking mountains and dense forest and a sense of Ecuadorian culture in fresh air—which is an unforgettable experience for many. The Chiva Express stops in forgotten villages along the tracks where tourists can see and buy locally made crafts, eat local food, and watch the villagers dance. Along the way, the Chiva Express comes to a part of the track known as the Devil's Nose, which is wildly popular among visitors because the bus goes backward down that portion of the track.

Selima Sultana

See also: Bamboo Trains (Cambodia); Chicken Buses (Central America); Jeepneys (Philippines); Mammy Wagons (West Africa).

Further Reading

Castro, Eric. 2015. "Chiva Express Train Tour." Streetdirectory.com. http://www.street directory.com/travel_guide/211442/adventure_travel/chiva_express_train_tour.html. Accessed July 2, 2015.

COAST STARLIGHT (UNITED STATES)

The Coast Starlight is one of Amtrak's long-distance passenger trains, running 1,377 miles between Los Angeles and Seattle in 36 hours. The train runs daily between Los Angeles's spectacular Union Station and Seattle's King Street Station, with stops at Santa Barbara, San Luis Obispo, San Jose, Oakland, Sacramento, Redding, Klamath Falls, Portland, and numerous smaller cities. Between Santa Barbara and San Luis Obispo the line runs along the coast. This section passes through Vandenberg Air Force Base, a busy satellite launch site. North of Oakland, California, the line runs north through California's Central Valley. Once past Redding, California, it enters the Sacramento River canyon. Near Dunsmuir it passes spectacular Mt. Shasta. Much of the route in Oregon passes through the Willamette Valley with views of Mt. Hood and other Cascade Range volcanoes. Crossing the Columbia River on a 1908 drawbridge, the train passes Kelso-Longview, Olympia, and Tacoma, Washington, before arriving in Seattle. Passengers traveling to and from San Diego, California, can use the Amtrak Surfliner train to connect to the Coast Starlight in Los Angeles, California. Union Station serves Amtrak's Southwest Chief and Sunset Limited trains as well as commuter trains and subways within Los Angeles. At the Seattle terminus of the Coast Starlight, travelers can make connections to the Empire Builder line that runs through the northern Rocky Mountains to Chicago, Illinois, and catch local commuter trains.

Joe Weber

See also: Amtrak (United States); Crescent, The (United States); Empire Builder, The (United States); Railroads.

Further Reading

Amtrak. "Amtrak Coast Starlight Route Guide." 2014. http://www.amtrak.com/ccurl/408/807
/Amtrak-Coast-Starlight-Train-Route-Guide-2014.pdf. Accessed August 2015.
Amtrak. "Coast Starlight." 2015. http://www.amtrak.com/coast-starlight-train. Accessed
August 2015.

COMMUTER TRAINS

Commuter trains provide passenger rail service between metropolitan centers and their surrounding suburbs. Commuters—travelers who ride the rail daily or on a regular basis—use trains to travel between home and work or school. Passengers riding commuter trains generally have a longer commute within a metropolitan area than those who ride buses or drive automobiles. Commuter trains—also known as commuter rails or suburban rails—travel at speeds of between 25 and 50 miles per hour along routes ranging from 10 to 125 miles long.

Passenger rail travel was born in September 1825, when the first paying riders boarded a steam locomotive, called the Locomotion, owned by Stockton & Darlington Railway in North East England. For decades, the railroad as a means of transportation and steam locomotives in general were the fastest way for passengers to traverse long distances. Berlin, Germany, introduced electric power to rails in 1879 and an urban rail system was born. As cities expanded toward the end of the 19th century, pushing populations farther outside city boundaries, a means of transportation that was a bridge between main-line passenger trains and intracity trains became necessary. Commuter trains that could run on main-line, standard gauge track and bring passengers to the city on shorter routes proved to be an efficient way to reconnect suburban dwellers with the urban centers in which they worked. In 1840, an estimated 1 million commuters rode the rails in England. That number jumped to 67 million just 10 years later and to 316 million in 1870. It was not uncommon for land developers in the late 1800s to entice people to move into their housing developments and ride the rail—oftentimes owned by the developer itself—into town.

Commuter trains differ from light-rail or rapid transit trains in several ways, namely commuter trains are larger and heavier, contain more seats, have fewer stops, tend to run on a schedule, serve lower-density suburban areas that feed into the city center, and share right-of-way and track with main-line regional trains. Commuter rail covers a broader area than metro rail or rapid transit lines but not as broad an area as regional rail—which transports freight and passengers between towns or cities—although the definition of each varies worldwide. Commuter trains are powered by locomotives running on diesel, electricity, or a combination of the two, or the cars are self-propelled and require no locomotive. Electric trains are supplied with power from a continuous conductor running along the track, either as an overhead catenary wire supplying alternating current (AC) or as a third rail installed on a standard track and supplying direct current (DC). Some trains

A commuter train at a quiet rural station. (Martin Brayley/Dreamstime.com)

run on both AC and DC current, such as the S-Bahn in Hamburg, Germany. Third rail trains—such as the Paris Métro in France and the Chicago "L" in Illinois—tend to accelerate quicker than AC trains, but anything that comes in contact with the third rail can be electrocuted, so low-level power (less than 1,500 volts) typically is used. Diesel multiple unit (DMU) trains, such as the WES in Portland, Oregon, the Adelaide Metro in Australia, and the Korail in South Korea, are powered by onboard diesel engines and require no locomotive.

The type of commuter rail installed depends greatly on the specifics of the existing system on which it is built, including where it is located in the world and the technology available at the time of construction. Most systems were installed up to 100 years ago, when electrification of rail lines first came into vogue, and therefore derive their power from electricity. Lines manufactured at the end of the 21st century have more sophisticated power electronics and microprocessors and the new technology is reflected in their design, manufacture, and operation.

Other than onboard batteries, third rail electrification is the oldest form of electric supply to commuter trains. It was created by Siemens and Halske in Germany in 1879. In this first iteration, the third rail ran between the track's two standard rails. Today the third rail generally is installed on the outer edge of the track. The first electric railway to use this system was the Gross Lichterfelde Tramway in a suburb of Berlin, Germany. The oldest operating electric railway in the world is the Volk's Electric Railway in Brighton, England, which started in 1883. It is mainly a seafront tourist attraction today.

Diesel multiple unit trains are powered by onboard diesel engines. They can be broken down further into three classifications according to the method the power is transmitted to the wheels: diesel-electric (DEMU), in which a diesel engine drives an electrical generator or alternator; diesel-mechanical (DMMU), in which rotating energy is transmitted by way of a gearbox and driveshaft, as in a car; and diesel-hydraulic (DHMU), which uses a torque converter transmission. The main benefits of DMUs are there is always enough motive power per number of passengers and the number of cars can be matched to the capacity of passengers, with more cars being added or removed along the route.

In terms of ridership, with more than 98 million annual commuters, the MTA Long Island Railroad in New York is the busiest commuter rail system in the United States. Some of its trains are electrified and some are diesel powered. It is the only American commuter railroad that runs nonstop. Worldwide, Japan's JR East, a line of the Japan Railways Group, has more than 17 million riders daily, totaling more than 6 billion riders annually.

Rosemarie Boucher Leenerts

See also: High-Speed Rail; Railroads; Rapid Transit.

Further Reading

Garrett, Mark. 2014. *Encyclopedia of Transportation: Social Science and Policy*. Thousand Oaks, CA: Sage.

Governing. 2013. "Commuter Rail Ridership Declining Despite Increase in Lines." www .governing.com/topics/transportation-infrastructure/col-commuter-rail-ridership -declining.html. Accessed September 8, 2015.

COMPLETE STREETS (UNITED STATES)

The term "complete streets" first was introduced in 2003 by Barbara McCann, a staff member of the advocacy group American Bikes, to convey the importance of accommodating the bicycle in transport planning. Although the term began to generate a lot of attention in the national media and newspapers such as *The New York Times*, *USA Today*, and *Time* magazine, understanding the concept of complete streets often has proved somewhat challenging. Generally, a complete street policy ensures that streets are designed and operated in a way that safeguards access for all users of all ages and abilities—including pedestrians, bicyclists, motorists, and transit riders. Today there are more than 700 agencies at the local, regional, and state levels that have adopted Complete Streets policies. The implementation of this concept is experiencing obstacles, however, as many communities still favor reliance on automobiles for mobility.

The idea of complete streets is not just about infrastructure, however; it is an integration process of creating livable communities for economic and social benefits. Complete streets not only can provide a feeling of safety and a sense of belonging, it also can give businesses the opportunity to cluster together, which makes travel easier for everyone. Similarly, it can reduce auto trips, road maintenance costs, and greenhouse gas emissions. It has been estimated that increasing bicycle trips from 1 percent to 1.5 percent of all travel could save 500 million gallons of gasoline per year in the United States. Most importantly, complete streets can provide freedom and independent mobility for everyone including children, the elderly, and the disabled, and not just for drivers. The concept of complete streets therefore is tied to larger land use and transportation planning efforts to consistently fund, plan, design, and construct community streets to accommodate all anticipated users.

Selima Sultana

See also: Bicycle Lanes; Bicycles and Tricycles; Sidewalks; Velomobiles.

Further Reading

Zavestoski, Stephen, and Julian Agyeman. 2015. *Incomplete Streets: Processes, Practice, and Possibilities*. New York: Routledge.

COVERED BRIDGES

Covered bridges usually are wooden truss bridges with a roof—and often walls—added to protect the structure from the weather. The covering does not serve as part of the structure holding up the bridge. Early covered bridges existed in Europe and China hundreds of years ago, but they are best known as a North American bridge type. They were a common form of bridges built in the Northeast and Midwest United States in the 18th and early 19th centuries, but can be found throughout the world wherever lumber for construction was available. They often were individually designed and built and thus differ widely in appearance and characteristics.

The first covered bridge in the United States was built in 1805, but most of those that remain were built in the later 19th and early 20th centuries, except in Oregon and parts of Canada where these were built into the 1950s. The oldest surviving covered bridge in the United States, the Hyde Hall Bridge (New York), was built in 1825. In the postwar era, the covered bridge came to be a treasured symbol of rural life in a more authentic age, and many were preserved and restored. Richard Sanders Allen (1917–2008) wrote many books about surviving covered bridges in the United States and helped popularize the topic. New covered bridges occasionally still are built, but often are not considered authentic as they could be steel or concrete bridges with a wooden cover.

Covered bridges are one of the most picturesque forms of bridges and the subject of considerable nostalgia. They are familiar to most people through their representation in film, TV, and literature. The six covered bridges in Madison County, Iowa, were made famous by the film *The Bridges of Madison County*, and many

Covered bridge at historic Roswell Mill, Roswell, Georgia. (Lloyd Smith/Dreamstime.com)

states publish guides to their surviving bridges. The National Society for the Preservation of Covered Bridges exists to preserve these structures. The Society produces a list of all known covered bridges in the world, along with their length, date of construction, and location. The organization estimates that there are 1,600 covered bridges around the world as of 2009, and 684 in the United States (down from 1,344 in 1959), though little is known about covered bridges in many parts of the world. China is known to have hundreds of covered bridges, but relatively few are known outside the country. In the United States, Pennsylvania has more covered bridges (213) than any other state. Midwestern and northeastern states have the most, but covered bridges can be found in all 50 states. In many states these bridges were constructed recently and might not be thought of as "authentic" covered bridges.

Today, many covered bridges are found in out-of-the-way locations along secondary roads. This location is not intentional; rather, it is a legacy of their age. At one time, many bridges were covered—including those on main roads and in big cities. Over time, these bridges were replaced with iron, steel, and concrete bridges. Those that were not replaced generally were the less important bridges. Those covered bridges that survive ironically are those that were of little importance. Many of these bridges no longer can be driven on. Many covered bridges have been moved, collapsed and been rebuilt, or reinforced with modern materials. The covered bridge in Germantown, Ohio, was built in 1870 and moved to its current location in 1911. It is now reinforced with steel beams to prevent another collapse.

Joe Weber

See also: Alpine Tunnels (Europe); Border Tunnels (Mexico and United States); Stormwater Management and Road Tunnel (SMART Tunnel) (Malaysia); Tünel (Turkey).

Further Reading

Allen, Richard Sanders. 1957. *Covered Bridges of the Northeast.* Brattleboro, VT: Stephen Greene Press.

Allen, Richard Sanders. 1959. *Covered Bridges of the Middle Atlantic States.* Brattleboro, VT: Stephen Greene Press.

Allen, Richard Sanders. 1970. *Covered Bridges of the Middle West.* Brattleboro, VT: Stephen Greene Press.

Allen, Richard Sanders. 1970. *Covered Bridges of the South.* Brattleboro, VT: Stephen Greene Press.

National Society for the Preservation of Covered Bridges. 2009. *World Guide to Covered Bridges.* National Society for the Preservation of Covered Bridges.

National Society for the Preservation of Covered Bridges. 2015. http://www.coveredbridge society.org/. Accessed August 4, 2015.

CRESCENT, THE (UNITED STATES)

The Crescent is an Amtrak passenger train currently connecting New Orleans and New York City. It runs 1,377 miles in about 30 hours. It has daily departures north and southbound, with stops in between at many cities including Birmingham, Alabama; Atlanta, Georgia; Greenville, South Carolina; Charlotte, North Carolina; Washington, DC; Baltimore, Maryland; and Philadelphia, Pennsylvania. At Washington, DC's Union Station, the diesel engines of northbound trains are replaced with electric and it becomes one of the high-speed trains of the Northeast Corridor. In New Orleans the Crescent connects with the Sunset Limited Train to Los Angeles and provides the city of New Orleans with service to Chicago.

From 1970 to 1979 the train was the Southern Crescent, operated by the Southern Railway between Washington, DC, and New Orleans. It was one of only two privately owned passenger trains in the country after Amtrak was created in 1971. Before that time, it was called the Crescent Limited, and portions of the route were operated under other names in the 19th century. The train was operated by various railroads as well.

Joe Weber

See also: Amtrak (United States); Northeast Corridor (United States); Railroads; Sunset Limited, The (United States).

Further Reading

Amtrak. "Crescent." 2015. http://www.amtrak.com/crescent-train. Accessed August 2015.
Amtrak. "Crescent Route Guide." 2015. http://www.amtrak.com/ccurl/680/728/Amtrak
-Crescent-Train-Route-Guide.pdf. Accessed August 2015.

CYKELSUPERSTIER (DENMARK)

The *Cykelsuperstier* (Cycle Super Highway) is a road network infrastructure specifically designed for bicycles in the greater Copenhagen area, in Denmark. Being the world's largest bicycle commuter city (it comprises about 35 percent of all commuters), Copenhagen already had a high-quality bicycle infrastructure. The cykelsuperstier, however, is a new innovation intended to draw attention to the benefits of bicycle commuting worldwide. This is designed to meet the needs of commuters, especially targeting the long-distance bicycle commuters—who commute between 5 and 15 miles. Research has shown that bicycle commutes drastically decline for destinations that are more than three miles away, hence this project will enhance the comfort and the experience of those commutes in a way that rarely existed before.

The routes of the cykelsuperstier are designed to connect Copenhagen by directly linking residential areas, workplaces, and schools. Synonymous with auto highways, these bicycle highways are built for increasing speed for bicyclists with minimal interruptions. Therefore, the routes are very direct with smooth pavement and synchronized traffic lights so that cyclists can maintain an average speed of 15 miles per hour and make as few stops as possible. Traffic lights usually are coordinated in favor of cars, but cykelsuperstier cyclists will have priority along the many main traffic arteries. Other facilities include bicycle pumps, footrests, automatic bicycle bumps, and countdown signals. The roads are maintained regularly and specially treated against heavy snowfall.

The realization of this project started in 2009 as Copenhagen wanted to be the first carbon-neutral European city by 2025. There are two phases of this project. In first phase, 28 routes with a total length of approximately 300 miles are planned to connect 26 municipalities with the center of Copenhagen. These routes will be built by upgrading existing bike paths as well as creating brand-new paths. The plan of these networks was adopted after thorough assessments of existing bike paths; locations of businesses and residences; and by feedback from stakeholders, users, and citizens. The main purpose of the plan was to find the best routes that could be easily accessible to bicycle commuters. To serve that purpose, the street network for automobiles was ignored whenever possible. A map of the entire network can be viewed online at www.supercykelstier.dk site. The total cost has been estimated at approximately US$127 million to complete, which is about $423,055 to build each 0.62 mile of the network. This cost is significantly less than the construction cost of highways in Denmark, which is about $11.7 million for building the same length of highway. There will be significant socioeconomic benefits for the country; and it will reduce greenhouse gas emissions and congestion as well as reduce automobile network maintenance and health care costs.

The first 11 miles of the cykelsuperstier, the Albertslund route, came into existence in 2012. This route links the towns of Copenhagen, Frederiksberg, Albertslund, and Rødovre and already has seen a 30 percent increase of cycle commuters. The 13-mile-long Farum route is constructed with a new type of lighting in the pavement that runs on solar energy. It has been open to the public since 2013. At this point, funding has been secured for nine more routes and should be completed by 2018. It is estimated that once these routes are completed, 52,000 commuters will be using these facilities each day, which is slightly less than 50 percent of current commuters in Copenhagen.

Selima Sultana

See also: Bicycle Lanes; Bicycle Libraries; Bicycles and Tricycles; Sidewalks.

Further Reading

Cycle Super Highways of Greater Copenhagen. 2014. *Cycle Super Highways: Capital Regions of Denmark.* http://www.supercykelstier.dk/sites/default/files/Cycle%20Superhighways _UK_maj%202014.pdf. Accessed July 22, 2015.

D

DARJEELING HIMALAYAN RAILWAY (DHR) (INDIA)

The Darjeeling Himalayan Railway (DHR) is one of the best-known mountain railways in the world. It passes through the beautiful mountainous landscapes of the Himalayas between New Jalpaiguri and Darjeeling, India (located in Northeast India). Known also as the "Toy Train," this 48-mile-long railroad runs from an elevation of 328 feet at New Jalpaiguri to 6,812 feet at Darjeeling, via a 7,407-foot summit at Ghoom. It first began operating in 1881, during the British colonial period of India and still is fully operational; many of its original features remain intact. The DHR in 1999 was listed as a world heritage site by the UNESCO World Heritage Committee as an example of one of the first mountain railways using an innovative engineering design representing multicultural values (local and colonial) and for its profound influence on social, cultural, and economic life of the region in the 19th century.

The DHR originally was introduced to connect Kolkata to Darjeeling in the foothills of the Himalayas. The beauty and temperate climate of Darjeeling was found to be an attractive place for British residents seeking to escape from the summer heat of the plains, and Darjeeling became a hill station. The idea of building a rail line between Kolkata and Darjeeling accelerated when the commercial cultivation of tea began in Darjeeling in the mid-19th century. In 1879, a broad gauge railway under the name of the Darjeeling Himalaya Tramway Company carried freight and regular mail between Kolkata and Siliguri. The line was connected to the Darjeeling main station in 1881, and the name of the company also changed to the DHR Company.

The DHR was a technological marvel in 1881. It used steam locomotives on each end of the line in turnaround service, but diesel locomotives later were used to haul trains from end to end. To increase efficiency, reverses and loops were introduced to ease the gradients of the steep line. As a result, by 1914 the DHR was carrying 250,000 passengers and 60,000 tons of freight annually, and World War I led to an increase to 300,000 passengers per year. After that war, two additional branch lines were connected to the DHR to meet the growing demands of the tea industry in Darjeeling: one to Kishenganj, southwest of Siliguri, and another to Kalimpong Road. During World War II the traffic on DHR increased significantly, as Darjeeling was being used as a "rest and recuperation" center for British soldiers. The DHR played a crucial role in transporting military equipment and personnel to various camps that were set up around Ghoom and Darjeeling. To meet the high traffic demand 39 working locomotives—including 5 ambulance trains—were added to the line. The DHR reached its peak capacity in 1947, when

The Darjeeling Himalayan Railway, nicknamed the "Toy Train," is a two-foot narrow-gauge railway from New Jalpaiguri to Darjeeling in West Bengal, India. (Samrat35/Dreamstime.com)

it had 45 locomotives, 139 passenger coaches, and 606 wagons.

Today, with 13 stations, the DHR has evolved not just as a mode of transportation linking the two worlds of mountains and plains lands, but also as a source of national pride for its major engineering features. For instance, the DHR tracks are built with zigzag reverses and loops with a ruling gradient of 1:31. Zigzag reverses are Z-shaped layout track where trains run forward first, then run in reverse backward up the slope, then proceed forward again at a higher elevation. There are six such reverses on the DHR, and the sharpest gradient is noted between Sukna and Rongtong. A loop is designed to enable a train to run in a full circle to reach higher elevations by crossing over itself. The DHR track consists of such three loops. The most famous loop is between Ghoom (the highest point) and Darjeeling, and is known as the "Batasia Loop," constructed in 1919. As it passes through terrain of great beauty—high waterfalls, green valleys, dense forest, and the snow-capped Kanchenjunga mountain range—at a speed of 13 miles per hour, the DHR brings a "sensuous immediacy" between nature and travelers.

Selima Sultana

See also: Alpine Tunnels (Europe); Channel Tunnel (France and United Kingdom); Chiva Express (Ecuador); Seikan Tunnel (Japan).

Further Reading

Roy, Sujama, and Kevin Hannam. 2013. "Embodying the Mobilities of the Darjeeling Himalayan Railway." *Mobilities* 8(4): 580–594.

United Nations Educational, Scientific, and Cultural Organization (UNESCO). 2015. "World Heritage List. Mountain Railways of India." http://whc.unesco.org/en/list/944. Accessed July 18, 2015.

DELHI METRO (INDIA)

The city of Delhi is within the National Capital Territory of Delhi, India. The urban area has a population of 11 million and the metropolitan area population is more than 16 million (second only to Mumbai). Urban expansion beyond the metropolitan area boosts the region's population to 25 million. Metropolitan growth was one reason why planning for India's third urban rail network (after Kolkata and Chennai) began in 1984—18 years before the first section of the Delhi Metro opened.

Globally, the Delhi Metro is the 15th longest urban rail network (in terms of route length), has the 12th greatest number of stations, and it is 18th in terms of ridership. The current network consists of five color-coded local lines and one airport express line. It has 121 route miles and 140 local stations (and six airport express stations). Daily ridership is around 2.4 million; annual ridership is 800 million.

The idea of Delhi having an urban mass transit system emerged from a traffic and travel characteristics study conducted in 1969. By 1984, the Delhi Development Authority and the Urban Arts Commission proposed an urban rail transport network to augment existing suburban rail services and urban road transport (including buses). Although studies and finance proposals to construct the metro were developing, the city's population kept increasing. The population doubled between 1981 and 1998 and the number of vehicles increased fivefold. Traffic congestion and air pollution choked the city as commuters abandoned the existing bus network in favor of private automobiles. When the state-run bus system was privatized in 1992, problems mounted as the new operators could not cope with overcrowding, poorly trained drivers, an aging fleet, and an unreliable timetable. By 1995, the government of India and the government of Delhi created a new company—the Delhi Metro Rail Corporation (DMRC)—to address the city's transport crisis. Since its inception, the DMRC has been a state-owned operation.

Construction began in October 1998; the DMRC had full control to oversee the project, which proceeded smoothly despite the arduous task of building an underground and ground-level rail network in a bustling city of millions. The first section of the metro opened on December 24, 2002. It was part of the first of four project phases. The first phase of the project was completed by October 2006; it included portions of three lines and consisted of 58 stations and 40 route miles—8 of the miles underground. It was completed three years ahead of schedule and on budget.

The second phase opened in various stages over a three-year period from 2008 and 2011. It added 77 route miles and 85 stations, including the airport express line. The third phase will open in 2016; when completed it will have added 65 route miles and 69 stations. The fourth and final phase should be completed by 2021. The completed network will be more than 257 route miles. Initially, the metro system's track would have been the same broad gauge (1,676 mm) scale as the rest of India's railway network, but a dispute led to a compromise. The result: The first three lines (Red, Yellow, and Blue) are broad gauge and the other three lines (Green, Violet, and the Orange–Airport Express) are standard (1,435 mm) gauge. All six lines are powered via overhead wires. The system runs from 5:30 a.m. until midnight with train intervals of three to five minutes. Overcrowding has led to increasing the number of cars per train as is possible.

The Delhi Metro has won accolades from organizations for environmentally friendly practices. The International Organization for Standardization (ISO) certified the system for environmentally friendly construction, and the Delhi Metro became the second urban railway in the world to receive this recognition. The United Nations' Clean Development Mechanism recognized Delhi as the first railway project in the world to earn carbon credits. The railway has earned 400,000 carbon credits by using regenerative braking systems on its trains. The railway corporation is also looking to harness solar energy and install solar panels at several stations in the system to reduce dependence on nonrenewable sources of energy. In addition to its environmental accomplishments, the Metro also has become an integral part of community infrastructure in a relatively short time. One example is the inclusion of artwork depicting the local way of life in Delhi. This artwork, created by students of different ages, is found at metro stations and along the concrete pillars of elevated sections.

The Delhi Metro has transformed the method of getting around one of the fastest-growing cities in the world in a little more than a decade. At the same time, the Metro contributes to cleaning up Delhi's polluted air, thus improving the quality of life for everyone.

Jason Greenberg

See also: Madrid Metro (Spain); Metro Railway, Kolkata (India); Mexico City Metro (Mexico); Shanghai Metro (China).

Further Reading

Advani, Mukti, and Geetam Tiwari. 2005. "Evaluation of Public Transport Systems: Case Study of Delhi Metro." Proceeding in START-2005 Conference, Kharagpur, India. http://tripp.iitd.ernet.in/publications/paper/planning/mukti_metro_kharagpur_05.pdf. Accessed July 28, 2015.

Murty, M. N., Kishore Kumar Dhavala, Meenakshi Ghosh, and Rashmi Singh. 2007. "Social Cost-Benefit Analysis of Delhi Metro." *Munich Personal RePEc Archive* 1658(7). http://mpra.ub.uni-muenchen.de/1658/1/MPRA_paper_1658.pdf. Accessed July 28, 2015.

Pucher, John, Nisha Korattyswaroopam, and Neenu Ittyerah. 2004. "The Crisis of Public Transport in India: Overwhelming Needs but Limited Resources." *Journal of Public Transportation* 7(4): 1–20. http://www.nctr.usf.edu/jpt/pdf/JPT%207-4%20Pucher.pdf. Accessed July 28, 2015.

DIRT ROADS

The term "dirt road" refers to any road that is not paved with asphalt, concrete, brick, or stone. A dirt road could be dirt, gravel, sand, or a similar substance. For many urban residents of the United States, a dirt road is seldom seen except in rural areas; in other countries dirt roads are more common. Originally all roads were dirt. In the 19th century engineers developed improved methods for constructing roads. These methods involved building a road on a firm foundation and making it higher than the surrounding land, and laying down gravel to ensure good drainage. John McAdam (1756–1836) was a Scottish engineer who popularized this design, and roads following his construction principles sometimes still are called "macadam" or "macadamized" roads.

The coming of the automobile caused tremendous problems for dirt roads, however well built. Rubber tires sucked up gravel, breaking down the road surface and generating tremendous amounts of dust. One solution was to spray oil on the road surface to keep down the dust. Eventually, oil was mixed with gravel before the gravel was applied, and the asphalt road was born. A range of classifications existed to describe different types of dirt roads, such as unimproved, improved, surfaced, and graveled. All of these were dirt roads; the terms distinguished different levels of construction.

In contrast to the elaborate construction of macadamized dirt roads, today dirt roads often simply are graded or bladed with a road grader or bulldozer. This removes the topsoil grading to create a road surface that usually is lower than the surrounding ground. Because graders push dirt to the side they can be lined with berms that can trap runoff, forming ponds in the road. A modern dirt road therefore is very different than a 19th-century dirt road.

Dirt roads are temporary and require frequent use and maintenance to remain. Heavy rain and washouts are problems, as is deep sand. Dust also can be a problem in populated areas. One of the most frustrating problems with dirt roads is "washboarding." This refers to a natural process whereby a dirt road surface becomes corrugated due to traffic. Over time the road becomes bumpier, producing a shaking that increases with vehicle speeds. The corrugated surface resembles a washboard (an item used before modern washing machines were invented). Regular grading is required to restore a smooth surface. If left undisturbed, however, dirt roads can last an impressive amount of time. In the Mojave Desert of California, dirt roads constructed in the late 19th or early 20th centuries often remain clearly visible and drivable if they are on ridgetops or other locations that are safe from flooding.

Some dirt roads consist of little more than two parallel tracks where tires have worn a path. These are "two-track" roads. These often have high centers and can require using high-clearance vehicles to keep rocks from scraping a vehicle's underside. Four-wheel drive might be required if the road is steep or rocky. These are not the four-wheel drive SUVs typically used in the city, and instead usually are high-clearance, short-wheelbase, lightweight vehicles. A winch is a helpful piece of equipment that could be necessary to pull the vehicle over obstacles. Regardless of whether a road is this extreme, dirt roads can be tourist attractions. National parks, such as Death Valley in California, offer many dirt roads, some suitable for a rental car and some legendary for their challenging conditions and the damage they in-flict on even the most rugged vehicles (off-road driving, however, is illegal). Al-though travel on dirt roads usually is much slower than on paved roads, it must be remembered that the fastest land vehicles actually were driven on a natural un-paved surface: the Bonneville Salt Flats.

Because dirt roads constantly change, their conditions can be unpredictable. Local soil conditions and climate can mean that the experience of driving on dirt roads varies tremendously from one locale to the next. Gathering local information is useful before attempting a long drive on a little-used dirt road. Not all GPS-based navigation devices distinguish the type of road surface ahead—this can be a prob-lem in isolated rural areas, and especially in the desert. A number of tourists visit-ing remote desert national parks in the southwestern United States have become lost and sometimes died when they followed computer-generated directions and ended up on rough or sandy dirt roads not suitable for their vehicles.

Joe Weber

See also: Bamboo Trains (Cambodia); Border Tunnels (Mexico and United States); Camel Trains (Mali); Sunset Limited, The (United States).

Further Reading

Blodgett, Peter J. (ed.). 2015. "Motoring West." *Automobile Pioneers, 1900–1909,* Vol. I. Nor-man, OK: University of Oklahoma Press.

Rehmeyer, Julie. 2007 (August 15). "Road Bumps: Why Dirt Roads Develop a Washboard Surface." *Science News.* https://www.sciencenews.org/article/road-bumps-why-dirt-roads-develop-washboard-surface. Accessed August 4, 2015.

DOG SLEDS

A dog sled is a sleigh or sled pulled by a single dog or a pack of dogs that travels over ice and through snow. Dog sledding originally was used as a means of trans-portation but has evolved into a sport and a tourist attraction as well.

Dog sledding dates back hundreds of years, to the ninth and 10th centuries CE. Arabian literature of the 10th century recorded the use of sled dogs in Siberia, and dog sledding appears in the writings of Venetian merchant and explorer Marco Polo in the 13th century, and of Italian nobleman Francesco da Collo in the 16th century. The 1675 edition of English explorer Martin Frobisher's book, *Historic Navigations*, contains an image of a dog pulling a sled shaped like a canoe. The first archaeological evidence of dog harnesses and related equipment was discovered in Canada and is thought to have belonged to the Thule people of coastal Canada and Alaska—the prehistoric ancestors of the Inuit—between 1000 and 1600 CE. Dog sledding increased the ability of the Thule to travel in search of food during the harsh winters. Dogs were the perfect form of transportation; horses could not sur-vive in the frigid cold, did not have the endurance of dogs, and could not master

Dog sled team racing on a beautiful, sunny winter day. (Kirkgeisler/Dreamstime.com)

the treacherous terrain and deep snow. Six dogs can pull between 500 and 700 pounds of weight on a sled and the Inuit made use of them to lead masters on a hunt and to help check traplines. They also were good transportation during migrations. Sleds of this period were made of wood and a basket that the Inuit called "*komatik*." When communities became more stable in the 19th century, dog sleds were used as a means of transportation for delivering mail and supplies and bringing news.

Early colonists in Canada adopted the use of dog sleds from the native peoples, and the sleds became a common way to travel by the 1700s. Additionally, dog teams were put into military service by French Canadians during the French and Indian War of 1754–1763. In the late 1700s and early 1800s, dog sleds were invaluable to fur traders of the Northwest. In trades at native camps, dog sleds carried furs, meat, firewood, and other goods and supplies, as well as messages between forts.

The dog sleds of this time had two basic designs: an open sledge and an enclosed cariole. Sledges resembled toboggans but with a higher curved front end. One to four dogs harnessed in single file would pull the sledge. Carioles resembled sledges but were partially enclosed and could carry a passenger. Several dog sleds typically would travel together in a "train" ("*traineau*"), with one sled leading the others. Men wore snowshoes and ran alongside the dogs. The trains traveled before dawn during the short winter days and would leave at night when the weather turned warmer so that the dogs would not have to travel in the soft daytime snow and tire more easily. The dogs used were large, long-legged animals. Sled dogs were trained to respond to two voice commands: "mush" (or "*marche*" in French), which would start the dogs running, and "whoa," to make them stop. A Hudson's Bay Company employee, William Miller, is reported to have taught dogs to turn right ("gee") or left ("haw").

In the 1800s, the Canadian Royal Mounted Police and U.S. police forces used dog sleds to patrol northern Canada and Alaska. After gold was discovered in the Yukon Territory in western Canada in the late 1800s, miners found dog sleds to be the best way to navigate the region. Explorers such as Richard Byrd, Robert Peary, and Roald Amundsen rode in dog sleds in their endeavors and brought attention to dogs of the Husky breeds, such as the Canadian Eskimo dog, which are strong runners with plenty of endurance. During World War I and World War II, dog sleds were used to haul heavy equipment over harsh terrain and used in search-and-rescue operations and to carry wounded soldiers.

Dog sledding also has become a sport. The first documented sled dog race took place in the mid-1800s. The race course was from Winnipeg, Manitoba, Canada, to St. Paul, Minnesota, in the United States. The first U.S. race was run in Ashton, Idaho, in 1917. The annual Iditarod race, first run in 1973, had its roots in a life-saving mission that occurred in 1925. That winter, residents of the town of Nome, Alaska, were besieged by the diphtheria virus, a lethal respiratory illness. Supplies for the antitoxin serum were running low and the closest available serum was in Anchorage, more than 1,000 miles away. The subzero temperatures made flying and dropping off supplies impossible, so the serum was shipped by train to Nenana, some 700 miles from Nome, where a relay of dog sleds was organized to deliver the drug. Known as the Great Race of Mercy, the five-and-a-half-day trek to deliver the drug sent the dogs across frozen waterways and rugged terrain in temperatures of 60 degrees below zero. The dog that led the final leg of the race, Balto, became legendary.

The first sled dogs were Husky dogs that bred with wolves and other dogs. The three most common sled dogs are Greenlands, from the island of Greenland; Siberian Huskies, from northeastern Siberia; and Alaskan Huskies, which descended from ancient dogs bred by nomadic hunter-gatherers crossing the Bering Strait into present-day Alaska. Today a variety of dogs is used in sledding. Dog sleds now are a form of recreation as well as transportation.

Rosemarie Boucher Leenerts

See also: Camels; Llamas; Ox/Bullock Carts.

Further Reading

Coppinger, Lorna. 1977. *The World of Sled Dogs: From Siberia to Sport Racing.* New York: Howell Book House.

Dogsled.com. "Dog Sledding's History and Rise." http://www.dogsled.com/dog-sleddings -history-and-rise/. Accessed June 5, 2015.

Gottfried, A. 2002. "Dog Sleds in the Northwest." *Northwest Journal* 4: 17–20. http://www
.northwestjournal.ca/IV5.htm. Accessed June 5, 2015.

DOUBLE-DECKER BUSES

Double-decker buses are those with two decks built upon a single, rigid chassis.
They have become a recognized symbol of London, England, where the majority
of buses are double-deckers. This bus style chiefly was created as a form of mass
transit and could double the number of passengers riding at one time. Being shorter
than conventional buses as well as articulated buses, which, like double-deckers,
can carry more passengers than traditional buses but are much longer, double-
deckers are easier to maneuver through London's narrow streets. Double-decker
buses measure about 31 feet to 36 feet long and are about 14 to 15 feet tall.

Besides in London, double-decker buses are used as public transportation in
cities in neighboring countries of the United Kingdom and in Europe, as well as in
Asia and other former British colonies. Some cities that outlaw long vehicles permit
double-deckers because of their shorter length as compared to traditional buses.
Double-deckers also are used as sightseeing vehicles in North America and elsewhere because of the high vantage point provided by the upper deck.

The first recorded bus—a horse-drawn vehicle called an "omnibus," meaning
"for all"—was operated in 1662 in France. Bus transportation did not flourish until
the 1820s, however, when commercial buses were widely introduced. Horse-
drawn double-decker buses first were produced in 1847 by Adams & Company of
Fairfield, England. This early bus featured a clerestory roof with built-in wooden
bench seats running horizontally along the length of the vehicle. These open-top
buses were run by the Economic Conveyance Company of London, which encouraged people to ride by offering half-price fare for those sitting on the top deck.
Initial ridership was low. In 1852, John Greenwood built a larger double-decker
that could transport 42 passengers and was pulled by three horses instead of
just two. Passengers used a ladder to reach the upper seats. Innovations such as
forward-facing seats and a staircase leading to the upper deck improved double-
decker ridership.

The introduction of the combustion engine in the early 1900s put an end to the
use of horse-drawn buses in London, but they still were used in more rural locations in the United Kingdom until the early 1930s. The first double-decker motorized bus was the NS Type, introduced in 1923. Built by the Associated Equipment
Company (AEC), it was considered a luxury ride, featuring upholstered seats and
a full, covered upper deck. Roofs had never been used on double-deckers before
the arrival of the NS Type because the high floor of earlier tall buses changed the
height-to-width ratio of the vehicle, causing an imbalance that could not withstand
the weight of a roof. The NS type quickly became popular—more than 2,400 were
built—and they drew riders away from the trams that had become a popular and
less-expensive alternative to buses. An advantage of the NS Type bus was it could

adapt to new routes more easily than trams and trains that required infrastructure to be built. This ease of access increased ridership as urban development prospered. The NS Type was in production until 1937.

The Routemaster—the iconic red double-decker vehicle, first built by AEC in 1954 and put into service in 1956—was a major development in double-decker transportation. Its unique construction included an alloy body shell and separate front and rear subframes, plus features such as power steering, independent front suspension, an automatic gearbox, and a power-hydraulic brake system—which, at the time, were considered advanced engineering. The Routemaster's design also was lighter and more fuel-efficient than rival double-decker vehicles. It soon became a symbol of London.

There were nearly 2,900 Routemasters built between 1957 and 1968. They were overhauled every five years, which kept the fleet well maintained. They survived the privatization of the public transit service, the Transport of London, and there still are 1,000 of these buses on the road today. The Routemasters were withdrawn from service on December 8, 2005, due to difficulties in accommodating disabled passengers. Two so-called heritage routes still are used in popular tourist areas of London. The New Routemaster—also called the New Bus for London—was developed in 2008 and entered service in time for the 2012 Summer Olympics.

As with many forms of public transportation, double-decker buses are undergoing changes to become clean, green vehicles. In 2007, the first hybrid-powered double-decker vehicle entered service in London. Since then, other manufacturers have built hybrids. The New Routemaster, a diesel and electric hybrid designed after the mayor of London called for a public competition to revamp the former style of the bus, emits less than half the carbon dioxide and nitrous oxide of diesel buses. Wrightbus of Ballymena, Northern Ireland, won the contest and was contracted to manufacture 600 hybrids between 2012 and 2016—a sum that represents 20 percent of all London buses. All-electric buses are planned to be phased in as well.

Rosemarie Boucher Leenerts

See also: Buses; Chicken Buses (Central America); Walking School Buses (United States).

Further Reading

Clean Fleets. 2014 (September 22). "The New Bus for London." http://www.clean-fleets.eu /fileadmin/New_Bus_for_London_Case_Study_for_Clean_Fleets_-_final.pdf. Accessed June 5, 2015.

London Bus Museum. 2015. "1923 AEC NS Bus—NS174." http://www.londonbusmuseum .com/museum-exhibits/double-deck-buses/aec-ns-ns-174/. Accessed June 5, 2015.

E

EL CAMINO DEL DIABLO (MEXICO AND UNITED STATES)

The Camino del Diablo ("Devil's Highway") is an old road that has returned to prominence. The name originally referred to a dangerous 250-mile road crossing the Sonoran Desert between northwest Mexico and Southern California, and running along the Arizona-Mexico border. It has almost certainly been used for hundreds of years, but the oldest documented crossing was in 1699. The route passed through sandy desert valleys and included rough mountain crossings and had no dependable water sources. The only water to be found was in *tinajas* ("tanks"), which are natural depressions in rocks that catch rainwater. Rainfall, however, is scarce, and summer temperatures are very hot. The graves of dozens of fallen travelers once were visible along the route. The construction of modern highways—including Mexico Route 2 and US 80—ended travel on the Camino del Diablo, as did the regulation of border-crossing traffic and the creation of a large bombing range on part of the route.

In the 1990s, however, this forgotten corner of the desert returned to prominence. Efforts by the U.S. Border Patrol to stop illegal border crossings in populated areas such as San Diego, California, and El Paso, Texas, caused border traffic to shift to rural areas, including the vast empty stretch of desert the Camino del Diablo crossed. Now, many people now cross this desert daily. They do not travel from east to west as people did before, but rather from south to north. The route remains dangerous. Water is scarce or nonexistent, the terrain and desert vegetation are difficult to walk through, and many illegal immigrants are completely unprepared for the experience. These immigrants originate from much farther south and are accustomed to tropical, not desert, conditions. They have paid significant sums of money to be guided through the desert by "coyotes"—professional guides who could just abandon them and keep their money. More than 2,200 deaths were recorded between 1999 and 2012, and many more victims have likely never been found. In 2004 alone 224 deaths were recorded. Were this desert passage a single highway it would be by far the deadliest road in the country.

Humanitarian groups in the United States—such as Humane Borders, Inc.—have tried to discourage travel by publishing maps showing walking distances from many border crossing points. These show that cities such as Phoenix are many days' walk from the border, not several hours as coyotes often promise. The groups also have left water and emergency radio beacons at designated spots along major migrant routes to help prevent the loss of life. The Mexican Ministry of

Foreign Relations published the "Guide for the Mexican Migrant," which provides tips for traveling through this area, including avoiding intense heat and recognizing the symptoms of dehydration.

Joe Weber

See also: Appalachian Trail (United States); Border Tunnels (Mexico and United States); Guoliang Tunnel (China).

Further Reading
Humane Borders, Inc. http://www.humaneborders.org/. Accessed June 5, 2015.

ELECTRIC VEHICLES

Electric vehicles (EVs) are a class of automobiles that utilize a battery and an electric motor for part or all of the power train of the vehicle. Several different configurations of battery electric vehicles (BEVs) currently exist, including conventional hybrid vehicles and plug-in hybrid electric vehicles, and plug-in electric vehicles.

Conventional hybrid vehicles are those that have both a traditional internal combustion engine (ICE) and an electric motor with a battery that stores electricity. These vehicles also can feature a regenerative braking system that uses the friction generated by applying the brakes to produce electric power to help recharge the battery. The most well known of these vehicles is the Toyota Prius. Introduced commercially first in Japan in 1997 and in the United States in 2000 with a starting price of less than $20,000, the Prius has become the most popular and most recognized non-ICE vehicle. Its distinctive design has contributed to people associating the Prius with "green," "environmentally friendly," and the standard-bearer for alternatives to using gasoline in cars. Virtually every major automaker in the world offers some version of its models with a conventional hybrid engine.

Plug-in hybrid electric vehicles (PHEVs) are like conventional hybrids that feature both an ICE and an electric motor with battery for electricity storage. They differ from hybrids in that the primary motor is the electric motor, and the battery can be charged independently from the ICE by plugging it into an external outlet. The ICE serves as a secondary source of power to use when the electric motor's battery is drained during operation. These vehicles also can feature a regenerative braking system to help recharge the battery during driving. The first commercially introduced version of this type of vehicle is the Chevrolet Volt, which debuted in the United States in 2011. Many major automakers are developing plug-in hybrid models of their vehicles, including Audi, BMW, Cadillac, Chevrolet, Ford, Honda, and Nissan.

Plug-in electric vehicles (PEVs) are vehicles that rely solely on an electric motor with a battery for electricity storage. These vehicles can be charged only by

Binary Power

Binary power is a radically new technology that in the coming years might change the future of transportation more rapidly than ever before. This technology's basic concept is to have two harmless beams of energy intersect at a point in space, creating a source of power. To clarify, imagine the following scenario. In a room, two invisible traveling beams intersect at a point in a room and glow like lights.

This binary power is an intense form of energy but does not create any greenhouse gas emissions. This binary power will substantially reduce the cost of power production, reduce the cost of owning a vehicle, and reduce utility costs. It also will significantly lessen both the weight and the cost of a vehicle. Experts predict that by 2030 the frictionless no-moving-parts flying vehicles will run on binary power, and by 2050 the entire automobile industry will use binary power as a principal source of power for vehicles. This currently still is a dream but, according to Google's research scientist visionaries, it most likely will come to fruition in the near future.

Selima Sultana

plugging into an external outlet, although some vehicles also feature a regenerative braking system. The Nissan Leaf was one of the first commercially available PEVs offered by a major traditional automaker. It was released in 2010 with a price of $28,000 and eligibility for a $7,500 tax credit (Plugincars.com). Tesla Motors represents the most prominent and successful of a series of start-up automakers that produce only electric vehicles. The company produced a sports car and a sedan that are among the industry leaders not only in performance but also in safety. In 2014, Tesla's Model S was able to boast being one of only three vehicles produced since 2011 to receive a five-star rating for both the U.S. National Highway Traffic Safety Administration (NHTSA) and the European New Car Assessment Programme (NCAP).

Electric vehicles differ from internal combustion engine (ICE) vehicles in several ways. One difference is that the driving range for a single full charge of the battery for most current BEVs is much less than that of comparable ICE models. Many BEVs manage between 30 to 100 miles on a charge, and the driving range for a typical car on a full tank of gasoline usually is between 250 and 500 miles. There are some exceptions to this—most notably the Tesla vehicles. Some Tesla models feature ranges of more than 300 miles for a single charge. Also, much work is under way to expand the range of electric vehicle batteries.

Another key difference is the refueling mechanism. Electric vehicles must recharge a battery by plugging into a power source. This differs from refueling an ICE with gasoline on several fronts. One is that the recharge time is much longer than the time needed to refuel an ICE. An electric vehicle can plug into a conventional 110-volt power outlet, but a complete recharge can take more than 12 hours.

Tesla Roadster Sport 2.5 on display during the AutoRAI motor show in Amsterdam, Netherlands, 2011. (VanderWolf Images/Dreamstime.com)

When plugging into a Level 2 Charger (a 240-volt outlet—similar to what a conventional dryer uses), the recharging time goes down to four to six hours. Development and introduction of "fast" charging stations also has increased. These stations
can recharge a depleted battery in approximately 30 to 60 minutes. Experimenta-
tion with "battery swap," in which a depleted battery is exchanged for a fully recharged one at a station, has begun. This process can take a few minutes or less but
presently is very experimental due to the wide variety of batteries used among
manufacturers.

Additionally, there are far fewer charging stations in public than there are gasoline stations. Electric vehicles, however, can be charged at homes that have access
to 110- or 240-volt outlets. Cities and states—as well as several automakers—are
working on investing in public charging stations both in urban areas as well as
along freeways. These can be free or paid for with a credit card at the site. Tesla has
completed a network of fast chargers that enable a user to drive one of Tesla's vehicles from the East Coast to the West Coast, and is looking to continue expanding
the network. Tesla's Supercharger Network includes the fastest charging stations
available, and currently has 388 locations across the world with 2,114 supercharg-
ers (Supercharger). The states of California, Oregon, and Washington likewise are

working on a network of charging stations that will connect the major cities in these states.

Electric vehicles currently are relatively expensive, costing up to two to three times as much as their gasoline counterparts. Many governments offer incentives in the form of tax rebates (worth up to $7,500 in the United States, and more in some countries) to purchase these vehicles, although the vehicles still are quite costly even after such rebates. Electric vehicles are offered in far fewer models of vehicles than their gasoline counterparts, but the number offered is increasing. Nearly every major automaker in the world offers at least one model as a hybrid—many are innovating PHEVs or PEVs into versions of existing models or completely new models, and car companies such as Tesla and numerous startups in China are engineering only electric vehicles.

Electric vehicles actually are not a "new" technology at all. For the first couple of decades of its existence, electricity and steam were the preferred choices of automotive power. Electric vehicles of the time performed similarly to some of the electric vehicles on the market today. The petroleum industry, however, grew with and helped shape the automobile industry when it was in its infancy; that fact—combined with benefits of petroleum that include its superior energy density, accessibility of supply, and ease of storage—helped make gasoline the fuel of choice for the automobile.

Alternatives to gasoline for the automobile were reconsidered seriously in the 1960s and 1970s, as problems with emissions from burning petroleum came to prominence and oil supplies were disrupted by the Organization of Petroleum Exporting Countries (OPEC) intentionally limiting supply. Additionally, the idea of Peak Oil and the recognition that petroleum is a finite, nonrenewable, polluting resource negated the long-term viability of gasoline and emphasized the need to look for an alternative fuel.

Electric vehicle research and development remained in fairly nascent stages through the 1970s and 1980s. The notion of global climate change arose to prominence in the 1990s and—somewhat in response to that news—General Motors began development of an electric vehicle called the EV-1. Initially drawing much attention and interest, it was scrapped under somewhat mysterious circumstances after a successful pilot program. After that, interest in electric vehicles subsided until the mid-2000s, when many countries—including the United States—began to set ambitious goals for market penetration of EVs and to introduce policies and incentives designed to encourage automakers to produce and consumers to buy electric vehicles. This has been done as much to promote economic competitiveness in the battery and automobile industries as to meet any environmental objectives.

The future of electric vehicles is at once promising but also very much uncertain. Transportation specialists generally regard using petroleum for gasoline as unsustainable, and EVs have made the most recent progress of any replacement. Electricity lacks the energy density to fully replace petroleum, however, and other alternatives also are available, including ethanol blends, natural gases, and hydrogen fuel cells. It remains to be seen what the fuel of the future will be for automotive transportation.

Bradley W. Lane

See also: Biofuel Vehicles; Hydrogen Vehicles.

Further Reading

Bowling, Clarke. 2013 (June 28). "Toyota Prius Turns 16: How the Hot Hybrid Has Changed Over the Years." *New York Daily News.* http://www.nydailynews.com/autos /toyota-prius-turns-16-hot-hybrid-changed-years-article-1.1385087. Accessed February 25, 2015.

Graham, John D. et al. 2011. *Plug-In Electric Vehicles: A Practical Plan for Progress. SPEA Insights.* Bloomington: Indiana University. https://spea.indiana.edu/doc/research/TEP _combined.pdf. Accessed December 7, 2015.

Lane, Bradley W., Natalie Messer, Devin Hartman, Sanya Carley, Rachel M. Krause, and John D. Graham. 2013. "Government Promotion of the Electric Car: Risk Management or Industrial Policy." *European Journal of Risk Regulation* 4(2): 224–41.

PlugInCars.com. "Nissan Leaf Review: Affordable All-Electric Car." http://www.plugincars .com/nissan-leaf. Accessed February 25, 2015.

ELEPHANTS

Asian elephants have served as a source of transportation and labor across Asia, the Middle East, and North Africa for millennia, due to their ability to carry heavy loads across rough terrain and their relative affinity toward humans. The use of working elephants has greatly declined in the past 100 years, however, due to the adoption of vehicles and heavy machinery, as well as a general decline in elephant populations due to poaching.

The earliest known use of elephants for human transportation dates to approximately 2000 BCE—the date of a Mesopotamian plaque depicting an Asian elephant with a man on its back. It is assumed that elephants were employed for transport at a much earlier date, however, because the plaque also shows a rope running between the elephant and the rider, implying a well-established elephant-riding culture. Ancient artifacts depicting elephant riding also have been found in Pakistan, and the use of elephants for transport in China has been traced back to 1100 BCE.

In addition to transporting humans, individual elephants can carry loads of several hundred pounds and are particularly adept at traveling across wetlands and through dense rain forests—landscapes that humans find virtually impenetrable during Asian monsoons. Elephants also were used in warfare in ancient India, Persia, and (later) the Mediterranean, serving essentially the same functions as modern transport trucks and tanks. They also have been used for logging—uprooting and transporting vast quantities of timber—in both ancient and modern times.

More recently, elephant rides have become popular at tourist destinations through-out much of Asia.

Despite the long-standing working relationship between elephants and humans, elephants never have been fully domesticated and continue to be regarded as un-predictable and sometimes aggressive in their behavior. Most working elephants are captured as adults from wild herds, because elephant calves take 15 years to mature and are virtually impossible for humans to raise from birth. In this context, elephant warfare researcher John M. Kistler writes in his book *War Elephants*: "The 'domestication' of elephants is simply a forced cooperation exchanged for food and social contact."

Even this tenuous taming of elephants in Asia proved impossible for most Afri-can elephants. Despite similar social structures in Asian and African elephant groups, African savannah elephants generally are more aggressive, more difficult to capture, and less amenable to human relationships than their Asian counterparts. The North African forest elephant subspecies (*Loxodonta africana pharaoensis*) might have been used in warfare before its extinction in Roman times, but evi-dence has proved inconclusive.

Today, the demand for working elephants has all but disappeared outside of their use as tourist attractions. An exception is the timber industry in Myanmar, where more than 5,000 elephants were still employed (Myanmar Timber Elephant Project 2016). Elephant advocates now worry that the traditional skills and knowledge of elephant drivers (also known as *mahouts*) could soon be lost forever. Additionally, as elephants become less valuable to humans as working animals, incentives to conserve and protect wild elephant herds decline and lead to increases in poaching—which already has devastated wild elephant popu-lations throughout Asia. To combat this trend, the United Nations Food and Agriculture Organization has suggested training *mahouts* in more sustainable income-generating activities related to their elephants, as well as registering and tracking all domestic elephants.

Terri Nichols

See also: Camels; Llamas; Ox/Bullock Carts.

Further Reading

Hart, Lynette A. 1994 (June). "The Asian Elephants-Driver Partnership: The Drivers' Per-spective." *Applied Animal Behaviour Science* 40(3–4): 297–312.

Iljas, Baker, and Masakazu Kashio (eds.). 2002. "Giants on Our Hands." Proceedings of the International Workshop on the Domesticated Asian Elephant. Bangkok, Thailand, February 5–10, 2001. Food and Agriculture Organization of the United Nations, Re-gional Office for Asia and the Pacific.

Kistler, John M. 2005. *War Elephants*, Westport, CT: Greenwood Publishing Group.

Myanmar Timber Elephant Project. 2016. Timber Elephant. http://myanmar-timber -elephant.group.shef.ac.uk/timber-elephants/. Accessed January, 4, 2016.

ELEVATORS

An elevator is a type of transportation equipment that carries passengers and goods vertically between floors of a multistory building or other structure. Outside of North America an elevator often is known as a "lift" and typically runs on electricity that moves cables or counterweights or forces a hydraulic lift (e.g., a jack). Elevators are responsible for helping change the landscape of metropolitan areas enabling the use of ever-taller buildings and providing access to all individuals—including disabled people—to buildings taller than one story.

Elevators are believed to have been in use as early as the third century BCE, when Greek mathematician, physicist, and inventor Archimedes devised a hoist using a rope-and-pulley device. The rope was wrapped around a capstan and the drum was turned by a lever operated by manpower. Emperor Nero had three such devices installed in his palace in ancient Rome. The first recorded use of an elevator was a hoist installed during the Middle Ages at a monastery in Greece that stood 200 feet above ground level. The hoist was used to lift people and cargo to the monastery. In the mid-1700s, King Louis XV of France had a counterweighted elevator installed in his personal chambers at the Palace of Versailles.

As industry boomed in the 19th century, the need to haul raw materials also ramped up. In turn, elevators were necessary to move items and passengers in both rural and metropolitan settings. These first elevators—which were similar to modern-day devices—operated using steam or hydraulic plungers. In the latter type, a cab was affixed to a hollow plunger that was lowered into an underground cylinder. The cab was elevated when pressure was created from water being injected into the cylinder. The cab would lower when water was released from the cylinder. Riders would turn valves inside the cab to manipulate the water flow. Other elevators in the era included one invented in 1833 that used reciprocating rods to raise and lower miners working in the Harz Mountains in Germany, as well as an elevator known as a Teagle, first installed in a factory in England in 1835. The Teagle was invented by Frost and Stutt and used counterbalances and a belt drive. The first elevator shaft was built in 1853 by Peter Cooper and installed in the Cooper Union Foundation building in New York in anticipation of the invention of a safe passenger elevator. The shaft was in the shape of a cylinder, which Cooper believed to be the most efficient design.

Elevators were powered by animals, humans, steam, gas, or water throughout history until electricity was invented in the mid-19th century, which coincided with the era of the first high-rise buildings being constructed in the United States. Because commonly used ropes—often made of hemp—broke easily over time, master mechanic Elisha Graves Otis devised a safety brake that would work automatically as soon as a rope gave way. Made of a spring meant to lock the lifting rope in place using ratchet bars attached to guiderails affixed to shaft walls, the brake was first tested in 1854 in what was considered a death-defying stunt. After boarding an elevator cab in New York's Crystal Palace, Otis cut the rope hoisting the cab he was in before quickly setting the brake. The brake stopped the cab—which impressed the audience. Otis's brake soon became a standard fixture in elevators of the time, along with Frost and Stutt's belt-driven mode of operation. Otis's first safety elevator was installed in the Haughwout Building at 488 Broadway in New York City and carried passengers for the first time on March 23, 1857. Standing just five stories, the building did not require an elevator, but the building's owner, E. V. Haughwout, thought the novelty of the elevator would draw customers to shop at his department store. By 1870, Otis elevators were installed in 2,000 buildings.

Advances in design improved elevator performance. German inventor Werner von Siemens created the first motorized elevator. Its motor was fitted to the bottom of the cab. The motorized gears allowed the cab to climb shaft walls fitted with racks. In the United States, Frank Sprague added improvements in safety and engineering that controlled the speed and acceleration of the cab, as well as adding multiple cars to the same shaft. Sprague sold his company to the Otis Elevator Company in 1895. In 1873, John W. Meeker of Michigan submitted his design to the U.S. Patent Office for elevator hatchway protectors that allowed doors to open and close safely, a system that was perfected by Alexander Miles of Minnesota in 1874. Miles invented a door that closed access to the elevator shaft.

As safety improved, perfections were made to modern elevators to improve convenience, cost, and efficiency. The direct-connected geared electric elevator, invented in 1889, sped up travel. Just 14 years later, this system was improved upon with a gearless traction electric design, opening the way for taller buildings to be built. Multispeed motors increased efficiency; and automatic leveling—invented in 1915—guided elevator cars to a precise position level with the floor passengers were to enter or exit. Doors also were motorized in this era. Improvements in speed enabled cars to travel 1,200 feet per minute when New York's Empire State Building was constructed in 1931. Speeds increased to 1,800 feet per minute when the John Hancock Building opened in Chicago, Illinois, in 1970.

Other improvements were electromagnetic technology, which replaced the ropes and switches, and push-button controls. A system called "collective operation" enabled elevators to stop at sequential floors when called. The hoistway door interlock was a safety feature that would not permit the car to move unless the outer door, or shaft, was closed and locked. Elevators became completely automatic by 1950, eliminating the need for human attendants. Double-decker elevators were first invented in 1932. They were made up of two cars, one mounted on top of the other. They moved as a unit and could serve two floors at one time. Several North American buildings had double-deckers installed in the early 1970s, including the Time-Life Building in Chicago, Illinois, and the Canadian Imperial Bank of Commerce in Toronto, Ontario, Canada.

Rosemarie Boucher Leenerts

See also: Carmelit (Israel); Funiculars; Ski Lifts; Stairways.

Further Reading

Columbia Elevator. 2015. "Elevator History." http://www.columbiaelevator.com/main /elevator-history/. Accessed June 5, 2015.

Mitsubishi Electric. 2015. "History of the Elevator." http://www.mitsubishielectric.com /elevator/overview/elevators/history.html. Accessed June 5, 2015.

Otis United Technologies. 2015. Products and Services. "Elevators." http://www.otisworld wide.com/k2-elevators.html. Accessed June 5, 2015.

EMPIRE BUILDER, THE (UNITED STATES)

The Empire Builder is a U.S. long-distance passenger train operated by Amtrak, the national passenger train agency. It runs over 2,200 miles between Chicago, Illinois, and both Portland, Oregon, and Seattle, Washington (the westbound train is split into two at Spokane, Washington, and eastbound trains are joined there). Passenger service has been offered over this route since 1929. It was operated by the Great Northern Railway and Burlington Northern Railway before Amtrak took over in 1971. There are daily departures each way, with the total journey taking about 45 hours.

The route is considered one of the most scenic in America, and the line has the greatest passenger ridership on Amtrak outside of the Northeast Corridor. The highlight of the trip is the journey through the Rocky Mountains, with stops at

Empire Builder passenger car moves behind railroad crossing signs near Brook Park, Minnesota, 2014. (Holly Kuchera/Dreamstime.com)

Glacier National Park in Montana. It crosses the Continental Divide at the 5,213-foot Marias Pass. The train has a special relationship with Glacier National Park, as that park was created in 1910 in large part due to efforts by the Great Northern Railway. The railway also built many of the park's lodges, including the Many Glacier Hotel. The Empire Builder schedule was set so that trains would pass through the spectacular scenery of the Rockies during the daytime, and Amtrak has continued this tradition.

Joe Weber

See also: Alaska Railroad (United States); Amtrak (United States); Northeast Corridor (United States); Trans-Siberian Railway (Russia).

Further Reading

Yenne, Bill. 2005. *Great Northern Empire Builder*. Minneapolis: MBI.

ESCALATORS

An escalator is a motorized staircase that moves passengers up or down an incline. Used worldwide to continuously transport large numbers of passengers, escalators most commonly are found in shopping centers, subway and other transit systems, airports, and hotels. They often are used in places where elevators are considered impractical, or where building designers want passengers to be able to see what is around them as they travel from one level to another.

The world's first escalator was invented in 1859—two years after the first passenger elevator—by American Nathan Ames, who that year filed a U.S. patent for a machine he called "Revolving Stairs." It consisted of an equilateral triangle of moving steps that required passengers to jump on and off. Ames died the following year, however, and his invention never actually was built.

It wasn't until 1892 that another type of escalator was invented—American Jesse W. Reno's "inclined elevator," which rose to a height of seven feet over an incline of 25 degrees. That machine was first used by the public in September 1896, when it was installed as a temporary carnival ride at New York's Coney Island amusement park. Reno's first inclined elevator design consisted of a set of revolving wooden, T-shaped cleats.

Also in the late 1890s, George H. Wheeler patented another type of moving stairway that included a moving handrail and flat steps—items that greatly increased safety. In 1898, Charles D. Seeberger purchased Wheeler's patent and then introduced it to the nascent Otis Elevator Company, which became the first business to develop the idea commercially. Seeberger coined the name "escalator," which comes from combining the word elevator with the Latin word *scala*, meaning "steps."

The escalator made its worldwide debut in 1900. That year, Reno improved his inclined wheel tread and had a cleat-type moving stairway installed in a New York

City elevated train station, and the Otis Elevator Company introduced a step-type moving stairway at the Paris Universal Exposition.

Soon after, escalators began to appear in department stores and public transit systems throughout the northeastern United States and Europe; some of the earliest commercial escalators were built in a Philadelphia department store, on the Brooklyn Elevated Railroad platform in Manhattan, and in the nascent Paris Metro transit system. In 1906, the London Underground transit system installed Reno's newest design—a spiral escalator—in some of its stations. In 1911, Otis Elevators purchased Reno's patents and became the primary manufacturer of escalators.

Aside from their practical use of efficiently moving a large number of people, escalators also have greatly influenced the design of many buildings and public transportation systems. Whether installed in upscale shopping malls or for efficiency in major subway systems, escalators have forever changed the spaces in which they operate, often dictating the design of the buildings that accommodate them.

Today the world's longest escalators exist in the Saint Petersburg Metro transit system, where three stations have escalators reaching 449 feet long and 225 feet high. The longest escalator in the Western Hemisphere—at 230 feet long and 115 feet high—is in the Washington, DC, Metro transit system's Wheaton station. The Washington Metro system also boasts the most escalators of any transit system on the planet—572.

Escalator safety has improved greatly since the earliest designs appeared at the turn of the 20th century, but a series of major escalator accidents in China in 2015 again raised concerns about the machines' potential dangers. Within one week, a woman was killed, a toddler's arm was mangled, and a man's leg was damaged beyond repair on separate escalators at different shopping malls around the country. Escalator engineers and manufacturers elsewhere in the world have been quick to blame old, poorly maintained escalators for the problem, stating that escalators are overwhelmingly safe with proper upkeep.

Terri Nichols

See also: Elevators; Stairways; Washington Metro (United States).

Further Reading

Goetz, Alisa. 2003 (September 1). *Up, Down, Across: Elevators, Escalators, and Moving Sidewalks.* London: Merrell Publishers.

Otis United Technologies. "About Escalators." http://www.otisworldwide.com/pdf/aboutesc alators.pdf. Accessed August 31, 2015.

"Reno, Jesse." 2015. Lehigh University Distinguished Alumni. http://www.lehigh.edu /engineering/about/alumni/reno.html. Accessed August 31, 2015.

F

FLYING CARS

Humanity has long dreamed of a flying car. For the vast majority of our species' history, our mobility has been exclusively tied to the ground, traveling either by land or by sea. The development of air travel freed humanity from the shackles of roads and path networks necessary for transportation. Early aviation innovators envisioned flight to be more personally oriented and less the more familiar high-capacity vessels. Flying cars might have reached their most prominent place in human psyche in the post–World War II United States. In an era of seemingly limitless scientific growth and possibility, the popular TV cartoon "The Jetsons" featured travel by flying cars that came to be idealized by the Baby Boomer generation.

Despite these powerful images and century-plus years of technological development, personal mobility and in-flight travel have yet to meaningfully combine. After the Wright brothers' success at flight, the first attempt at building a flying car was in 1917 by the disputed father of the flying car, Glenn Curtiss, whose "auto-plane" sported three wings and a four-bladed propeller and managed a few hops off of the ground. Accidents and misgivings kept getting in the way of the development of the flying car. "The Airphibian" created in 1946 by Robert Fulton was the first airborne vehicle that was certified by the Civil Aeronautics Administration, the predecessor of the Federal Aviation Administration. The Airphibian had wings that were detachable and a propeller that could be stored in the fuselage. Although the vehicle could fly at 150 miles per hour and drive 50 miles per hour, a lack of funding killed the project (Bonsor 2000). In 1947, American airplane manufacturer Consolidated Vultee Aircraft Corporation planned to build 160,000 Convair Model 18 vehicles—a lightweight version of an automobile with wings that could be removed—until crashes during testing derailed the initiative (Cummings 2013). Many other similar false starts have plagued development of the flying car dream.

Flying cars come from the Vertical/Short Takeoff and Landing (V/STOL) family of aircraft. The advantage of these craft is the lack of runway space needed to get the craft off and on the ground. The most widely known example of this is the helicopter; there have also been military-grade examples of jet-propelled V/STOL aircraft. For V/STOL craft to function, the main requirement is an extremely high thrust-to-weight ratio to achieve vertical takeoff and landing directly against the force of gravity. This is easier to achieve the lighter the vehicle, and makes the power plant of the aircraft the most important component in achieving V/STOL. Most aircraft engines—especially jet engines—have plenty of thrust available

Accompanied out the door by their children, American inventor Robert Edison Fulton Jr. (1909–2004) and his wife, Florence, walk to their Fulton FA-2 Airphibian plane/car hybrid, Connecticut, 1949. (Bernard Hoffman/The LIFE Picture Collection/Getty Images)

to achieve this ratio; the main problem becomes directing the thrust downward. In aircraft that achieve a high enough thrust-weight ratio while also directing thrust exhaust downward, an additional issue arises in the placement of an engine of that size, which makes the aircraft's cruising thrust extremely inefficient. Vehicles that have separate lift and cruise engines partially address this issue, but additional weight from the extra engines also contributes to thrust inefficiency. The shorter the takeoff distance required and the heavier the aircraft (as well as the higher the altitude and the hotter the climate), the bigger the thrust of the engine required. The most common feature of propulsion system so far on the (very limited) flying cars that have been produced are ducted fans or shrouded propellers (as opposed to open propellers on civilian personal aircraft or commercial turboprop planes), or the much heavier jet engine. This type of engine produces the necessary thrust, but is somewhat inefficient at higher cruising speeds and also greatly limits the size and weight that the vehicle can be. To address these issues with physics, flying cars are low speed, lightweight, and short range. This also partly explains why flying cars that are currently deployed largely consist of close-surveillance vehicles (such as drones) or exclusive single or two-seat personal transport vehicles.

There are many issues with developing a flying car from laboratory potential and prototype testing to real-world adoption. One has been cost; most attempts so far have been far too expensive to be considered for mass commercialization, with price tags in the hundreds of thousands of dollars. Additionally, V/STOL vehicles are notoriously less safe than their long-runway airborne or ground vehicular counterparts. The high weight-to-thrust ratio makes it difficult for these vehicles to sustain lift, which leads to a dramatically increased accident rate. Helicopter crash rates are a notorious example of this.

There also are many regulations that a flying car must meet from both the aviation and automotive worlds that hold back commercialization. One is the licensing procedure for piloting a flying car. If the parallel is the licensing required for piloting a private airplane, then an individual interested in operating a flying

Fully Automated Navigation System

This is a navigation system equipped with electronics that is capable of determining the location and the direction a car is moving by collecting and processing information from satellites. Advances in electronics, computers, communications, controls, and sensor technologies enable scientists to develop a fully automated navigation system. This system will use high-tech electronics to monitor surrounding environments (e.g., roads, parking lots, buildings) and control the movement of a vehicle. This system also will have the capability to communicate with other vehicles, as well as with the roads—including the cameras, radar sensors, and traffic signals—to detect objects around the vehicle itself.

Automotive companies are focusing on building a fully automated navigation system and autonomous cars by 2020. A fully automated navigation system would need little to no human interaction to navigate roads and parking lots. This system also could sense when a vehicle's driver is getting drowsy and could awaken the driver or assume control of the vehicle. It also will enable people who otherwise are not able to drive a vehicle to be able to do so, for example, the elderly and persons with physical, visual, and mental disabilities. Similarly, because flying cars must be able to land and take off safely within highly congested urban settings, a fully automated navigation system must be able to handle the driving before "flying cars" become a reality.

Selima Sultana

car is facing a much longer, more difficult, and more expensive process than that required to meet the requirements for an automobile. A second is the regulation of the travel of flying cars. Again, the parallel is drawn with private and commercial air travel. Proliferation of flying cars would add demand to an air traffic control network that is already complex to manage. This complexity increases if those cars are added to urban air traffic networks, which already balance commercial activity and fixed flight plans and paths with the less regulated, more free-flowing civilian air traffic.

To date, there are more than 80 patents on file for various types of flying cars, and several currently in what is described as "serious development." The most immediate likelihood for proliferation is militarized use. Military applications for flying cars involve machines designed to get to hard-to-reach sensitive battlefield areas and deposit (and pick up) highly trained units (such as U.S. Navy SEALs) into difficult-to-access locations, or picking up personnel, cargo, or medical supplies. The U.S. Department of Defense, for example, currently is working with Lockheed Martin and Advanced Tactics on the production of such vehicles. The most prominent commercial civilian production initiative currently appears to be coming from Terrafugia, which is producing a road plane called the "Transition" that can cover 600 miles on the ground or 400 miles in the air, at a cost of $279,000 per vehicle. Other applications include vehicles as alternatives to helicopters for such tasks as herding cattle, short business or leisure travel, or medical emergency applications.

Although the Federal Aviation Administration (FAA) regulations and categorizations have loosened, the Jetson-like dream of everyone owning their own flying vehicle seems to have died. Most seem to agree that flying cars have a civilian future beyond extremely niche tasks and markets only with automation. Drones could fall into the realm of flying cars. There already are numerous military and surveillance applications. Additionally, a proliferation of civilian applications might be on the horizon. Amazon has reported experimenting with a drone delivery mechanism for smaller packages, and individual citizens already have access to and use personal drones in a variety of capacities.

Bradley W. Lane

See also: Autonomous Vehicles; String Transportation System; Vacuumed Trains.

Further Reading

Bonsor, Kevin. 2000 (December 1). "How Flying Cars Will Work." HowStuffWorks.com. http://auto.howstuffworks.com/flying-car.htm. Accessed February 5, 2015.
Cummings, Mary. 2013. "A Drone in Every Driveway." *Scientific American* 308(1): 28–29.
Hsu, Jeremy. 2014. "When Cars Fly." *Scientific American* 310(4): 28.
Kohler, Nicholas. 2012. "Back to the Future." *Maclean's* 125(41): 86–89.
Rusting, Ricki. 2009. "Flying Car." *Scientific American* 301(3): 71.
Saeed, B., and G. B. Gratton. 2010. "An Evaluation of the Historical Issues Associated with Achieving Non-Helicopter V/STOL Capability and the Search for the Flying Car." *The Aeronautical Journal* 114(1152): 91–102.
"What Happened to the Flying Car?" 2012 (March 3). *The Economist* 402 (8774): S.3–S.4.

FREEWAYS

A freeway is a type of highway with multiple lanes in each direction, is physically separated from oncoming traffic, and access is restricted except at on- and off-ramps. Crossing roads go over or under the freeway and connect only at interchanges. Speed limits generally are higher than on other roads, as are traffic volumes. Despite their high speeds, accident rates are lesser for these highways than for city streets and rural highways.

These roads were developed gradually in the early 20th century, but freeways also require special laws. A public road normally is open to all who live or own property alongside it, but a freeway requires that it can be entered or exited only at the interchanges. Early solutions were to build parkways, or highways running through long linear parks. The Merritt Parkway, which opened in Connecticut in 1938, is an early example of this type of road. Another type is a turnpike, which is a road that a company or agency has the right to operate a toll road. This right includes preventing access except at designated points. The Pennsylvania Turnpike, which opened in 1940, is a well-known and influential early example. An elevated road or skyway makes access impossible. Eventually a legal solution was found by passing laws in each state allowing the legal limitation of access to a road; however, not all highways with legal access control are freeways. The Blue Ridge Parkway is a spectacular two-lane scenic road but definitely is not a freeway. Freeways therefore are both a type of road design and a legal concept.

The term "freeway" first was applied to the Arroyo Seco Parkway (now called the Pasadena Freeway) in Los Angeles in 1939. The term referred to the absence of tolls on the highway, which were very common on similar roads in the northeast United States at the time. This term caught on and became a generic term, although turnpike and parkway still are used. Other countries use different terms, including "autobahn," "motorway," "autostrada," or "autopista."

Most large American metropolitan areas now have extensive freeway systems. Freeways are popularly associated with Los Angeles, but New York City was the first American city to have a substantial network thanks to the parkways built under the control of Robert Moses. The 1950s and 1960s were the golden age of freeway construction. Freeways are enormous and costly highways, and they have had tremendous unintended consequences in cities. In the 1960s, many urban residents became opposed to the massive construction required by new freeways. Many neighborhoods were demolished to make way for these highways. Across the country people began to protest new freeways to save their neighborhoods. This has been called the "freeway revolt" and achieved notable success in San Francisco and New Orleans. In most other cities the highway builders won. In recent years, a "teardown" movement has emerged, calling for older freeways to be removed in place of parks or landscaped streets. Conversely, many U.S. freeways now are old enough to be considered of historic significance. Although transit, cycling, and pedestrian travel are heavily promoted in metropolitan transport planning, the construction of freeways continues. Large, fast-growing cities such as Phoenix, Arizona, continue to build them, as do smaller cities that are growing into metropolitan status.

Many countries around the world have built freeways. Some are building large national networks, such as the National Trunk Highway System of China. This has many amazing bridges in rural areas and big cities such as Shanghai. Other countries have built one or more freeways on heavily traveled routes. Mexican Autopistas and Moroccan Autoroutes are examples of these. Freeways in some countries are toll routes and can be built by either the national government or private corporations. Almost all of these freeways were built by state highway departments and most are free of tolls.

There now are freeways in every state of the United States, totaling nearly 60,000 miles of roads. Nearly 43,000 miles of these are part of the Interstate Highway System. They provide the framework for travel in almost every city in the country and throughout much of rural America. They have seen considerable change over their nearly century-long existence, and will remain vital for the foreseeable future—although the types of vehicles driven on them in the might be very different from what we are used to today.

Joe Weber

See also: Autobahn (Germany); Autostrade (Italy); Interstate Highway System (United States); National Trunk Highway System (China).

Further Reading

Brodsly, David. 1981. *L.A. Freeway: An Appreciative Essay.* Berkeley: University of California Press.

Lewis, Tom. 1997. *Divided Highways: Building the Interstate Highways, Transforming American Life.* New York: Penguin.

Swift, Early. 2011. *The Big Roads: The Untold Story of the Engineers, Visionaries, and Trailblazers Who Created the American Superhighways.* Boston: Houghton Mifflin Harcourt.

FUNICULARS

A funicular is a form of transportation that combines aspects of both an elevator and a railroad to propel cars on rails up and down steep slopes, with the ascending and descending vehicles counterbalancing one another. The country of Turkey operates one of the oldest funiculars in the world, the Tünel, which began operations in 1875. It carries passengers between the former districts of Pera and Galata. The Tünel is the oldest underground funicular and one of the oldest underground railways, second only to England's London Underground, built in 1863. Turkey also operates a second funicular known as Kabataş-Taksim, which connects Kabataş with Taksim Square in Istanbul. This funicular commenced operations on June 29, 2006.

Funiculars have a unique mode of operation. Two cars, one ascending a slope and the other descending, are attached to each other by a cable. The cable runs through a pulley at the top of the slope. When one car ascends the slope, the other descends, and their weights counterbalance each other. The car heading down the slope minimizes the energy needed to lift the ascending car. An electronic winch turns the pulley. Water is another counterbalancing device used to propel some funiculars. The water tanks under the floors of the cars are filled and then emptied to achieve the perfect imbalance that enables the cars to move.

The invention of the funicular is credited to Swiss engineer Carl Roman Abt, who constructed 72 mountain railways worldwide. Early-day funiculars ran on two parallel tracks, with the tracks containing a sufficient space in between them for the cars to pass midway along the route. Modern-day tracks require less width.

The underground railway line in Istanbul, Turkey (the Tünel), has two stations and connects the quarters of Karaköy with Beyoğlu. It travels along the northern shore of the Golden Horn, a major urban waterway and the primary inlet of the

Funicular in Budapest, Hungary, 2009. (Stavrositu Iuliana/Dreamstime.com)

Bosphorus Strait. The Tünel's starting point is close to sea level and the Tünel travels uphill about one-third of a mile.

French engineer Eugène-Henri Gavand designed the Tünel in 1867 to connect the neighborhoods of Galata (now Karaköy), at sea level, with Pera (now Beyoğlu), about 200 feet up a steep hill. Galata was the business center of Constantinople (today's Istanbul) and many people working in that part of the city lived in Pera. At the time, the most common way to reach Pera was on foot along the Yüksek Kaldirim, a narrow street on which 40,000 people traveled each day. Gavand presented his railway plan to connect the two districts to Sultan Abdülaziz of the Ottoman Empire. The sultan granted Gavand permission to build the Tünel on June 10, 1869, and construction commenced on June 30, 1871. Construction was delayed several times because of conflicts between the company Gavand formed, the Metropolitan Railway of Constantinople, and property owners along the route. The Tünel finally was completed on December 5, 1874. The two trains were powered by two 150-horsepower steam engines. To prove its safety, the Tünel's very first trains only carried animals. After several successful test runs, the first trains carrying people were inaugurated on January 17, 1875.

Today, the publicly run Istanbul Elektric Tramway ve Tünel (IETT) owns and operates the Tünel. There are still only two stations, one at Karaköy and the other at Beyoğlu. The original trains were made of wood and each train contained two cars, one reserved for passengers (who were separated by class and sex), and the other transporting goods and animals. In contrast, today's electrified steel cars run on pneumatic tires, which replaced the wooden cars in 1971. They cruise at just 16 miles per hour, taking 1.5 minutes to go from one station to the other. A second funicular was built in 2006 and connects Kabatas to Taksim. The trains run every 5 minutes and the trip between the stations takes just 150 seconds. Each day an average of 30,000 people ride this funicular, and the train makes 195 trips.

Rosemarie Boucher Leenerts

See also: Aerial Tramways; Cable Cars; Elevators: Rack Railways; Railroads.

Further Reading

Istanbul Trails. Transportation. 2009. "Getting Around in Istanbul by Metro, Tram and Funicular." http://www.istanbultrails.com/2009/08/getting-around-in-istanbul-by-metro-tram-and-funicular/. Accessed June 5, 2015.

Istanbul Trails. Transportation. 2009. "Tünel, the Shortest and Third-Oldest Passenger Underground in the World." http://www.istanbultrails.com/2009/01/tunel-the-shortest-and-third-oldest-passenger-underground-in-the-world/. Accessed June 5, 2015.

G

GHAN TRAIN, THE (AUSTRALIA)

The Ghan is a passenger train that crosses the great Australian Outback between Adelaide and Darwin on a 1,851-mile-long line. It is a very popular tourist attraction in Australia. The train started operations on August 4, 1929. At an average speed of 53 miles per hour, the Ghan can make a one-way journey in just about 54 hours (two and a half days) with a four-hour stopover in Alice Springs. The Ghan Train is an abbreviated version of its original name "The Afghan Express," named after pioneer camel drivers that arrived in the 19th century from Afghanistan. These were collectively referred to as "Afghans" or "Ghans." The railway was named when an Afghan passenger bolted off the train to face Mecca and recite his evening prayers. A rail worker joked that if that man had been the only passenger, the train should be called the Afghan Express, and since then this name has remained attached to the Ghan.

The Ghan Train's original line followed the route of explorer John McDouall Stuart, traveling from Adelaide to the town of Stuart (now Alice Springs). In its initial run, the train carried supplies and about 100 passengers to this remote town. Soon, the Ghan would travel back and forth between these two towns and assist in the growth of Alice Springs. The original line, despite its efficiency, faced some hardships in the Outback. Intense heat and flash flooding led to very irregular services. Flash floods are quite common in the Outback, especially in the low-lying plains the train ran through. Floods were notorious for blocking paths and even lifting rails from the ground. Because of these issues, the old track was abandoned in 1980 and a new track was set in place a few miles west of the original. A new standard gauge rail line also was set in place, complete with termite-proof concrete sleepers. The new Ghan train embarked on its first transcontinental journey from Adelaide to Darwin on February 1, 2004.

Selima Sultana

See also: Alaska Railroad (United States); Bangladesh Railway; Qinghai-Tibet Railway (China).

Further Reading

Great Southern Railway. "The Ghan." http://www.greatsouthernrail.com.au trains/the
    _ghan/. Accessed July 8, 2015.

GOING TO THE SUN ROAD (UNITED STATES)

The Going to the Sun Road is one of the most famous and spectacular scenic roads in the U.S. national park system. It is located in Glacier National Park in Montana and is named after Going-to-the-Sun Mountain, a very prominent peak along the road. It is 53 miles long and is the only road that spans the entire width of the park. It crosses the Continental Divide at Logan Pass, the highest point along the road. Its endpoints are both lakes—Lake MacDonald on the west side and Saint Mary Lake on the east side. The road climbs above timberline and provides many views of glacier-carved mountains. There are many overlooks along the road and numerous points of access to a variety of trails that lead into the mountainous wilderness of the park.

Construction took more than 10 years, and the road was completed in 1932. It has proved to be the most popular attraction in Glacier National Park and one of the most scenic drives in any national park. The road was built in a rustic style, and local stone was used to build retaining walls and guardrails. Today the road is considered a model of such design, and now is itself a historic structure.

Because of its elevation and latitude the road receives very heavy snowfall and must be closed each winter. The road can be buried under 80 feet of snow, which takes many weeks to clear. The road over Logan Pass usually does not open completely for the summer tourist season until June or July. The lower sections of the road at either end usually are open year-round.

Joe Weber

See also: Blue Ridge Parkway (United States); Natchez Trace Parkway (United States); National Scenic Byways (United States); Parkways.

Further Reading

Library of Congress. Prints and Photographs. Historic American Buildings Survey. "Going-to-the-Sun Road, West Glacier, Flathead County, MT." http://loc.gov/pictures/item/mt0242/. Accessed August 5, 2015.

National Park Service. U.S. Department of the Interior. 2015. "Glacier National Park." http://www.nps.gov/glac/index.htm. Accessed December 19, 2015.

GRAND CANYON MULE TRAIN (UNITED STATES)

Transport is difficult in the vast and rugged Grand Canyon of Arizona. To reach the bottom of the canyon requires a long hike or the assistance of a sturdy four-footed animal. Although elsewhere a horse or camel is used, at the Grand Canyon the preferred animal always been the mule—the offspring of a horse and donkey. Since the earliest days of tourist activity at the canyon, mule trains have been used to carry visitors to the bottom of the canyon and then back up again the next day. The ride from the South Rim to Phantom Ranch (at the bottom of the canyon) descends 4,800 feet over 10.5 miles. This route takes more than five hours going down and seven hours climbing out of the Grand Canyon. The trips are very popular (and expensive) and must be scheduled well in advance. Mule trains also haul supplies and mail to Phantom Ranch (the only place in the United States where mail is carried using animals). This form of transport has remained unchanged for more than a century and will likely continue unchanged for many years to come.

Joe Weber

See also: Camels; Llamas.

Further Reading

Henry, Marguerite. 1991. *Brighty of the Grand Canyon.* New York: Aladdin.
National Park Service. 2015. "Plan Your Visit. Things to Do. Mule Trips." http://www.nps .gov/grca/planyourvisit/mule_trips.htm. Accessed December 7, 2015.

GRAND CANYON RAILWAY (UNITED STATES)

In 1901, visitors first reached Arizona's Grand Canyon by train when the Atchison, Topeka and Santa Fe (ATSF) railroad completed a branch line northward from its main transcontinental line. Travelers on the ATSF could take a scenic detour to see the Grand Canyon, staying at the luxurious El Tovar hotel—owned by the

Vintage Grand Canyon Railway tourist train engine on curved track. (Scott Griessel/Dreams time.com)

ATSF—on the very rim of the canyon. The Grand Canyon attracted more and more visitors and, over time, they increasingly came by car. The passenger trains faced financial hardships. The line ceased carrying passengers in 1968 and was shut down entirely a few years later. The rails likely would have been taken up and sold for scrap, but in 1988 it was bought by private investors. The tracks were restored and passenger trains again began running to the Grand Canyon the following year. This new service, the Grand Canyon Railway, runs 64 miles from the town of Williams—where Amtrak passenger trains still run on the old ATSF transcontinental line—to the south rim of the canyon. The original ATSF train stations at both ends were restored and serve passengers today. Both steam and diesel engines are used by the railway, and many special excursions are offered.

Joe Weber

See also: Amtrak (United States); Grand Canyon Mule Train (United States); Railroads.

Further Reading

Grand Canyon Railway Depot. 2008. http://grandcanyonhistory.clas.asu.edu/sites_southrim _railwaydepot.html. Accessed August 2015.
"Welcome to the Grand Canyon Railway." 2015. http://www.thetrain.com/. Accessed August 2015.

GREAT BELT FIXED LINK (DENMARK)

The Great Belt Fixed Link spans the 11-mile-long Danish *Storebælt* (Great Belt) strait via two bridges and a tunnel. It connects the Danish islands of Zealand and Funen and, by extension, all of Scandinavia. When the multibillion-dollar project was completed in June 1998, it was Denmark's largest-ever engineering venture— and arguably its most economically important. The Fixed Link for the first time enabled trains and vehicles to travel between Denmark's Jutland Peninsula on the European mainland and Denmark's capital city, Copenhagen. In a larger context, it provided the final land-based connection between southern Spain and Norway, leaving Ireland as the only European nation not physically connected to the mainland.

The Great Belt Fixed Link uses the tiny, uninhabited island of Sprogø to make the land link between Zealand and Funen possible. Sprogø lies about halfway across the Great Belt strait and previously was slightly more than 90 acres in size. To accommodate connections for the east and west sections of the Fixed Link, Sprogø's landmass was expanded to about 380 acres. The original 90 acres have been set aside as a protected wildlife area, and the rest lies within the domain of the Fixed Link.

Connecting Sprogø to Zealand is the East Span, consisting of a 4.2-mile-long bridge for vehicles and two parallel, 5-mile-long tunnels for trains. The East

Bridge boasts one of the longest free, or suspension, spans in the world—5,328 feet—between two pylons. At 833 feet tall, the pylons are the highest points in Denmark (the country's highest natural point rises to just 561 feet). On either side of its free span, the East Bridge extends across another 23 smaller approach spans (14 leading to Zealand and 9 leading to Sprogø). Connecting Sprogø to Funen is the West Span—a double bridge accommodating vehicles on one side and trains on the other. The 4.1-mile-long West Bridge is composed of 63 separate spans.

The two tunnels comprising the East Tunnel each accommodate one set of train tracks, allowing rail traffic to travel east in one tunnel and west in the other. Excavated using four boring machines that were specially built just for the Fixed Link project, the tunnels run between 40 feet and 131 feet below the seabed for most of their length and reach 246 feet below sea level at their deepest point.

Construction of the Fixed Link began in 1988. The rail portion opened in June 1997, and the road portion opened in June 1998. The Fixed Link's impacts on travel and transport were immediate, reducing travel time across the Great Belt strait from one hour via ferry to seven minutes via car. As the first land-based link from the Baltic Sea to the North Sea, and from mainland Europe to the bulk of Scandinavia, its influence on trade also has been significant. Within just five years, Danish authorities had recorded 34 million separate trips across the Fixed Link—an increase of 77 percent when compared with trips across the Great Belt by ferry in the previous five years. Commuter business was estimated to have increased up to fifteenfold.

Terri Nichols

See also: Channel Tunnel (France and United Kingdom); Railroads; Suspension Bridges.

Further Reading

Railway-Technology.com. 2015. http://www.railway-technology.com/projects/denmark/. Accessed September 3, 2015.

Storebælt website. 2015. http://www.storebaelt.dk/English. Accessed September 3, 2015.

GREENWAYS

A greenway is a pathway for nonmotorized travel set within a linear, open space corridor. Greenways are "green" in that they are located within vegetated park like settings and offer numerous environmental benefits. Greenways are "ways" or routes that go places and connect to destinations. Greenways are designed to provide a continuous path with sufficient width and an appropriate surface for safe use. They often follow linear natural features such as rivers, shorelines, or ridgelines. Greenways have become popular in recent decades as a way of creating safe and enjoyable routes for nonmotorized travel, promoting exercise and healthy lifestyles, improving the health of local ecosystems, and raising local property values. These areas give users everyday access to the natural world, enabling them to move through the landscape using a self-propelled, nonpolluting mode of travel. They commonly are used by walkers, runners, bicyclists, and in-line skaters. Many greenways have been designed to avoid barriers and obstructions so they can be used by individuals in wheelchairs. Depending upon local conditions, some greenways are open to other modes of nonmotorized travel including snowmobiling, cross-country skiing, and horseback riding.

The concept of a greenway was inspired by both parkways and greenbelts. Parkways were a 19th-century innovation that placed scenic roadways within generously landscaped open spaces. As automobile traffic increased on parkways, there arose a desire for separate greenways for bicyclists and pedestrians. A greenbelt is a continuous expanse of protected open space surrounding a city. American regional planner Benton MacKaye (1879–1975) proposed combining a greenbelt around the perimeter of a city with spokes of recreational open space reaching inward toward the center of the urban area. MacKaye's model has inspired the design of many metropolitan regional greenway networks. In 1921, MacKaye also proposed what grew to be America's greatest greenway, the 2,100-mile Appalachian National Scenic Trail.

Acquiring land for a greenway is complicated because of the many properties a path potentially could cross. Thus, many greenways are created from abandoned railroad lines, former canal towpaths, or along existing transportation or utility corridors. Other greenways are created in floodplains where permanent buildings are not allowed. In Denver, a greenway built along the polluted and flood-prone South Platte River contributed to the restoration of the river. Greenways can improve water quality by filtering sediment and pollutants out of storm water runoff and allowing rainwater to infiltrate and replenish aquifers. Greenways also function as ecological corridors that reduce habitat fragmentation, facilitate wildlife movement, and increase biodiversity.

The growth of bicycling is a primary driver for the growth of greenways. When bicycles first became popular in the 1890s, the urgent need was for smooth roads on which to ride a bicycle. Today, most cities have an abundance of paved streets, roads, and highways built for motorized vehicles; and the urgent need is for bicycle infrastructure that protects bicyclists from the hazards of automobiles, buses, and trucks. In every European city with high rates of bicycle use there is an extensive bicycle-route system with bicycle paths separated from motor vehicle traffic. Hannover, Germany, for example, built a 50-mile greenbelt encircling the city with biking and hiking trails that cross hills and river valleys. In addition to its many bike routes on city streets, Copenhagen, Denmark, has added greenways for longer commutes and recreational rides.

North American cities where a significant percentage of trips are made by bicycle are those with extensive greenway and bicycle path systems such as Davis, California; Boulder, Colorado; Victoria, British Columbia; Portland, Oregon; and

Minneapolis, Minnesota. Located at the edge of the Rocky Mountains, Boulder has protected its natural resources and encouraged outdoor recreation through a greenbelt and a network of greenways along Boulder Creek and its tributaries. Portland benefits from the state-funded and state-operated 255-mile Willamette River Greenway. In Minneapolis, a sunken railroad corridor was turned into the Midtown Greenway, which connects inner-city neighborhoods with the city's parkway system.

Louisville, Kentucky, has used greenways to expand a 19th-century parkway system into a countywide network of interconnected parks, open space, and trails. New York City turned a disused elevated railway into the beautifully landscaped High Line greenway, which attracts tourists and residents alike. In the Pearl River Delta region of China, plans for six extensive regional greenways are a part of attempts to improve the livability of this industrialized region. In summary, greenways are popular around the world because they promote environmental sustainability and give people the opportunity to transform their everyday travel into healthy exercise and contact with the natural world.

Mark Bjelland

See also: Appalachian Trail (United States); Bicycle Lanes; Bicycles and Tricycles; Parkways; Wheelchairs.

Further Reading

Flink, Charles A., and Robert M. Searns. 1993. *Greenways: A Guide to Planning, Design, and Development.* Edited by Loring Schwarz. Washington, DC: Island Press.

GUOLIANG TUNNEL (CHINA)

Construction of China's unique Guoliang Tunnel began in 1972 to provide an alternate mode of transportation accessibility to the village of Guoliang, located at the top of Taihang Mountain. Guoliang is a very small, isolated settlement that at one time had a communication and trade barrier separating it from the rest of the world. This village originally was accessible only by steps or a sky ladder, until 13 villagers took on the immense task of constructing a 0.75-mile-long tunnel so that cars could reach the village. It took the 13 strong-willed villagers—slowly carving away at the mountainside with hand tools—five years to accomplish this task. The workers created windows from the tunnel to throw debris and rubble off the side of the cliff while carving, which in turn created wide outlooks for those travelling through the tunnel. At 12 feet wide, this is virtually a one-way road carved into the facade of a towering rock face in the mountains and is one of the most famous and most spectacular tunnels in the world.

Selima Sultana

A minivan travels along the Guoliang Tunnel in the Taihang Mountains in central China's Henan Province, 2012. The Guoliang Tunnel was built by local villagers and opened in 1977. (Imaginechina/Corbis)

See also: Border Tunnels (Mexico and United States); Seikan Tunnel (Japan); Stormwater Management and Road Tunnel (SMART Tunnel) (Malaysia); Tünel (Turkey).

Further Reading

LosApos.com. Road Trip Routes. Road Trips in Asia. "China's Guoliang Tunnel Road." http://www.losapos.com/guoliang_tunnel_road. Accessed August 5, 2015.

H

HAMBURG S-BAHN (GERMANY)

Serving the Hamburg Metropolitan Region in Germany, the Hamburg S-Bahn is a rapid transit railway system that is part of the Hamburg Public Transport Network (known as HVV for Hamburger Verkehrsverbund GmbH), which also includes the underground rapid transit system (Hamburg U-Bahn), the AKN Railway commuter and freight trains (A-Bahn), and the regional rail (R-Bahn). The Hamburg S-Bahn is noteworthy among German transit systems for its use of both direct current (DC) supplied by a third rail and alternating current (AC) supplied by overhead lines. It is also known for its dense schedule and wide-ranging coverage of the metropolitan area. Unlike transit systems in Berlin and Hanover, the S-Bahn's network runs entirely within city limits except for one 20-mile stretch into the Lower Saxony region that was added in 2007.

The Hamburg Public Transport Network is a subsidiary of DB Regio and part of the Deutsche Bahn short- and medium-distance commuter train services. The S-Bahn runs six lines that serve 68 stations along 91 miles of track. Ten of the stations are completely underground. Nearly 600,000 commuters ride the S-Bahn on any given workday, which equals a total of approximately 220 million riders annually. Since January 2010, Hamburg's S-Bahn network is supplied with carbon-dioxide-free electrical energy that is generated from hydroelectric power plants.

"S-Bahn" is an abbreviation of the German word "*Stadtschnellbahn*," meaning "city rapid rail." Today, many cities and metropolitan areas in Germany operate S-Bahns, all of which had branched off from main-line railways operating steam locomotives. Unlike the locomotives, the new electric trains provide service within metropolitan areas and from suburban to urban areas. They have proven to be a quieter and cleaner form of transportation than steam-powered trains—an important aspect in heavily populated sections of cities. Berlin, in 1891, was the first city to run an S-Bahn, although the term was not coined until 1930; Hamburg was second. Hamburg's S-Bahn spawned from the rail system that was in place since 1866. To accommodate the new electric system, two separate tracks were laid in 1906. The first electric trains boarded passengers on October 1, 1907. The first exclusively electric line ran between Blankenese and Ohlsdorf stations, a 13-mile route.

The network expanded in 1924, when a local company that had built and run a railway in the district of Stormarn went bankrupt and was taken over by the German Imperial Railway—the Deutsche Reichsbahn—the national agency formed after World War I. The German Imperial Railway turned its new purchase into an electric rail and extended the line, which was designated an S-Bahn in 1934.

Beginning in 1939 and continuing for a decade and a half, the trains and infrastructure underwent an overhaul. Part of the renovation included converting the overhead AC electric lines to third-rail DC power. To allow for improved acceleration over the Berlin trains that ran at 750 volts, the Hamburg S-Bahn trains' third-rail power was set at 1,200 volts. The train carriages also were replaced at this time. The AC trains had been composed of two articulated carriages on six axles and with doors on each side of the compartments. The new DC trains consisted of three carriages on four axles, with four sliding double doors on each side of the carriage.

In the 1940s, the S-Bahn network expanded along the main line to Berlin, but many wartime disruptions to the Berlin system resulted in several sections of the route being closed. After World War II, the division of Germany between communist East Germany (German Democratic Republic) and free West Germany (Federal Republic of Germany) caused further disruptions to main-line train travel between Hamburg and Berlin. To extend service to commuters living in the suburbs of Hamburg in the direction of Berlin, an S-Bahn line to Bergedorf was added in 1958. This was the first section in which the S-Bahn and main-line trains shared tracks. (Service was not fully restored to Berlin until the early 1990s, after the fall of the Berlin Wall separating communist East Berlin and free West Berlin and the reunification of Germany.) The Hamburg S-Bahn has experienced several more extensions over the years, most notably in 1969, 1979, 1983, and 1984. Recent work includes a four-mile northwestern extension of the S3 line to Stade and a two-mile extension from Ohlsdorf to the Hamburg airport, completed in December 2008.

The German Imperial Railway teamed up with two other local transport companies to implement a common tariff system for bus and U-Bahn travel in 1965, and the S-Bahn joined the system in 1966. At that time, the rail lines were numbered, with the Hamburg S-Bahn first carrying the designations of S1 through S6. Currently its six lines are labeled S1, S21, S3, and S31 for the main lines, along with S11 and S2. Cities outside Hamburg, in other parts of Germany, as well as in Austria and Switzerland followed suit, labeling their S-Bahn networks with a letter "S" and U-Bahn networks with a letter "U." Currently, the HVV has a one-ticket, one-price ticketing system among its more than 30 public transport operators, which include bus, ferry, rapid transit, and regional rail lines. Ticketing is on the honor system, whereby no ticket need be presented upon entering the vehicle. Tickets are checked, however, and if a rider has not purchased a ticket in advance then the rider is fined on the spot.

Rosemarie Boucher Leenerts

See also: Berlin U-Bahn/S-Bahn (Germany); Rapid Transit.

Further Reading

Hamburg. UrbanRail.Net. 2004. www.urbanrail.net/eu/de/hh/hamburg.htm. Accessed September 3, 2015.

Piekalkiewicz, Janusz. 2008. *The German National Railway in World War II*. Atglen, PA: Schiffer.

Public Transport in Germany. 2015. www.german-way.com/travel-and-tourism/public -transport-in-germany/. Accessed September 3, 2015.

HEADLOADING (SUB-SAHARAN AFRICA)

Many areas in Sub-Saharan Africa (SSA), particularly the rural areas, are characterized by poor spatial accessibility. Spatial accessibility here refers to the ease by which one location can be reached from another. In fact, SSA is considered as the least-accessible region in the world. The region has a total road network (the dominant transport system) of two miles per 1,000 people, as compared with a world average of five miles. In the rural areas, reliable or navigable roads form a small proportion (about 34 percent) of the entire transport network and the rest are in deplorable conditions (African Development Bank 2010).

The limited accessibility in SSA results in a high cost of transportation. Such conditions discourage potential investors from investing in the transport sector, leading to acute shortages that in turn result in prohibitive transport charges. Consequently, the significant cost of transportation coupled with limited transport supply and pervasive high poverty levels compel a majority of the locals, particularly women, to resort to cheaper and readily available means of transportation. The most common is headloading.

Headloading is an alternative means of transportation characterized by walking and carrying of goods on the head. It is popular in the developing world, and particularly in SSA, even in the midst of new innovations in transport systems. It is believed to be well suited to the precarious nature of most rural terrain—unpaved, rugged, and flood-prone roads. Although it is practiced by both sexes, in SSA headloading culturally is regarded as a female activity. Women—oftentimes with their children—use this mode in their daily activities, such as for water collection, fuel harvesting, and transporting farm produce to market centers. Even in places where there is considerable vehicular operation, some women still head-load their goods to the nearest paved roads in an attempt to reduce costs.

Headloading also is practiced in urban centers. In Ghana, paid headloading—popularly called *kayayei*—is popular in major cities and market centers. It is tagged as a low-status occupation and often is undertaken by young female migrants from the less-developed northern regions of the country. Also, in urban areas where the water supply is irregular, residents often harvest water using headloading.

Many studies have reported that locals often carry loads of up to 20 percent of their body mass when headloading but expend no extra energy (Maloiy et al. 1986). Although these findings defy the norm, some scholars have concluded that years of practice have led to the development such remarkable physical

African woman head-loading a bag of goods, 2005. (Nlphotos/Dreamstime.com)

capabilities. Headloading, however, has been found to be associated with some health complications, including backache, headache, chest pain, and deformation of the spine. Other reported health issues include miscarriage and the malformation of unborn fetuses. Nonetheless, there is very little empirical evidence backing these claims and they have not been conclusively established. Headloading also is associated with negative energy balance, which is attributed to malnourishment of individuals practicing headloading. People who perform this task tend to have low purchasing power, which constrains them from obtaining enough food. Such conditions eventually could result in other health issues such as stunted growth and low cognitive functioning.

Headloading remains the most common and affordable means of transportation in SSA. Though economically efficient, headloading is comparatively less efficient in terms of time and load capacity—many hours that could be devoted to other productive ventures instead are spent walking. Consequently, some scholars have recommended the adoption of Intermediate Means of Transport (IMTs); a simple, nonmotorized, and cheaper form of transportation. The lack of operation skills, however, as well as poverty has limited the full adoption of this system. The construction of culverts and bridges over water-crossing points to help ensure uninterrupted movement of vehicles also has been suggested. At present, however, the deficient transport services still prevalent coupled with widespread economic poverty mean that headloading is expected to remain popular in the region.

Nathaniel Dede-Bamfo

See also: Bicycles and Tricycles; Rickshaws (Bangladesh).

Further Reading

African Development Bank (ADB). 2010. "Infrastructure." http://www.afdb.org/en/topics -sectors/sectors/infrastructure/. Accessed May 13, 2014.

Dweck, Jessica. 2010 (August 27). "Head Case: The Art and Science of Carrying on Your Head." *Slate*. http://www.slate.com/articles/news_and_politics/explainer/2010/08/head _case.html. Accessed April 13, 2012.

Maloiy G. M., N. C. Heglund, L. M. Prager, G. A. Cavagna, and C. R. Taylor. 1986. "Energetic Cost of Carrying Loads: Have African Women Discovered an Economic Way?" *Nature* 319: 66869.

Mwase, N. R. L. 1989. "Role of Transport in Rural Development in Africa." *Transport Reviews* 9(3): 235–53.

HIGH-OCCUPANCY TOLL (HOT) LANES

See High-Occupancy Vehicle (HOV) Lanes (United States).

HIGH-OCCUPANCY VEHICLE (HOV) LANES (UNITED STATES)

High-occupancy vehicle (HOV) lanes are specially designated highway lanes that are reserved for use by carpools, vanpools, buses, and sometimes motorcycles. Also known as "carpool lanes" or "diamond lanes," HOV lanes most commonly are found in urban areas that experience significant levels of vehicle traffic and that have several lanes on the roadways. The HOV lanes are intended to reduce traffic congestion and air pollution by providing an incentive for travelers to join carpools or to take buses. Vehicles using HOV lanes usually move more quickly through traffic, although in some regions HOV lane popularity has decreased their effectiveness.

Each state has its own laws regarding the use of HOV lanes but, in general, these corridors are located in the far-left lane and are separated visually from the rest of the highway or freeway by the use of specialized pavement markings, electronic signs, and (in some cases) raised barriers. In some states HOV lanes serve their primary function 24 hours a day, in others they operate only during designated times when traffic levels are highest and could be open to general traffic the rest of the day and at night. In places where traffic congestion is worse in one direction in the morning and in the opposite direction in the evening, some regions reverse the direction of the HOV lanes depending upon the time of day.

In some places, particularly where use of HOV lanes is low, HOV lanes have been converted to "HOT" (high-occupancy/toll) lanes. This designation allows vehicles carrying fewer than the required number of passengers to use the HOT lanes for a fee. Elsewhere, HOV lanes are so well utilized that the number of passengers required for a vehicle to use them has been increased. In all states that have HOV lanes, there are fines in place to punish drivers who use these facilities without having enough passengers.

The designation of certain lanes for vehicles carrying more than one passenger began in the early 1970s, as the number of cars on U.S. roadways increased along with concerns about reduced air quality due to vehicle emissions. The first such lanes—on Northern Virginia and Washington, DC's Shirley Highway, and along the entrance to New York and New Jersey's Lincoln Tunnel—were for buses only. By the 1980s, however, most HOV lanes were used primarily by passenger cars and trucks carrying multiple travelers.

Today there are more than 150 HOV lane projects in metropolitan areas across the nation, and federal transportation grants strongly support such facilities. Many HOV lane projects include park-and-ride parking lots at major intersections; these give travelers a safe place to leave extra vehicles when carpooling. In some regions, HOV lanes are estimated to be carrying as much as one-third of all travelers along routes where they are used.

As the number of vehicles on U.S. roadways and the number of miles traveled continues to increase, however, HOV lanes alone are not always sufficient to reduce traffic congestion. According to the U.S. Department of Transportation's Federal Highway Administration, the number of miles of urban roadways nationwide has increased by about 60 percent since 1980. At the same time, however, the number of miles traveled by vehicles on the nation's roadways has increased by approximately 120 percent.

Terri Nichols

See also: Autobahn (Germany); Car Sharing; Freeways; Interstate Highway System (United States).

Further Reading

Collier, Tina, and Ginger Goodin. 2004 (November). "Managed Lanes: A Cross-Cutting Study." Federal Highway Administration. FHWA-OP-05-037. Texas Transportation Institute, Texas A&M University System, College Station, TX.

U.S. Department of Transportation. 2002. *2002 Status of the Nation's Highways, Bridges, and Transit: Conditions & Performance, Report to Congress.* Washington, DC.

U.S. Department of Transportation. Federal Highway Administration. 2007 (July). "Active Traffic Management: The Next Step in Congestion Management." http://international .fhwa.dot.gov/pubs/pl07012/. Accessed September 1, 2015.

U.S. Department of Transportation. Federal Highway Administration. 2012 (November). "Federal-Aid Highway Program Guidance on High Occupancy Vehicle (HOV) Lanes." http://ops.fhwa.dot.gov/freewaymgmt/hovguidance/index.htm. Accessed September 1, 2015.

HIGH-SPEED RAIL

High-speed rail (HSR) refers to a new type of railway with advanced technologies for fast-speed operation. The HSR's fast speed requires complex operation systems, such as an upgraded signal and control system; high-quality tracks supported by tunnels and viaducts; and high-performance rolling stocks using tilting trainsets, aerodynamics, air brakes, regenerative braking, engine technology, and dynamic

High-speed rail hub station in Beijing, 2015. (Linqong/Dreamstime.com)

weight shifting. The term "HSR" covers these complex components of operating and developing the HSR system.

The criterion of high speed varies with the different operators. The European Union defines HSR as railway with the designed speed of 150 miles per hour on new high-speed tracks or 124 miles per hour on upgraded conventional tracks. High-speed rail has four different operation models: dedicated HSR, mixed high-speed, mixed conventional, and fully mixed. The dedicated HSR model involves high-speed trains on dedicated HSR tracks separated from conventional railways (e.g., the Japanese Shinkansen). The mixed high-speed model operates high-speed trains on conventional tracks as well as on dedicated HSR tracks (e.g., France's Train à Grande Vitesse (TGV)), which enable expanding HSR service coverage beyond HSR tracks. Some high-speed trains use different operation models when crossing borders (e.g., Spain's Alta Velocidad Española [AVE]). In Germany, HSR tracks are offered for all types of trains, such as ICE (high-speed train), conventional passenger trains, and freight trains.

Since beginning operation of the Tokaido Shinkansen between Tokyo and Osaka in 1964, HSR has been developed in many countries in Europe and East Asia. France launched TGV Sud-Est, serving Paris to Lyon in 1981 and improved the maximum operation speed to 186 miles per hour in 1989. Germany began Intercity-Express (ICE) service connecting Hamburg to Frankfurt and Munich in 1991. Spain opened AVE service between Madrid and Seville in 1992. Later, HSR began operation in England, South Korea, Switzerland, Taiwan, the Netherlands, and China.

Hyperloop Trains

Presently only a concept, the proposed Hyperloop train is a superfast, affordable, and more energy-efficient form of high-speed rail than any ground-transportation modes that currently exist. The concept was proposed publicly in 2013 by Elon Musk, CEO of Tesla Motors and SpaceX, in a paper published on the SpaceX website. Using hyperloop transportation technology, the train promises to transport passengers in floating pods inside low-pressure tubes at speeds of 800 miles per hour using magnets. Powered by solar panels along the tube, the system also would have a set of fans attached to the pods to enable the train to rest on a cushion air. Currently, the five-mile-long world's first hyperloop test track is proposed to be built in central California along the I-5 to be used to test the potential of this train, according to Hyperloop Transportation Technologies Inc. If the Hyperloop train passes the test and the company can alleviate all other technological concerns, then this train could transport passengers from Los Angeles to San Francisco in only 30 minutes—which is much faster and quieter than a maglev train.

Selima Sultana

Today, 14 countries operate HSR service, with a combined length of 13,670 miles. Additionally, more than 8,000 miles of HSR tracks were under construction by 2015. Since the introduction of the TGV, the HSR network in Europe has been extending rapidly to connect rail networks and facilitate integration between European countries. As a result, international express trains have operated on various HSR and conventional tracks crossing the borders of European countries (e.g., Brussels to London and Paris, Madrid to Barcelona and Marseille, Barcelona to Lyon or Paris). The European HSR network reaches 4,568 miles and will extend to 13,108 miles by 2025.

East Asian countries also have been extending their HSR networks. China has rapidly developed its HSR network since 2003. The Chinese government has invested considerable funding in HSR projects to develop four north-south lines and four east-west lines. As a result, China has the world's longest HSR network of 6,917 miles and plans to extend it to 13,969 miles. Along with the growth of HSR networks, the various technologies integrated with HSR—such as construction, operation, and train manufacturing—also have been growing in China.

HSR has various economic and environmental benefits as an efficient mass-transportation system. Its fast speed—surpassing other ground transportation modes—significantly reduces travel time. This increases the turnover ratio, which consequently raises the service capacity. HSR systems also contribute to economic development by promoting interaction between cities and creating jobs and investments with improved technology levels. HSR provides alternatives for passengers with its safety and reliability—avoiding traffic congestion, car accidents, and the effects of bad weather conditions. HSR has an advantage of accessibility to city centers as compared to the outer-city location of many airports. The competition between transportation modes upgrades service quality, especially in fare systems and convenience of traveling, which impacts the growth of low-cost airlines. The efficiency of HSR also contributes to reducing greenhouse gas emissions by using electricity and by carrying more passengers.

Based on these benefits, HSR has shown competitiveness with other transportation methods in many countries. In Japan, HSR serves four times more passengers between Tokyo and Osaka. The TGV Sud-Est in France showed increased market share after it began operation, increasing its share from 40 percent to 70 percent. Forty-nine percent of the Eurostar (connecting London, Brussels, and Paris) passengers switched from airplanes to rail. Sixty percent of passengers between Madrid and Barcelona have switched from airplanes. HSR has more competitiveness within 250 miles of travel, however; thus, there are some perspectives highlighting the complementary relationship between HSR and airplanes supporting long-distance travel.

HSR proves to be an innovative transportation method that changes interactions between cities and changes people's lives. The HSR network is growing in many countries and high-speed trains are getting even faster. A TGV test train recorded 357 miles per hour on conventional HSR tracks in 2007. Moreover, a maglev system has been developed to offer faster connection. The maglev system has been operating in Shanghai and is planned to serve Tokyo to Nagoya and Osaka (Chuo Shinkansen), with a maximum speed of 311 miles per hour.

Hyojin Kim

See also: High-Speed Rail in the United States; Maglevs; Shinkansen (Japan).

Further Reading

Gutiérrez, Javier. 2001. "Location, Economic Potential and Daily Accessibility: An Analysis of the Accessibility Impact of the High-Speed Line Madrid-Barcelona-French Border." *Journal of Transport Geography* 9(4): 229–242.

International Union of Railroads (UIC). 2014. http://www.uic.org. Accessed June 5, 2015.

Shaw, Shih-Lung, Zhixiang Fang, Shiwei Lu, and Ran Tao. 2014. "Impacts of High Speed Rail on Railroad Network Accessibility in China." *Journal of Transport Geography* (40): 112–122.

Todorovich, Petra, Daniel Schned, and Robert Lane. 2011. *High Speed Rail: International Lessons for US Policy Makers*. Cambridge, MA: Lincoln Institute of Land Policy.

HIGH-SPEED RAIL IN THE UNITED STATES

High-speed rail (HSR) has been developed in many European and East Asian countries since the first HSR was launched in Japan in 1965 as an alternative transportation for reducing greenhouse gas emissions. In the United States, however, the current passenger railway system lags behind other modes of transportation, such as cars and planes. Amtrak, the government-owned rail system, has been continually operating at a budget deficit; hence, infrastructure investments have not been available. Amtrak still serves intercity connections, but the frequency of trains is less than 10 times per day in most major stations except for the northeast corridor (NEC). Additionally, train speed remains relatively low due to curves and single-track sections. Additionally, speed upgrades for rail transportation in the United States have not made much progress. Acela Express is the only HSR line along the northeast coast of the United States, but its maximum speed in the NEC remains 150 miles per hour, which is close to the semi-HSR standards in other countries.

The introduction of HSR systems has been actively discussed in the United States, however, since the 2009 official government announcement of an HSR network plan with goals of reducing intercity travel time and expanding employment. Following the announcement, the Federal Railroad Administration (FRA) announced the High-Speed Rail Strategic Plan, which was funded by the American Recovery and Reinvestment Act. The plan, called High-Speed Intercity Passenger Rail (HSIPR), focuses on creating fast connections between cities through a nationwide HSR network. More than $10 billion was allocated to upgrade and improve current railways in 33 states and the District of Columbia.

The HSIPR mainly focuses on conventional railway upgrades. There are three strategies included in the proposed HSIPR to adjust rail speed according to the population density of a service area: core express, regional service, and emerging services. The FRA applies the "hub" concept to central cities in its plans for the HSIPR network. The core express corridor, supported by the fastest speed of 125–250 miles per hour, is planned for connections between dense areas such as New York, Los Angeles, and Chicago, which are referred to as "cores" or "hubs." The regional corridor, with a speed of 90–125 miles per hour, covers travel between a dense city and low-density cities, which would be an improvement to current rail speed. The emerging corridor, reaching speeds up to 90 miles per hour, supports connections to the core express and regional corridors.

The HSIPR has been progressing in the following sections: Boston to Brunswick; Chicago to various destinations such as Detroit, Omaha, St. Louis, and Minneapolis; New York to Buffalo; New Haven to St. Albans; Washington DC to Boston (NEC); Philadelphia to Pittsburgh; St. Louis to Kansas City; Eugene to Vancouver; Charlotte to Washington, DC (Piedmont); San Diego to Los Angeles and San Francisco (California HSR); and Dallas to Houston and Austin. The NEC will increase train speeds to more than 125 miles per hour between Boston and Washington, DC, which will save an anticipated 30 minutes of travel time, and serve connections with upgraded regional routes to cities such as Albany, Rochester, Montreal, and New Haven. Meanwhile, speed upgrades of as much as 110 miles per hour will enhance the Chicago hub's role in the railway network by providing improved accessibility. The Piedmont corridor along major cities in North Carolina will connect with the northeast corridor. The California HSR—the most prominent corridor in the United States—will serve Los Angeles and San Francisco with a maximum speed of 200 miles per hour by 2029. The network later will be connected to San Diego and Sacramento. In Texas, another HSR project has been under development by a private company that plans to operate an HSR line between Dallas and Houston with a 90-minute travel time.

Although aircraft and automobile are the principal modes of transportation in the United States, HSR has the potential to be competitive for several reasons. Most rail stations in the United States are located in city centers, providing the option of convenient travel between cities (i.e., avoiding additional travel to the airport, check-in, layover, danger of accidents, and stress from driving). Additionally, cities in low-access areas that fall under the hub-and-spoke system can benefit from the travel advantages of HSR if service is available, especially as road congestion continues to worsen and overcrowded airports experience increases in costs due to delays. Thus, the implementation of the HSIPR in the United States is expected to change competition among modes of transportation and to improve the spatial accessibility of cities.

The progress of U.S. HSR, however, faces obstacles. Although California has been making excellent progress on its HSR project, Florida, Ohio, and Wisconsin have canceled HSR funding. Despite the economic and sociological benefits associated with HSR, the most important issue is the construction cost, which is approximately $50 million per mile. Additionally, the ability to construct high-speed tracks actually is somewhat far off in the future. As a result, the train speed in many corridors will remain at 90–110 miles per hour, which is far slower than the typical speed of HSR trains in other countries after the HSIPR completion. Nevertheless, although the speed of trains in many corridors is not compatible with the speed available in the northeastern or California corridors, it still is faster than that of cars even without traffic congestion.

Hyojin Kim

See also: Amtrak (United States); High-Speed Rail; Korea Train eXpress (KTX) (South Korea); Northeast Corridor (United States).

Further Reading

Button, Kenneth. 2012. "Is There Any Economic Justification for High-Speed Railways in the United States?" *Journal of Transport Geography* 22: 300–302.

Todorovich, Petra, Daniel Schned, and Robert Lane. 2011. *High Speed Rail: International Lessons for US Policy Makers.* Cambridge, MA: Lincoln Institute of Land Policy.

United States Department of Transportation. April 2009. *Vision for High-Speed Rail in America.*

Williams, Michael J., Dawna L. Rhoades, and Thomas A. Simms. 2013. "High-Speed Rail: Will It Change the Dynamics of US Intercity Passenger Travel?" *World Review of Intermodal Transportation Research* 4(1): 73–95.

HITCHING

Hitching or hitchhiking is the act of soliciting a ride from passing drivers while situated along the roadside—usually for free. Customarily, a hand gesture is used to indicate the need for a ride, although a sign can also be used to specify a destination. In North America, extending a fist with the thumb up is the standard gesture but gestures vary worldwide.

Hitchhiker at the city limits of Waco, Texas, 1939. (Library of Congress)

In most areas of the United States, hitchhiking is legal if the person is standing off the pavement. Several states do outlaw hitchhiking in all cases, however, and it is fair to say that it is discouraged nationwide. The most frequent argument against it is safety, as it is increasingly considered a risky behavior, both for the hitchhiker and the driver. Tales of hitchhikers being picked up by those who might harm them or of drivers picking up crazed criminals along the roadside abound—and seemingly are a common occurrence in horror movies. In other countries hitching is more common and widely accepted.

Joe Weber

See also: Car Sharing; Slugging (United States); Uber.

Further Reading

"The Hitch Hikers Handbook." 2014. http://hitchhikershandbook.com/. Accessed July 17, 2015.

HOT LANES

See High-Occupancy Vehicle (HOV) Lanes (United States).

HOV LANES

See High-Occupancy Vehicle (HOV) Lanes (United States).

HYDRAIL

Hydrail (hydrogen fuel cell battery hybrid rail vehicle systems) technology is electric railway traction that is powered using hydrogen energy carried on board, rather than electricity supplied from an external overhead wire or third rail. Fuel cells convert hydrogen and oxygen into electricity, water, and a little heat. Batteries store and modulate the electricity to the traction motors and capture stopping energy to reaccelerate the rail vehicle economically after stops. By the middle of the 21st century hydrail will be the most common form of railway propulsion. This holds true on all scales, from very small rail systems in mines to freight and high-speed intercity passenger trains. Mines and streetcars will transition first; energy-intensive freight trains and high-speed passenger services will be last.

Hydrail technology was created in the United States to replace battery power in mining vehicles for which recharging downtime was costly. Soon it was considered for subway vehicles and, in 2000, a full-scale switching locomotive was built—and able to double as a mobile, self-propelled power plant. Soon after 2000, two single-unit hydrail passenger railcars were demonstrated in Japan but they never entered regular service. Experts believe if rapid deployment had followed, the Fukushima nuclear accident would not have stopped that portion of the transit network instantly.

The term "hydrail" was coined by retired planner Stan Thompson. The word was chosen as a search-engine target to aid scientists and others around the world to find and build on published work about the technology. In 2005, with the help of Jason W. Hoyle of Appalachian State University, North Carolina, Thompson invited hydrail experts from around the world to Charlotte, North Carolina, for the First International Hydrail Conference (1IHC). It brought together scientists, engineers, and industries' tech experts to expedite worldwide replacement of a major technology for societal—rather than commercial—reasons. Among the presenters was Canada's Dr. Alistair I. Miller of Atomic Energy of Canada, Ltd., whose 1999 paper was one of two that prompted Mooresville's hydrail initiative. Dr. Miller's paper showed that railways and marine craft were the forms of transportation most readily adapted to hydrogen fuel cells. The other was a paper by Dr. Max Wyman that showed the strategic value of hydrogen as an alternative to petroleum importation. Since the first conference, International Hydrail Conferences have been held each year in cities around the world, and organizations from nearly 20 countries have given presentations.

The first business to manufacture passenger hydrail vehicles commercially is believed to be the Los Angeles, California, firm TIG/m LLC. Its first system (using battery operation at the time of this writing) now operates in the Caribbean Netherlands island of Aruba. Aruba has almost unlimited wind energy and is committed to becoming—by 2020—the first country entirely powered by sustainable, zero-carbon energy. North Germany also has abundant wind resources. The North German states expect to have 40 hydrogen multiple unit (HMU) trainsets in service before the end of 2020.

Selima Sultana

See also: Alternative Fuels; Hydrogen Vehicles.

Further Reading

Krivit, S. B. (ed.). 2010. *Nuclear Energy Encyclopedia: Science, Technology, and Applications.* Hoboken, NJ: John Wiley & Sons.

Marin, G. D., G. F. Naterer, and K. Gabriel. 2010. "Rail Transportation by Hydrogen vs. Electrification Case Study for Ontario, Canada, II: Energy Supply and Distribution." *International Journal of Hydrogen Energy* 35: 6097–6107. http://ac.els-cdn.com /S0360319910005768/1-s2.0-S0360319910005768-main.pdf?_tid=4645aa84-4852 -11e5-b210-00000aacb35f&acdnat=1440195571_1517e20ed1750563aaa5bdd380 c0a429. Accessed January 11, 2016.

Thompson, S. 2015. "The Mooresville, NC Hydrail Initiative: A Small-Town Environmental Transit Project Has Reached around the World." http://hydrail.org/node/44. Accessed August 8, 2015.

HYDROGEN TRAINS

See Hydrail.

HYDROGEN VEHICLES

Hydrogen vehicles are viewed by many experts as a candidate to replace conventional gasoline- and diesel-powered vehicles in the coming decades. Worldwide, the current transportation system relies on the use of petroleum, thus leaving economies vulnerable to petroleum price changes, geopolitical instability in petroleum-producing regions, and uncertain future supplies. For these reasons, many major automobile manufacturers began investing in hydrogen vehicle technology during the late 1990s. Hydrogen, which is an energy carrier, in the United States is

Toyota Motor's new fuel cell vehicle (FCV) on display at the company's showroom in Tokyo, 2014. Buoyed by its success with electric-gasoline hybrid vehicles, Toyota is betting that drivers will embrace hydrogen fuel cells, an even cleaner technology that runs on the energy created by an electrochemical reaction when oxygen combines with hydrogen. (AP Photo/Koji Sasahara)

classified as an alternative fuel by the Energy Policy Act of 1992. Hydrogen is the most abundant element on the planet and enough exists to easily meet current transportation fuel demands from a supply standpoint. The increased use of hydrogen vehicles also has the potential to drastically reduce global carbon and greenhouse gas emissions.

In contrast to a conventional vehicle with an internal combustion engine that relies on the burning of gasoline or diesel fuel, hydrogen vehicles employ a fuel cell and an electric motor. In most hydrogen vehicles, the fuel cell has a polymer electrolyte membrane (PEM) placed between a positively charged electrode and negative electrode. Hydrogen enters the negatively charged side of the fuel cell as an H_2 molecule, and oxygen is introduced into the positively charged side. When hydrogen passes through the fuel cell an electrical current is generated. This electricity is used to power the vehicle's motor—a process similar to the manner in which battery electric vehicles operate. In addition to the electricity generated, the hydrogen atoms combine with the oxygen molecules in the fuel cell, creating water (H_2O).

Because fuel cells produce only water and electricity, carbon dioxide emissions would be greatly reduced by the use of hydrogen as a transportation fuel instead of gasoline or diesel. Fuel cell vehicles also are two to three times more efficient at converting hydrogen fuel to electrical energy than are internal combustion engines. Onboard storage of hydrogen in the vehicle requires a fuel tank, and due to

Hydrogen Fuel Cell Forklifts

Forklifts are an essential—if often overlooked—mode of freight transportation. They are used to load and unload trucks in warehouses and retail stores, and for many other applications. Forklifts are available in a wide range of sizes and with different capabilities. A recent development is using hydrogen fuel cells in place of gasoline engines or batteries in these vehicles. These cells use oxygen and compressed hydrogen (carried in an onboard tank) to generate electricity. Cells can operate longer than batteries can and they are not affected by temperature. They can be refueled in minutes simply by changing a hydrogen tank. They also do not generate air pollution and can be used indoors without the need for extra ventilation.

Joe Weber

its low density, hydrogen must be compressed into a tank capable of withstanding high pressure. The U.S. Department of Energy (DOE) is developing hydrogen storage technology to enable vehicles to carry enough hydrogen to drive approximately 300 miles, which is competitive with current gasoline-vehicle ranges.

A number of challenges face the widespread adoption of hydrogen vehicles. Although fuel cells have been in operation for decades—including in applications for the military and for space exploration—their use in vehicle propulsion was unproven until recently. Given the novelty of the technology and the expenses involved in production, hydrogen vehicles presently have a much greater cost to consumers than do conventional vehicles. Although costs should be reduced as production increases, hydrogen vehicles currently are only affordable to relatively wealthy people. In 2014, Hyundai began leasing its Tucson Fuel Cell model to about 200 individuals in Southern California, and Honda and Toyota began selling hydrogen fuel cell vehicles on the U.S. market in 2015.

Another challenge to adopting hydrogen vehicles is the production of the fuel. Though hydrogen is abundant, its atomic structure is such that it almost always is bound to other elements. Separating the hydrogen from the molecular bonds requires some form of energy. At present, the most common way to obtain hydrogen is through steam reformation of natural gas. Producing hydrogen from natural gas, coal, or other nonrenewable resources would decrease the emissions savings gained by using hydrogen as a transportation fuel instead of gasoline. Hydrogen also can be produced from steam reformation of biofuels, or through electrolysis using electricity derived from renewable sources such as wind or solar energy, but these methods currently are more expensive than steam re-formation from natural gas.

One of the most substantial hurdles facing the adoption of hydrogen vehicles is the lack of a refueling infrastructure. Unlike electric vehicles that can be recharged at home or work, hydrogen must be dispensed at public refueling stations.

A limited number of hydrogen refueling stations are in operation in Southern California, but most of the country lacks even one station within a reasonable driving range for any potential buyer, which makes consumers less likely to purchase hydrogen vehicles. This phenomenon is referred to as the "chicken and egg" problem, because consumers are hesitant to purchase hydrogen vehicles with so few existing refueling stations, and station developers are unwilling to build more stations without enough hydrogen vehicles on roadways. Many experts suggest that the government should subsidize the construction costs of initial refueling stations, placing them in such a way as to encourage adoption and ensuring that station developers have enough of a customer base to operate. A pipeline network also would have to be built to move hydrogen from production facilities to refueling stations. Estimates vary, but the total cost of constructing a hydrogen infrastructure could exceed $500 billion.

Scott Kelley

See also: Biofuel Vehicles; Electric Vehicles; Plug-In Battery Electric Vehicles (BEVs); Plug-In Hybrids.

Further Reading

DeLuchi, M. A. 1989. "Hydrogen Vehicles: An Evaluation of Fuel Storage, Performance, Safety, Environmental Impacts, and Cost." *International Journal of Hydrogen Energy* 14(2): 81–130.

Tollefson, J. 2010. "Hydrogen Vehicles: Fuel of the Future?" *Nature News* 464: 1262–1264. doi:10.1038/4641262a. http://www.nature.com/news/2010/100428/pdf/4641262a.pdf. Accessed August 8, 2015.

INDIAN RAILWAY SYSTEM (INDIA)

The vast Indian railway system is a heritage of British rule on the subcontinent. Most of its lines are controlled by Indian Railways, a state-owned division of the Ministry of Railways. In the early 21st century, the system is the fifth largest in the world.

Britain had extended control by means of the East India Company over most of India by the 1820s. The colonizers built several short rail lines over the next few decades for the transport of materials, but the first passenger line in the country—and the first in Asia—was a section of what would be known as the Great Indian Peninsula Railway (GIPR). It was on this 21-mile route between the western Indian port of Mumbai (then known as Bombay) and the smaller city of Thane (Tannah) that a train made a trial run in 1852. The railway was financed by investors who had been granted concessions by the East India Company; and it began offering public service the following year.

Another line—the East Indian Railway (EIR)—running out of Kolkata (Calcutta) in eastern India was completed in 1854. More followed in a short time, including the Madras Railway, the Sind and Punjab Railway, the East Bengal Railway, and the Great Southern Railway. Although some of the new railroads were state owned, most were run by private companies operating under the concession system, which granted them 99-year leases of land and guarantees of reasonable return on investment. Additionally, the wealthy rulers of some semi-independent Indian states built their own lines.

Although publicly extolled by the East India Company's administrators as instruments of progress and development, the railways also allowed authorities to maintain what they believed was better control of their Indian subjects. The disastrous Indian Rebellion of 1857 revealed serious shortcomings in the company's rule, however, and it was taken over by the British government, which continued to expand the colony's rail system. In 1869 there were about 4,000 miles of track in India, and within two years it had become possible to cross the entire country from west to east on the GIPR. In 1880 the country could boast some 9,000 miles of track, and by 1901 the number had grown to nearly 25,000 miles. Just a short time before, Indian factories had begun building their own locomotives modeled on British designs.

The government exercised its option under the concession system by taking over the GIPR and the EIR in 1925, and followed suit with other major lines over the coming years. Builders were adding more than 1,000 miles of track each year,

Indian Railways has a total state monopoly on India's rail transport. It is one of the largest and busiest rail networks in the world, transporting 18 million passengers daily and more than 2 million tons of freight daily. (Samrat35/Dreamstime.com)

and while the country had about 43,000 miles of track by 1931, the global depression brought expansion to a standstill. During World War II (1939–1945) much of India's rolling stock was transferred outside the country and its facilities converted to producing materiel for the war effort.

India became independent in 1947 and it subsequently was partitioned into the nations of India and Pakistan. Several of the major railway routes now lay in Pakistan, and other routes were disrupted. In the years that followed, the newly independent Indian government consolidated the GIPR and several other government-owned railways on the west-central coast of India as the Central Railway. The government also divided the EIR into two systems, Northern Railways and Eastern Railways. Other realignments followed, and by 2014 Indian Railways operated 16 zonal (regional) systems as well as Kolkata Metro, the country's first underground metro railway. The system has links with other railways in the neighboring countries of Pakistan, Bangladesh (once East Pakistan), and Nepal.

As of 2013, Indian Railways ran more than 9,950 locomotives and 660,000 units of rolling stock over 55,500 miles of track. Most of its tracks are 5-foot 6-inch broad ("Indian") gauge, but some are 3-foot 3 3/8-inch gauge, and a few are 2-foot 6-inch and even 2-foot narrow gauge. The agency employs 1,307,000 workers and operates about 11,000 trains every day, some 7,000 of them passenger carriers.

Indian Railways also manufactures much of its rolling stock and exports to other countries.

Over the years, Indian trains have suffered several major disasters, the worst of which occurred in 1981, when a train crossing a bridge in the eastern state of Bihar was blown into a river by a cyclone, killing some 800 passengers and crew members.

Grove Koger

See also: Metro Railway, Kolkata (India); Railroads.

Further Reading

Indian Railways Fan Club. 2010. "IR History." http://www.irfca.org/faq/faq-hist.html. Accessed December 8, 2015.

Kerr, Ian J. 2001. *Railways in Modern India.* Oxford: Oxford University Press.

Kerr, Ian J. 2006. *Engines of Change: The Railroads that Made India.* Westport, CT: Praeger.

Pet, Paul C., Geoffrey Moorhouse, and J. B. Hollingsworth. 1986. *Rail Across India: A Photographic Journey.* New York: Abbeville.

INTELLIGENT TRANSPORTATION SYSTEMS

Intelligent Transportation Systems (ITS) refer to the set of new and developing solutions for making transportation safer and using technology more efficiently. Advancements in computer and information technology—such as GPS devices, sensors, and data storage—have been used in a broad range of applications to combat growing concerns caused by increased transportation use in urban areas. These concerns include economic, social, and environmental issues that potentially can be reduced using technological solutions. ITS programs in Japan, Germany, and the United States have been in development since the 1960s. Programs were not formalized as a strategic plan in the United States until the late-1990s. Since then, ITS research has contributed to a broad range of applications in the areas of safety and environmental impact, as well as public and private transportation efficiency.

Modern transportation has been greatly influenced by two persistent factors over the course of the last century. One is that exponential population growth is creating a growing demand for automotive vehicles. At the same time, those growing populations are using their means of mobility to increasingly migrate to urban areas, thus producing dense population clusters. Due to this urbanization, transportation systems have been plagued by traffic accidents, vehicle congestion, and pollution generated by fuel emissions. The objective of ITS is to monitor the status of the current systems and use this information to improve land use and system functionality, and to reduce environmental impact.

The trajectory of interest in ITS potential began in the 1960s with Japanese researchers' attempts to implement route guidance and traffic-control systems, and

Vehicle-to-Passenger (V2P) Communications Technology

Vehicle-to-passenger (V2P) communications technology is a component of what often are called "intelligent transportation systems" (ITS). The goal of V2P is to enable the vehicle to share with the driver any information the vehicle obtains through wireless contact with other vehicles and through roadside sensors. This could include changing speed limits on the route, a light about to turn red at an upcoming intersection, a traffic accident just around the corner, a car wandering out of its lane, and many other—sometimes dangerous—situations. Information about alternate routes also can be displayed on a screen. Many cars have sensors providing information such as warning of slowed traffic on the route, informing the driver of lane changes to be made, and navigation instructions. To be more effective, however, this technology should be integrated with vehicle-to-vehicle and vehicle-to-infrastructure communications technology.

Joe Weber

the development of an electronic toll-collection system that requires no human contact. Today, the Vehicle, Road, and Traffic Information Society (VERTIS) exists to facilitate research among private, government, and academic institutions to improve infrastructure, as well as to collaborate internationally. In Europe, ITS has been driven strongly by Germany since the mid-1970s, where the research agenda aimed at reducing congestion. The primary focus throughout the 1980s and 1990s for the European Union was on road transport informatics, which pairs transportation with information technology to better understand traffic conditions. The United States did not get involved in ITS until the mid-1980s, when the Partners for Advanced Transit and Highway (PATH) program was started. The PATH program was dedicated to finding means by which to increase highway capacity and safety and concurrently reduce highway congestion, pollution, and energy uses. Intelligent Transportation Systems of America is the federal advisory committee currently overseeing the program.

ITS America laid out a strategic plan during the mid-1990s in anticipation of implementing aspects of the ITS agenda. From 1997 to 1999, the project was in the "Era of Travel Information and Fleet Management." This beginning stage was intended to primarily be occupied with establishing an information infrastructure for private, government, and academic entities to communicate with each other to share data, information, and ideas. The stage was also to include introductory intelligent vehicle systems, such as cruise control and route guidance. From 2000 to 2005, the "Era of Transportation Management" was to use that information to connect vehicles to the infrastructure. "Smart travelers" would be able to use vehicle-to-roadside communication and electronic payment systems for tolls and parking to make travel easier and more efficient. The final—and current—stage was set to begin in 2010 and continue into the future as the "Era of the Enhanced Vehicle." This stage is the transition from smart traveler to smart vehicle. It includes advancements in automated braking, steering, and space control. The goal is the development of an automated highway system, with every vehicle operating at maximum efficiency.

In 1995, the U.S. Intelligent Transportation Systems Congressional Caucus laid out seven categories of potential applications for using ITS technologies. Electronic payment systems were some of the earliest applications and have seen great improvements in changing the way tolls, parking fees, and fines are collected. Travel and transportation management applications use ITS solutions and traffic information to improve travel flows for efficient operations. Similarly, travel demand management focuses specifically on single-rider vehicles to encourage ridesharing and advanced travel planning. Public transportation and commercial vehicle operations focus on the issues and activities particular to each avenue of transportation with the aim of streamlining and automating feasible aspects, such as ridership demand for public transport or weight inspection for private freight transportation. ITS information technology also is being directed toward emergency management, with the goal of improving response time using traffic and infrastructure activity data. Lastly, advanced vehicle control and safety systems include applications dedicated to developing vehicle enhancements that can contribute to safer transportation conditions. The safety concerns are not only for the driver, but also for other drivers, pedestrians, and cyclists.

Amanda Richard

See also: Autonomous Vehicles; Smart Cars.

Further Reading

Branscomb, Lewis M., and James H. Keller (eds.). 1996. *Converging Infrastructures: Intelligent Transportation and the National Information Infrastructure.* Cambridge, MA: MIT Press.

Figueiredo, Lino, Isabel Jesus, J. A. Tenreiro Machado, and L. L. Martins de Carvalho. 2001 (August). *Towards the Development of Intelligent Transportation Systems.* Presented at Intelligent Transportation Systems. 1206–1211.

McQueen, Bob, and Judy McQueen. 1999. *Intelligent Transportation Systems Architecture.* Norwoord, MA: Artech House.

Miller, Harvey J., and Shih-Lung Shaw. 2001. *Geographic Information Systems for Transportation: Principles and Applications.* New York: Oxford University Press.

U.S. Congress. 1995. *High-Tech Highways: Intelligent Transportation Systems and Policy.* Washington, DC: Congressional Budget Office.

INTERSTATE HIGHWAY SYSTEM (UNITED STATES)

The interstate highway system is a national freeway system and the most heavily used system of highways in the United States. It makes up only 4 percent of U.S. road mileage but carries 20 percent of the nation's highway traffic.

Interstate highway signs show the intersection of Interstate 70, 64, and 55 in St. Louis, Missouri, 2006. (Americanspirit/Dreamstime.com)

The interstate system originated in planning efforts dating back to 1938 when President Franklin Roosevelt (1933–1935) became interested in the possibilities of long-distance highways. Originally to be called "interregional highways," the interstate highway system was created by the Federal Aid Highway Act of 1944. This law specified that the system should be no more than 40,000 miles and should connect major cities. President Eisenhower (1952–1960) usually is given credit for the system, as large-scale construction began during his term. This was due to the Federal Aid Highway Act of 1956, which created the Highway Trust Fund, a collection of taxes on gasoline, oil, and tires. The Highway Trust Fund provided an enormous amount of money to pay for the expensive new highways, and the federal government paid at least 90 percent of the cost of each mile. Despite the importance of federal funding, individual state highway departments built and own the interstate highways. The 1956 Act also changed the name to the Interstate and Defense Highway System, although defense needs have had very little to do with the system.

The highway system was to be completed by June 30, 1972, but it was only about 80 percent finished at that time. New freeways continued to be added to the system and rising costs slowed down construction. In 1992, most of the system had been constructed and it was officially declared complete. The interstate system now contains 42,793 miles of freeways, or about three-quarters of the 60,000 miles of freeways in the United States. New freeways continue to be added to the interstate system, thus the system never truly will be finished. Many of the routes that have been added are urban, and today more than one-third of the system is urban.

The interstate system is a grid of highways connecting the major cities of the country. All interstate routes are numbered in a consistent system. Routes running east-west use even numbers, with smaller numbers in the south and greater numbers farther north. Interstate 10 is the southernmost highway running coast to coast, and I-90 is the northernmost. North-south running highways use odd numbers, with smaller numbers on the West Coast and greater numbers farther east. Interstate 5 is the westernmost highway, and I-95 the easternmost. Before 1944 the term "interstate highway" was used to refer to any long-distance highway. Many interstate highways exist in only one state, thus the term is misleading. Technically, there are interstate highways in Alaska and Hawaii, although those in Alaska are not freeways. This came about because the funding plan from the interstate system was extended to these states once they were admitted to the union.

In addition to the network connecting cities, many urban freeways were made part of the system. These were designed to allow traffic to bypass the congestion of big cities. Larger cities have spurs, bypasses, or beltways. These are given three-digit numbers, with the second and third corresponding to the main highway from which they branch off. These numbers can be repeated in different states. For example, there is an I-110 in Los Angeles, California; El Paso, Texas; Baton Rouge, Louisiana; Biloxi, Mississippi; and Pensacola, Florida. Along the Gulf and Atlantic coasts many interstate routes are part of hurricane evacuation routes. These have been modified to allow contraflow travel, or wrong way driving, to make use of all lanes for evacuation. Crossover lanes, additional signs, and gates on on-ramps have been provided to aid in evacuation.

The interstate system made travel across the country much faster and easier than before. Being on an interstate highway is an important factor in the growth of many small towns and cities. It has reshaped American cities, as suburban interchanges often are the sites of large malls and office complexes. These "edge cities" can rival the old downtown in size and in the number of high-rise buildings located there.

Many other countries have built their own version of the interstate system. The newest and largest is National Trunk Highway System being constructed in China. This is designed to be a 52,000-mile system of toll roads. The first freeway in China was built in 1988, which means that creating the system is a massive construction project. Another example is the Malaysian Expressway System, which will consist of at least 1,253 miles when it is complete.

Joe Weber

See also: Autobahn (Germany); Autostrade (Italy); National Trunk Highway System (China).

Further Reading

Weber, Joe. 2011. " 'America's New Design for Living': The Interstate Highway System and the Spatial Transformation of the U.S." In *Engineering Earth: The Impacts of Megaengineering Projects*. Edited by Stanley D. Brunn. 553–567. Dordrecht: Springer.

J

JEEPNEYS (PHILIPPINES)

The jeepney is a good example of an unusual form of transportation and is extremely popular in Philippines. This vehicle is called the "jeepney" because it originally was made from American jeeps left behind in the Philippines after World War II. Instead of using these vehicles for combat, the Filipinos decided to use them for more effective ways of transporting people within cities and across regions. Modifications were made to jeeps to make the jeepney serve its new purpose. The Filipinos altered the jeep by extending it by 6.6 feet and adding two long seats on each side, which can hold up to 16 passengers. This capacity is less than that of a bus but more than a taxi. Since then jeepneys have been used as an affordable and faster form of mass transport for people and goods in the Philippines.

Jeepneys are not just a mode of transportation, they also are a cultural icon and include elaborate colors, religious and nationalistic artwork, and vibrant ornaments (such as a horse or a trumpet). They all are built in the Philippines using a labor-intensive process, but parts such as engines and transmissions are imported from Japan. The art and bodies of the vehicles are original and make each jeepney visually attractive and unique. Two automotive manufacturing companies in the Philippines—Sarao Motors and Francisco Motors—are major jeepney manufacturers and their sentimental statements about the jeepneys are that

> the vehicle represents the multi-culture history of the Philippines. There is a bit of Spanish and Mexican traits incorporated by the vivid colors and fiesta-like feel. There is a little bit of American because it evolved from the jeep. There is as little bit of Japan because of the Japanese engine. But it was built by Filipino hands. (Syed 2013)

The operation of jeepneys is flexible and most are run by private owners. License plates are required from the board of transportation to run jeepneys on a designated route. Many jeepneys run without license plates, however—especially in rural areas—even though operators could get caught by police. Operators of jeepneys follow the fare rate determined by the government. Jeepneys run on all major streets in the Philippines but their routes usually are designated. The route names are painted clearly on each side and on the front below the windshield of the vehicles. The designated route name usually consists of three things: The location of two terminal points where service starts and ends, and the major street name that the route follows. Drivers are required to pick up and drop off passengers only at the stops designated along the routes. Occasionally, however, short detours might be made for an additional payment.

The front seats next to the driver can hold two passengers and are the most comfortable and desirable seats in a typical jeepney. The atmosphere and interaction of individuals inside the vehicles make jeepneys a friendly means of transportation, and some of the many reasons that jeepneys are so popular. Passengers usually engage in casual conversations with each other, asking about destinations and trying to get to know one another at personal level. Jeepneys, however, are known to be overcrowded—so overcrowded that it is not unusual to see passengers hanging from the outside or riding on top of the vehicle, especially during the rush hours. Another deficiency of the jeepney service is that no seating or covered waiting areas are provided for passengers.

Since the 1950s, jeepneys have represented Philippine pride and are labeled the "King of the Road" in the Philippines. Diesel-operated jeepneys, however, are the cause of significant air pollution throughout the Philippines. Diesel jeepneys are being replaced with more fuel-efficient transportation options, such as the electric jeepneys (eJeepneys). The Institute for Climate and Sustainable Cities (ICSC) currently is working to address cost issues related to eJeepneys in an attempt to make jeepney costs reasonable for owners.

Selima Sultana

See also: Bamboo Trains (Cambodia); Chicken Buses (Central America); Chiva Express (Ecuador).

Further Reading

Grava, Sigurd. 1972. "The Jeepneys of Manila." *Traffic Quarterly* 26(4): 465–485.
Otsuka, Keijiro, Masao Kikuchi, and Yujiro Hyami. 1986. "Community and Market in Contract Choice: The Jeepney in the Philippines." *Economic Development and Cultural Change* 34(32): 279–298.
Syed, S. 2013. "End of the Road for Jeepneys in the Philippines?" http://www.bbc.com/news/business-23352851. Accessed August 3, 2015.

KARAKORUM HIGHWAY (PAKISTAN)

Mountain ranges long have been barriers to land transportation, and there is no greater mountain range than the Himalayas. This range—which includes Mt. Everest, the world's highest peak—runs from Burma to Afghanistan, and separates South Asia from China. The western end of this range includes the Karakorum Mountains, which lie between Pakistan and China. This range includes K2—the second highest peak on earth—among many other tall mountains, most of which have glaciers. Both Pakistan and China desired to open up the mountainous parts of their countries and saw a road between them as a potential economic stimulus. Construction started on this road—the Karakorum Highway (KKH)—in 1959, following an agreement between the two countries, and it finally opened in 1979.

The southern terminus is Abbottabad, in the northern foothills of Pakistan, at an elevation of about 4,000 feet. The road ascends the Hunza River valley into the mountains. Switchbacks are sometimes necessary to gain elevation. Earthquakes and landslides are very common along the road. In fact, a 2011 landslide created a large lake high up in the Hunza Valley that flooded the highway and destroyed several villages. A new road has yet to be built and ferry boats are needed to be able complete the journey. The boundary between Pakistan and China is at Khunjerab Pass, said to be the highest elevation paved-road border crossing in the world. This is 15,397 feet above sea level, higher than any road in the United States but still thousands of feet lower than the surrounding peaks. The pass is closed in the winter due to snow. Across the border, the highway descends through river valleys into the Tarim Basin. The city of Kashgar, located at approximately 4,200 feet, is the northern terminus of the Karakorum Highway. It is located on the old Silk Road trade route and is a hub for travel in western China.

The total length of the road is approximately 800 miles and located mostly in Pakistan. Its spectacular mountain scenery has made the road a major tourist attraction in Pakistan, and also has produced significant economic transformations in the mountains through which it passes. Many villagers in formerly self-reliant areas have left their homes to seek work in the cities to the south, and goods are shipped by truck into remote communities that once produced everything they needed. Improvements to the highway are expected on both sides of the border, making travel in the mountains and between these two countries easier.

Joe Weber

Bicycle tourist on the Karakorum Highway riding toward Shandur Pass in northern Pakistan. (Paweł Opaska/Dreamstime.com)

See also: Alaska Highway (Canada and United States); Alpine Tunnels (Europe); National Trunk Highway System (China).

Further Reading

European Council on Foreign Relations. 2015. China's Silk Road to Nowhere? http://www.ecfr.eu/article/commentary_chinas_silk_road_to_nowhere3025. Accessed August 31, 2015.

Pamir Times. 2012. "Misery in Hunza Area." http://pamirtimes.net/tag/karakoram-high way/. Accessed August 31, 2015.

KARLSRUHE MODEL (GERMANY)

The Karlsruhe model is a dual-mode system of transportation in Germany that combines tram lines within the city of Karlsruhe with light rail and other commuter and regional train lines in the nearby countryside. It serves the entire region of the middle-upper Rhine Valley and is the first track-sharing system of its kind. It is considered a high-quality, highly patronized, flexible system of public transportation that is used as a model worldwide.

The city of Karlsruhe is located in the Baden-Württemberg region of southwest Germany, near the border of France, and has a population of nearly 300,000. Because of its relatively small size as compared to cities such as Berlin and Hamburg, which have populations more than 3 million and 1 million, respectively, Karlsruhe in the mid-20th century did not warrant a freestanding S-Bahn network of electric light rail trains. The city, however, desired a system of transportation that would link urban and rural areas using available locomotive infrastructure. The solution was a combined light and heavy rail system.

The Karlsruhe model stems from the 1957 merger of the regional light rail and dual-mode operator Albtal-Verkehrs-Gesellschaft mbH (AVG) with the Verkehrs-betriebe Karlsruhe GmbH (VBK), the urban tramway and bus system. Since 1897, the Albtalhahn, run by the Deutsche Eisenbahn Betriebsgesellschaft (DEBG)—the precursor to the AVG—had transported passengers along its narrow gauge track from Karlsruhe to Bad Herrenalb in the Black Forest region. By the mid-20th century the Albtalbahn experienced economic troubles and underwent a financial restructuring. The German Federal Railway (or Deutsche Bundeshahn, DB), estab-lished in 1949 by the new Federal Republic of Germany, refused to accept the Albtalbahn into the national railway system and refused to substitute buses in place of the rail line in a section from Bad Herrenalb to Karlsruhe, claiming that buses could not accommodate the vast number of riders. The DB suggested con-necting the VBK tramway system with the Albtalbahn and operating the tramway vehicles on the heavy rail track. Trains were modified to run on the tramway net-work, and track was changed from narrow gauge to standard gauge. The heavy rail line was electrified with 750 direct current (DC) volts to be compatible with trams in the Karlsruhe network. These modifications were the first steps in developing the Karlsruhe model. The city of Karlsruhe then established the AVG to operate the newly merged rail line.

The 1960s and 1970s ushered in more changes and expansions due to increased ridership on trams versus buses. Tram lines to Waldstadt and Knielingen were ex-panded and the local transit authority, the Karlsruhe Transport Authority (Karlsruher Verkehrsverbund, KVV), extended the urban tramway to a new hous-ing development in Nordweststadt. This phase was significant because it was the first instance of track sharing with the DB.

The mixed-use network meant passengers no longer had to transfer trains, and this caused ridership to increase. To accommodate the greater number of passen-gers, more connections were needed. Technical changes also were necessary, such as making power upgrades to the electrical system and correcting the difficulties encountered when connecting lighter trams to the existing heavy rail line. The solution was a dual-mode light rail vehicle known as a "tram-train," which was created by the Siemens Corporation. The tram-trains can operate in the street as a tram (or streetcar) and then connect to standard railway lines, sharing the tracks with freight and passenger trains.

The first tram-train line of the new Karlsruhe Stadtbahn system ran about 19 miles from Karlsruhe to Bretten on September 25, 1992. Between Karlsruhe Marktplatz and Grötzingen, the tram-train operates as a streetcar. At Grötzingen, the tram converts from DC to AC voltage and then operates as a heavy rail train to Bretten. Before the Stadtbahn, commuters along this route had to transfer from a tram to a regional train and then to a bus, making for a 15-minute-longer commute. The new system implemented shorter intervals between trains (one hour at off-peak and 30 minutes at peak times), more stops, a single-fare structure, the integration of the local bus network, and the availability of park-and-ride facilities for automobile drivers and bicyclists.

After just six months, weekly ridership in the system increased from 533,660 to 2,554,976. The Stadtbahn cites the elimination of transfers, the increase in the number of stops, comfortable seating in bright and clean vehicles with restrooms, and its accessibility to commuters in outlying areas as the reasons for the dramatic bump in ridership. Because of the tram-train's success, improved funding under a new financing commission was established in 1996 to expand the tram-train network even further. By 2004, when the network introduced its 100th tram-train, it occupied 292 miles of track, with the AVG owning 162 miles and Deutsche Bahn (the private company that succeeded the merged Deutsche Bundesbahn and Deutsche Reichsbahn) owning the remaining 130. The latest phase of construction, a tunnel built to relieve some of the tram-train traffic between Marktplatz and Augartenstrasse, was proposed in 2010 in hopes of reaching a 2015 completion date, in time for Karlsruhe's 300th anniversary, but the project has been delayed.

United States rail services that incorporated the Karlsruhe model include UTA Trax in Salt Lake City, Utah; the River Line in New Jersey; and Capital MetroRail in Austin, Texas. Worldwide, the Karlsruhe model has been adopted by the Regio-Tram in Kassel, Germany; the Mulhouse Tramway in France; and the Randstad Rail in the Netherlands.

Rosemarie Boucher Leenerts

See also: Light Rail Transit Systems; Rapid Transit.

Further Reading

Karlsruhe Model. Genesis. www.karlsruher-modell.de/en/index.html. Accessed September 3, 2015.
Vuchic, Vukan R. 2007. *Urban Transit Systems and Technology*. Hoboken, NJ: Wiley.

KOREA TRAIN EXPRESS (KTX) (SOUTH KOREA)

The Korea Train eXpress (KTX) is the South Korean high-speed rail service. The KTX began operation along the Gyeongbu Line between Seoul, the capital city, and Busan via the Gyeongbu high-speed railway (HSR) on April 1, 2004, with a speed of 190 miles per hour between Seoul and Daegu. At that time, the travel time between Seoul and Busan was cut from 4 hours and 10 minutes by conventional train to 2 hours and 40 minutes by KTX train. Today, after the completion of Gyeongbu HSR from Daegu to Busan, the travel time has been reduced to 2 hours and

16 minutes. The HSR tracks have extended from 139 miles in 2004, to 215 miles in 2010, to 329 miles and 42 stations in 2015.

The decision to build HSR in South Korea was made in the 1980s when the country was suffering from severe congestion due to its rapid economic growth (starting in the 1960s) along the major corridor of Seoul to Busan. This corridor alone contained about 80 percent of the country's manufacturing facilities and 73 percent of the national population and carried 66 percent of all rail passengers every day. Therefore, relieving congestion between Seoul and Busan became the most important part of the economic development strategy in the country in the 1980s. The construction of the HSR Gyeongbu corridor progressed in two steps: from Seoul to Daegu in 2004, and from Daegu to Busan in 2010. Additionally, the construction of Honam HSR between Cheongju and Gwangju was completed in 2015.

The KTX connects major cities of South Korea with 920 miles of track. The Gyeongbu and Honam HSR serve three major routes connecting Seoul to Busan (Gyeongbu KTX), Seoul to Mokpo (Honam KTX), and Seoul to Yeosu (Jeonra KTX) and three branch routes connecting Seoul to Changwon and Jinju (Gyeongjeon KTX) and Seoul to Pohang (Donghae KTX). Some KTX trains also run via the conventional railways, serving cities not included in the HSR coverage in three routes connecting Seoul to Suwon and Busan, Seoul to Miryang and Busan, and Seoul to Daejeon and Iksan. Additionally, some KTX trains offer through service from Seoul to Incheon International Airport using the Incheon Airport Line.

The reduced travel time provided by KTX changed people's concept of traveling. A round trip within a day became possible between Seoul and distant cities such as Busan, Ulsan, Daegu, Gwangju, Yeosu, and Mokpo by KTX. This promotes interaction between cities and contributes to local economic growth. Many cities prepare for tour programs linked with the KTX, which could be a convenient option for travelers. KTX stations act as regional hubs connecting conventional railways, subways, and bus networks. Thus, many KTX stations located in the city center have been reconstructed into multipurpose complexes serving travelers and local residents. KTX station areas also have become attractive locations for both residences and offices due to the improved accessibility to other cities. As a result, some new KTX stations located far from the city center have been developed into multiuse complexes.

The operation of the KTX changes the competitive relation between transportation modes. After the first KTX operation, the demand for airplanes between Seoul and Busan decreased by half by 2007, and many airlines consequently reduced or ceased domestic service in routes competing with the KTX. Conversely, demand promotes the growth of low-cost airlines that compete with the KTX for travel between Seoul and Busan.

The KTX service effectively covers the population, with a limited budget for construction costs. The service was expanded from 9 percent of the country's area in 2004 to 22 percent of the total area in 2011, and from 45 percent of the population in 2004 to 56 percent of the total population in 2011. Additional regions still wait to receive KTX service. Therefore, the Korean railway authority will upgrade conventional lines to serve as semi-HSRs by 2018 as an alternative for the KTX. This considers both spatial equity in services and efficiency in budget expenditure. The semi-HSRs reaches speeds of 112–155 miles per hour and will connect Seoul to Jecheon and Andong (Jungang Line), Seoul to Wonju and Gangneung (Wonju-Gangneung Line via Jungang Line), and Gangneung to Pohang (Donghae) through the current poor KTX service region. At the same time, the Capital Area HSR Line will be added from Pyeongtaek on Gyeongbu HSR and southeastern Seoul as a branch of Gyeongbu HSR to improve the capacity of KTX trains near Seoul. This plan will increase the frequency of Gyeongbu and Honam KTX trains and improve accessibility of KTX stations on the new HSR. The KTX service upgraded people's lives in South Korea and successfully covers the heavy traffic of Gyeongbu corridor.

Hyojin Kim

See also: High-Speed Rail; Railroads.

Further Reading

Chun-Hwan, Kim. 2005. "Transportation Revolution: The Korean High-Speed Railway." *Japan Railway and Transport Review* 40: 8–13.

Chung, I., and S. Lee. 2011. "The Effects of KTX on Population Distribution Between 2004 and 2009." [In Korean.] *Journal of the Korean Regional Science Association* 27(3): 121–38.

Korean Statistical Information Service (KOSIS). http://kosis.kr/statisticsList/statisticsList_01List.jsp?vwcd=MT_ZTITLE&parentId=A. Accessed June 5, 2015.

MOLIT, Ministry of Land, Infrastructure, and Transport of South Korea. 2011. "The Nation Railway Network Plan" [in Korean]. http://www.molit.go.kr/USR/I0204/m_45/dtl.jsp?idx=8132. Accessed June 5, 2015.

Mun, Jin Su, and Deok Kwang Kim, 2012. "Construction of High-Speed Rail in Korea." *The KDI School of Public Policy and Management*.

LAS VEGAS BOULEVARD

See Las Vegas Strip (United States).

LAS VEGAS MONORAIL (UNITED STATES)

Las Vegas is famous for the Strip, several miles of street that lined by huge casinos. Once a rural highway, the Strip now is a congested urban street with both heavy vehicle and heavy pedestrian traffic. A form of transport that could move large numbers of people along this linear corridor, preferably without being stuck in traffic, became an important goal by the late 1990s. The solution, the Las Vegas Monorail, opened in July 2004. It runs along a single route from the MGM Grand hotel to the SLS Las Vegas Hotel (formerly the Sahara), a distance of 3.9 miles. Along the route there are stops at Bally's, the Flamingo, Las Vegas Convention Center, and Harrah's Casino. It is an elevated design, originally using vehicles from Disneyworld for a smaller monorail route between the MGM Grand and Bally's.

Monorail arriving at the station on the Las Vegas Strip, 2011. (Vacclav/Dreamstime.com)

Ridership has not been strong enough to avoid financial troubles, partly due to the changing fortunes of casinos along the route as well as the fact that it does not run directly along the strip but runs parallel to it, sometimes requiring lengthy walks between stations and casinos. A number of extensions to the line have been proposed, including northward to downtown Las Vegas, southward to UNLV and McCarran International Airport, or to the Mandalay Bay resort. In 2015, local transport planners unveiled the idea of a light rail line running underneath the strip, with connections to the airport. This will likely compete against the monorail should it ever be built.

Joe Weber

See also: Las Vegas Strip (United States); Monorails.

Further Reading

Las Vegas Monorail Corporation. 2015. "Las Vegas Monorail." http://www.lvmonorail .com/. Accessed August 31, 2015.

Shine, Conor. 2015. "Report: Future of Las Vegas Transportation Includes Light Rail under Strip, Monorail Extension." http://lasvegassun.com/news/2015/may/27/future-las -vegas-transportation/. Accessed August 31, 2015.

LAS VEGAS STRIP (UNITED STATES)

The Las Vegas Strip, or simply "the Strip," is a four-mile section of Las Vegas Boulevard at the core of the Las Vegas tourism, gaming, and entertainment industry. It is defined by a dense cluster of casino-hotels stretching from the Stratosphere north of Sahara Avenue to Mandalay Bay near Russell Road on the south. As an agglomeration of resorts, the Strip is one of the world's most recognizable tourist attractions. It is visited by tens of millions of people every year who use a variety of transportation strategies to see sites and patronize casino-resorts along the Strip's flanks.

Las Vegas Boulevard is part of the city's historic commercial strip along US Highway 91, which, in the pre-Interstate era, was the gateway for tourists and the only road connecting Vegas with cities in Utah and California. Open spaces and an absence of burdensome taxes and regulations south of city boundaries along Highway 91 appealed to the California hotelier Thomas Hull, who opened El Rancho Vegas in 1941 at the southwest corner of the Boulevard and (what is today) Sahara Avenue. If Hull is considered "founder" of the Strip, mobster Benjamin "Bugsy" Siegel made the road famous and flashy through his own infamy and colorful design choices at the Flamingo resort, which opened in 1946. In time, another casino operator, Guy McAfee, gave the gambling destination its name; the emerging resort corridor reminded him of the Sunset Strip in Los Angeles. The development of the Strip has influenced the growth of the Las Vegas metropolitan area, in spatial scope (early housing developments followed the general trajectory of the Strip's expansion to the south downtown), in politics (the Strip remains outside city boundaries, a continual source of mild tension between county and city), and in economics (jobs in the casino industry and construction are a main driver for local incomes and industries).

Visitors move about the "Strip" using various modes of transportation, and such diversity of mobility options makes the Strip a unique case in transportation geography. The street is notoriously congested with vehicular traffic, especially after dark. The city's proximity to large tourist markets within driving distance, the convenience of spacious and free parking garages at every casino, and the desire for freedom that automobiles afford make a personal (or rented) vehicle an appealing option for many visitors. Fleets of taxis escort many tourists to and from the airport and between clubs and casinos. A double-decker bus line, aptly called "the Deuce," is an affordable sightseeing option that attempts to ease (but also adds to) congestion on the road. A number of casino properties also offer shuttle-bus service between casinos or to the airport. A monorail connects resorts between Tropicana Boulevard and Sahara Avenue, running behind the casinos a block east of the Strip. Lastly, several resorts offer tram service between neighboring properties. Walking still is a preferred mode of transportation for tourists who desire a more personal experience and who appreciate the ability to step in and out of different properties with ease. Installed over the last 25 years escalators and raised walkways at each major intersection have brought increased efficiency and safety to the movement of both pedestrians and motorists.

As the most prominent feature in the Las Vegas metropolitan area, the Strip serves as a symbolic landscape with different meanings for different groups. For tourists, it represents a weekend away; the vacation of a lifetime; or an experience of extravagance, sin, and vice not available at home. As the headquarters of the world's largest and most prominent gaming corporations, the Las Vegas Strip also is at the center of the gambling world. For the 2,000,000 local residents of Las Vegas, the Strip takes on additional meaning. In the past, when Las Vegas was a much smaller city, the Boulevard was akin to a city's main street and served as a gathering spot for entertainment, gambling, or dining. Most residents don't visit the tourist core as frequently anymore, except for specific purposes such as a fundraiser, a special night out, or to accompany visiting family or friends. For Las Vegans today, Las Vegas Boulevard represents the engine that drives the local economy, a literal barrier that must be crossed to get from one side of town to the other or, more symbolically, the uniqueness of their life in a tourist town.

Rex J. Rowley

See also: Boulevards; Double-Decker Buses; Las Vegas Monorail (United States); Monorails.

Further Reading

Moehring, Eugene P. 2000. *Resort City in the Sunbelt: Las Vegas 1930–2000*. 2nd ed. Reno: University of Nevada Press.

Roske, Ralph J. 1986. *Las Vegas: A Desert Paradise.* Tulsa, OK: Continental Heritage Press.
Rowley, Rex J. 2013. *Everyday Las Vegas: Local Life in a Tourist Town.* Reno: University of Nevada Press.

LIGHT RAIL TRANSIT SYSTEMS

Light rail transit (LRT) systems have been in existence since the 1970s, following the decline of the post–World War II tramway era. The light rail concept originally was coined in the United States by the U.S. Urban Mass Transportation Administration (precursor to the Federal Transit Administration) in 1972 and was quickly adopted by European countries to represent a subset of rail transit designated specifically to carry light loads at fast speeds (comparable to those of freight rail). Light rail was intended to function within cities and carry people in a mode alternative to personal vehicles. The systems were to operate within set right-of-ways and operate singly or in tandem at street grade with other vehicles, so as to minimize construction of separate guideways or excessive infrastructure changes.

In many cases, light rail came about as a transit mode to replace out-of-date or degrading passenger railway systems with transit systems that used existing infrastructure with more advanced and cost-effective technology. This is especially true for England, where much of the tram infrastructure was discarded in the 1950s to be replaced by the automobile.

Light rail commuter train at a station. (Americanspirit/Dreamstime.com)

Other concepts brought back into the foreground with the emergence of light rail were the integration methods of LRT into the urban framework. This is done by having at grade systems, where trains are in the road side-by-side with automobiles and buses, or above grade where the LRT stops and stations typically are elevated to near highway level or regional train level. Above-grade stations typically do not reside within cities, but more likely are on the edges of the primary city—such as some of the Portland, Oregon, area stations—and often are underserved in those locations. On top of the grade itself, the stations where passengers board can be at the same grade as the rail, or the passengers might have to step up to board the train (e.g., in the city of Tampa). If not managed and kept in check, this difference in grade of boarding could impact the efficiency of passenger boarding times and tip the cost-effectiveness of the system.

The first LRT plans in Europe began under the supervision of the Tyneside Public Transit Authority, an authority first established by the Transport Act 1968. The Transport Act 1968 was an Act of Parliament and described as "An Act to make further provision with respect to transport and related matters" (legislation.co.uk). The TPTA was authorized under the Act to take over run-down local rail lines and develop interchange infrastructure to connect the lines for a large-scale network of electric suburban lines. The TPTA's planned system included bus-feeder interchanges to account for the vast integration of bus transportation already present within the United Kingdom and many countries at that time.

As mentioned above, the concept of LRT came about as the need to upgrade and improve upon old tramway systems where new stock of cars became more important. An additional practice that came later was the segregation of LRT lines and intersection restrictions in an attempt to improve average travel speeds in many other European countries, such as Germany, Austria, and Belgium (LRTA). Following the successes of these Euro LRT systems, cities that had benefitted from tram interdependence in the United States sought to learn from other Western nations how to improve their existing networks. Many U.S. cities, such as Boston and San Francisco, sought to adapt light rail vehicle technology to existing tramways, with the help of Boeing. Unfortunately, this vehicle concept failed, as Boeing attempted to start from scratch and the prototype became too expensive to be profitable.

Edmonton, Alberta, Canada, provided an alternative means to North American LRT upgrades by using imported German light rail vehicles and a combination of realignment of underused railway and subway tracks. This method was an immediate success, with a rail model that was mimicked by San Diego, California, and Calgary, Alberta, Canada.

With many successes in the 1980s, light rail transit systems have spread to other countries, such as China, which has an LRT system that has been in operation since 1988 with 112 million passengers per year in the suburban township of Tuen Mun. Japan also has worked to improve its automobile-dominated country, where congestion and auto use are higher than that of the United States. Over the years, Japan has been working to improve its infrastructure, upgrading existing tramway lines and creating segregated street-grade LRT lines. The current goal of the Japanese government is to have 10 light rail lines by 2016.

Dylan Coolbaugh

See also: Buses; Cable Cars (United States); Commuter Trains; Rapid Transit.

Further Reading

Department of Transportation and Communications. Light Rail Transit Authority. http://lrta.gov.ph/. Accessed November 3, 2014.

Dixon, Scott. 2014 (August 24). "Japanese Companies Ride U.S. Light Rail Boom." *The Japan Times*. http://www.japantimes.co.jp/news/2014/08/24/national/japanese-compa nies-ride-u-s-light-rail-boom/#.VXnJ0WdFDIU. Accessed November 3, 2014.

Experience, Economics, and Evolution from Starter Lines to Growing. 2003. "Proceedings from 9th National Light Rail Transit Conference, November 16–18, 2003, Portland, Oregon. Transportation Research Board." http://onlinepubs.trb.org/onlinepubs/circulars/ec058/01_Front_Matter.pdf.

Taplin, Michael. 1998. "The History of Tramways and Evolution of Light Rail." Light Rail Transit Association. http://www.lrta.org/mrthistory.html. Accessed November 3, 2014.

LLAMAS

The llama, a long-necked, thick-furred South American ruminant, was domesticated and first used for transportation in the Incan Empire of Peru, as long ago as 4000 BCE, although the breed dates back 40 million years to the central plains of North America. They are of the genus *Lama*, relatives of the camel, and are native to the higher Andes Mountains. They are closely related to alpaca, guanaco, and vicuña, a camelid group collectively known as lamoids. The average height of a llama is 47 inches at the shoulder and they weigh about 250 pounds.

The Inca used the llama in a multitude of ways and, because of their importance, the state llama herders known as *llama-michis* controlled their breeding. The hunting of llama and the smaller alpaca also was forbidden. In religious ceremonies, male llamas were sacrificed as well as female llamas that were unable to reproduce. Llamas have a double coat of earth-toned fur that has long been used in weaving. The common people wove a coarse fiber called *aluascay*, and weavers for the nobility wove a finer alpaca and vicuña fiber. The llama also provided food, and parts of the animal's digestive tract were made into medicine.

The most common use of llamas from earliest domestication was as a beast of burden. Llama caravans would transport farm goods from Lake Titicaca to the capital of the Incan Empire, Tiwanaku, in present-day Bolivia. Not only was food transported but so was the Ecuadoran *Spondylus* shell, which was coveted by the wealthy as a symbol of fertility and power. Llamas also are believed to have carried construction materials to build a huge ceremonial structure at Tiwanaku between 500 and 200 BCE, as the population of llamas increased sharply at this time. Many

Peruvian women with llamas near Arequipa, Peru. (Noamfein/Dreamstime.com)

llamas died during a great drought in Tiwanaku in 562 CE, partially from starvation and partially from a decreased resistance to disease. Also, fewer llamas were needed for transportation at this time and many were moved to lower elevations where food was more plentiful. During the Spanish conquest of South America in the early 1500s, many llamas were killed by the Spaniards and replaced with domestic European species, such as sheep. Those llama and alpaca that did survive were forced into the higher regions.

Most herds of llamas still live in the Andean region of Bolivia, Peru, Ecuador, Chile, and Argentina. There are an estimated 7 million llama in South America. A small minority of llamas has been exported to North America, New Zealand, Australia, England, and Europe. There currently are 150,000 llamas in North America. In addition to being a source of transportation, llamas provide food, fiber, tallow, and fuel. Llamas are used as pack animals and sometimes travel in trains of several hundred. Like camels, llamas can cover a great distance of land without water and they are durable animals, traveling an average of 15 to 20 miles per day and able to carry a load of up to 132 pounds. Also, like camels, they have a cleft upper lip that is prehensile, as well as large, sharp teeth, which help them forage for fine or small vegetation that other animals might pass up. Their feet, which have two toes and are not hooves, are cloven and padded and have a large surface area, allowing them to traverse steep, hilly terrain easily. The llama's bloodstream has a high hemoglobin content, and oval-shaped red blood cells help llamas adapt to oxygen-poor, high-altitude environments. They derive water from the vegetation they eat, but they also need fresh water to survive. They are ruminants with a three-compartment stomach, and they regurgitate their food before eating it again, which helps them fully digest it.

Llamas are social animals that interact in herds and will bond with other animal species if removed from the herd. Llamas have an easygoing nature but, unlike camels, are known to spit or hiss when provoked or overburdened. The spitting or spraying is a defense mechanism usually reserved for other llamas. Llamas are gaining popularity as a pack stock in recreational wilderness and backcountry travel in the United States and South America. Llamas also are sometimes used as guard animals, known to protect domestic stock such as sheep from predatory animals such as wolves and coyotes.

Rosemarie Boucher Leenerts

See also: Camel Trains (Mali); Dog Sleds; Ox/Bullock Carts.

Further Reading

Buckhorn Llama Co., Inc. 2015. "Llamas." http://www.llama-llama.com/text/physical.html. Accessed June 5, 2015.
National Geographic. 2015. "Llama: Lama Glama." http://animals.nationalgeographic.com /animals/mammals/llama/. Accessed June 5, 2015.

LONDON UNDERGROUND/THE "TUBE" (UNITED KINGDOM)

The London Underground is the oldest, one of the largest, and one of the busiest urban rail networks in the world. It began operation in 1863 using steam locomotives. In 1890, it became the first system in the world to adopt electric traction. Today the system has 270 stations on 250 route miles, half of which actually is underground. It has 11 lines that carry 1.2 billion passengers a year; thus, it is the 12th busiest urban rail operation in the world. It operates throughout Greater London, although extensions of the system carry it beyond the city. It is a major component of London's complex transportation network, which includes buses, light rail, and commuter rail lines.

The "Tube" is one of the most recognized aspects of London, and two features have long marked its existence. First is the 100-year-old Underground symbol called the "Roundel." It consists of a red circle with a white center crossed in the middle horizontally by the name of a station, line, or merely "Underground." The second feature is Harry Beck's diagrammatic tube map, which first appeared in 1933 and still is in use. Despite a different color identifying each line on the map, Londoners refer to each line by its name.

The Metropolitan Railway built the first section of the railway in 1863 and employed the "cut-and-cover" method of construction. Today, the Metropolitan, Hammersmith and City, District, and Circle Lines make up the portion of the Underground that utilized this technology. These lines are "sub-surface" lines. The seven deep-level or "tube" lines (giving reference to the shape of the tunnels) were built deep into London's clay. The seven lines are the Bakerloo, Central, Jubilee, Northern, Piccadilly, Victoria, and the Waterloo & City. Subsurface trains are larger than tube trains, and although the two sets of lines do intersect, subsurface trains do not operate in tube sections. Typically, subsurface lines are close to street level, and tube lines necessitate the use of more than 400 long escalators (and more than 150 elevators or "lifts") to reach the platform level.

Historically, private companies built the railway. As the lines extended further from central London, the railway companies also developed the land along the routes into new housing areas. World War I (1914–1918) briefly interrupted expansion. By 1933, London's various forms of public transportation merged to become the London Passenger Transport Board. Expansion of the system continued during the 1930s, but halted again during World War II (1940–1945). London was heavily bombed in 1940 and damage occurred to parts of the network. Some tube stations became air-raid shelters. After World War II, the Underground received lower priority toward expansion in favor of rebuilding the damaged British railway network.

By the 1960s, however, expansion of the network continued. The Victoria Line opened in the late 1960s with automated operation (one person on board still operated the doors). In 1977, the Piccadilly Line reached Heathrow Airport, thus linking one of the busiest airports in the world with the Underground (and London). By the year 2000, a brand-new section of the network (the Jubilee Line) opened. Other recent developments include a new London "Overground" network, created by combining commuter rail lines and the East London (subsurface) line. At the same time, maintenance and refurbishment of the network is a constant factor for riders who deal with closures of stations or lines; the Underground is in a perpetual state of repair. For years—until new trains were put in service in 2000—many passengers on the Northern Line called it "The Misery Line," referring to the quality of their commute.

Politics have been a part of London's transport for decades, especially more recently. In 1981, when the Greater London Council controlled the transport network, London Transport introduced a system of fare zones for buses and underground trains that cut average fares for passengers. By 1984, however, control shifted and the network—now known as London Regional Transport—reported directly to Britain's Secretary of State for Transport and not to the City of London. By the year 2000, "Transport for London" became the new name for the network, with the Greater London Authority (GLA) and the Mayor of London assuming considerable control and responsibility for the operation of the entire network. Since its beginnings, labor unions have been an integrated part of the Underground; as a result, industrial actions (strikes) are common occurrences for the network and can cause major disruption in this city of 9 million people.

Jason Greenberg

See also: New York City Subway (United States); Paris Metro (France).

Further Reading

Bownes, David, et al. 2012. *Underground: How the Tube Shaped London.* London: Allan Lane.

Glover, John. 2010. *London's Underground.* 11th ed. Shepperton, UK: Ian Allan Publishing.

Green, Oliver. 2012. *The Tube: Station to Station on the London Underground.* Colchester, UK: Shire General.

LOS ANGELES FREEWAYS (UNITED STATES)

Among the most famous highways in the world are those that comprise the freeway network of the Los Angeles metropolitan area. Although Los Angeles is widely thought of as being a sprawling city because of its freeway system, this is not the case. The decentralized nature of the city is due to its earlier streetcar system, the largest in the country. These lines connected the scattered towns and cities that would coalesce to create the vast metropolitan area. Automobile travel also was important at an early date, and building highways became an important part of the region's transportation system. In 1940, the Arroyo Seco Parkway opened. Later known as the Pasadena Freeway it was the first freeway in the city. A freeway plan for the region was created in 1947 that largely was followed by later construction. Construction was rapid in the 1950s and 1960s but slowed after that due to the increased cost of construction, inflation, and opposition to more freeways. Many freeways planned decades ago were never built and a number of new routes still might be built.

Many of these freeways are part of the Interstate Highway System, including I-5 and I-10. In addition to their numbers they also have names by which they are more commonly known. Interstate 5 is known as the Golden State Freeway north of downtown and Santa Ana southwards, and I-10 is called the Santa Monica Freeway west of downtown and is called the San Bernardino Freeway to the east. The Hollywood Freeway, or US 101, runs from downtown Los Angeles north to the San Fernando Valley. It was built between 1940 and 1968. The Century Freeway, or I-105, is one of the latest, completed in 1993 after a lengthy legal and political battle. Other major freeways include the Foothills Freeway (I-210), The San Diego Freeway (I-405 and the busiest freeway in the United States), Long Beach Freeway (I-710), San Gabriel Freeway (I-605), Harbor Freeway (I-110), and Gardena and Artesia Freeway (California route 91, and the first freeway in Southern California to have toll lanes). The interchange between the Hollywood, Harbor, Santa Ana, and Pasadena freeways was built in 1953 and was the first fully directional freeway interchange in the world. It is known as the "Four Level Interchange" and is a common symbol of the freeways of Los Angeles. Many other interchanges also have their own names. The Newhall Pass Interchange on the Golden State Freeway at the north end of the San Fernando Valley is infamous for having collapsed after major earthquakes in 1971 and 1994.

Downtown Los Angeles is the hub of the freeway network and is completely encircled by freeways. Due to the region's mountains, connections to other cities and parts of the state must negotiate mountain passes. The Golden State Freeway to Northern California passes through Newhall and Tejon passes, I-15 (known by several names) climbs up Cajon Pass on its way to the high desert and Las Vegas, and the San Bernardino freeway runs east through the wide San Gorgonio pass to Palm Springs and beyond. Only the Santa Ana Freeway south to San Diego avoids mountains (though an observant traveler might catch a glimpse of the snow-covered Matterhorn peak at Disneyland), and is one of the freeways in Los Angeles that offers a view of the beach.

The Los Angeles freeways are a famous part of the city's life and the Southern California experience. Reyner Banham's book *Los Angeles: The Architecture of Four Ecologies* gives the highways equal standing with the beaches, mountains, and vast flatlands of the metropolitan area. Much of the world's population has experienced Los Angeles freeways through their frequent appearances in TV shows and movies, among them the TV series *CHiPs* and the movie *Speed*. This filming was relatively easy at a time when many freeways were under construction as the empty freeway could be rented out to a studio. The slowdown in construction in Los Angeles has prompted filmmakers to seek out freeways in other cities that are still actively building new highways, such as Las Vegas and Phoenix. Car chases and spectacular crashes are common in such appearances. In reality, however, freeways are the safest type of roads in the United States. Despite the reputation of Los Angeles as a freeway-dominated sprawling city, it actually ranks quite low in freeway miles per person and has the highest population density of any urban area in the country.

Joe Weber

See also: Autobahn (Germany); Freeways; Interstate Highway System (United States); Streetcars.

Further Reading

Banham, Reyner. 1971. *Los Angeles: The Architecture of Four Ecologies*. Harper and Row.

Brodsly, David. 1981. *L.A. Freeway: An Appreciative Essay*. Berkeley: University of California Press.

Rizzo, David. 2006. *Survive the Drive: How to Beat Freeway Traffic in Southern California*. Lorikeet Express Publications.

LYFT (UNITED STATES)

Lyft is an American ridesharing transportation company founded by Logan Green and John Zimmer that officially launched in San Francisco, California, in 2012 to provide an affordable transportation solution for people. The company has created a mobile phone app for users; it can be downloaded to any smartphone that has an operating system iOS 8.0 or later. Users who are interested in finding a ride from Lyft drivers create an account on the Lyft website (https://www.lyft.com/) and download the Lyft App. By using this app anyone can request a ride with drivers who have passed required driving background checks conducted by Lyft. When a ride is requested, a nearby registered Lyft driver confirms the ride. Users can then trace the driver's name, photos, and the car that will pick them up. Users also can read comments and ratings submitted by past passengers that used that particular driver for a lift.

Like Uber, the Lyft company does require personal information such as name, cell phone number, e-mail address, and credit card numbers to create an account. Uber, however, provides services within the United States as well as worldwide, and Lyft only provides services within the United States. Additionally, drivers who are registered under Lyft are required to put a furry pink mustache on the front of their vehicles. Unlike Uber drivers, Lyft drivers are allowed to collect tips.

Lyft faces criticism from taxi companies for ignoring existing regulations designed for the taxi industry. In particular, critics believe that the Lyft on-demand ridesharing program is less safe than traditional taxi service because the background check Lyft performs is not as rigorous as that for traditional taxi companies. Anyone can drive under Lyft company rules with a minimum background check. There also are concerns about Lyft drivers who do not carry commercial insurance. This raises a legitimate question of how passengers are insured in case of accident and property damage. Lyft also faces criticism from other ridesharing transportation companies such as Uber. Uber claims that Lyft is not a transportation service company, rather it connects drivers with people who just need a lift and care less about safety and security. In response to Uber's criticisms, Lyft claims that the primary goal of the company is to serve the community with affordable transportation services.

Despite these concerns, it is estimated that Lyft promises to generate $600 million in revenue by the end of 2015. Lyft currently provides services in 65 large cities within the United States though its long-term vision is to provide services worldwide.

Selima Sultana

See also: Car Sharing; Uber; Yellow Cab, New York City (United States).

Further Reading

Feeney, M. 2015. "Is Ridesharing Safe?" *Policy Analysis at the Cato Institute* 767. http://www
    .memphistn.gov/Portals/0/pdf_forms/CATO.pdf. Accessed August 31, 2015.
Hoge, Patrick. 2013 (November 18). "Lyft and Sidecar Replace Voluntary Donations with
    Set Prices." *San Francisco Business Times*. American City Business Journals. Accessed
    September 2, 2015.

M

MADRID METRO (SPAIN)

Madrid Metro is the subway system serving Spain's capital city of Madrid. Composed of nearly 300 individual stations and measuring 182 miles in length, the Madrid Metro is easily the largest metro system is Spain. Officially known as the "*Metro de Madrid*," the Madrid Metro had its beginnings in 1919, and subsequent construction focused on lines in and near the historic city center. Since the middle of the 20th century Madrid's Metro has undergone extensive expansion. Today, new lines reach far beyond Madrid proper. The system currently is one of the longest metros in Europe, second only to the London Tube.

The Madrid Metro has high ridership for a city its size; the metro lines alone accounted for more than 550 million riders during 2013. Fares vary among city zones and usage of light rail, but a typical fare ranges between 1.5 Euros per trip in the city center and 3 Euros per trip system wide. Currently the system encompasses metro lines, light rail lines, and nodal connections to other forms of transportation such as buses. There are 13 mostly underground metro lines (*lineas*). Lines 1 through 12 are numbered and color-coded, and an additional short *Ramal* branch line links Opera station and Príncipe Pío station in the city center.

Most metro lines radiate out from the historic city center of Madrid to outer suburbs, forming a system of outward-running spokes. An exception to the radial spoke system includes the Gray Line 6 and the Dark Green Line 12. Line 6 is the *Circular*, which forms a large oval around the main neighborhoods of the city of Madrid, and follows a rough approximation of Madrid's Calle 30 ring road in many areas. Because of its circumferential pattern, the *Circular* usually is the most crowded metro line in Madrid.

Another exception to the linear spoke pattern is the Dark Green Line 12. Line 12 is the MetroSur, which follows a looped perimeter that encircles the suburban towns just southeast of Madrid. All other metro lines radiate out from central Madrid. For example, Line 1, Line 2, and Line 3 converge at the *Puerta del Sol* (or *Sol*) station, which is located a short distance from the famous Plaza Mayor space in downtown Madrid. In recent years, the expansion of lines has corresponded to large development projects beyond the city. The Pink Line 8 connects Madrid to Spain's largest airport, Barajas Airport. Metro stations serving the airport have all been completed within the last 15 years.

The 95-year-old metro system has certain antiquated characteristics. Trains in the Madrid Metro run on the left instead of the right—a trait unique among Spanish subway systems—and rail track gauge is a bit wider than most international standards. Many older metro stations are entirely inaccessible to persons with disabilities, although newer stations are mandated to have more accessible facilities. Conversely, the metro system also has incorporated 21st-century elements into its business. To raise revenues, for example, the Madrid Metro allowed the telecommunications firm Vodaphone Spain to purchase naming rights in 2013, and the Metro renamed its *Linea 2* (Line 2) to "Linea 2 Vodaphone."

The wider Madrid Metro system encompasses more than subway lines. Madrid's Metro system also includes a nascent *Metro Ligero* (light rail) system, composed of three numbered and mostly aboveground light rail lines. Two lines located west of Madrid, beyond Casa de Campo, and another line is northeast of the city. Madrid's Metro system also is well integrated into other transportation modes. The Avenida de America station at the northeastern edge of downtown is one of many metro stations that also houses a bus station, with service reaching across Spain. Other metro stations include links to the extensive *Cercanias Madrid* commuter train system, which links the city of Madrid to suburbs and towns throughout the community of Madrid. (Madrid's Cercanias commuter train system became tragically well known across the globe in the aftermath of a 2004 terrorist bombing in four aboveground Cercanias trains.)

After a transportation *Consorcio* (consortium) was established for the Madrid area in 1993, rapid expansion of the Metro followed. Examples of this recent wave of construction include the enormous MetroSur Line 12 southeastern loop. Ambitious construction projects continue, including future construction plans for entirely new lines to link Madrid with towns further out in the community of Madrid region. A new light rail line linking Madrid to the upscale Las Rozas municipality in the northeast already is under construction. Additionally, the metro system enacted several forward-thinking environmentally friendly measures during the 2000s, including an improved cleaning system for metro cars, and additional measures to work toward increased sustainability.

Patrick D. Hagge

See also: Busan Subway (South Korea); London Underground/The "Tube" (United Kingdom); New York City Subway (United States); Seoul Metropolitan Subway (South Korea).

Further Reading

Metro De Madrid. 2009. "Metro De Madrid." http://www.metromadrid.es/en/index.html. Accessed November 1, 2014.
Ovenden, Mark. 2007. *Transit Maps of the World.* New York: Penguin Books.

MAGLEVS

Maglev (derived from "magnetic levitation") is a railway system that uses magnetic force to propel a levitated the train above a guideway. Maglev trains are faster than trains that use conventional tracks due to minimized friction between the train and

The Shanghai maglev train, November 12, 2012. (Yinan Zhang/Dreamstime.com)

tracks. The idea of maglev appeared in the early 20th century, and researchers developed principles and were awarded patents for establishing this new type of train system. After the development of the magnetic linear motor in 1940, the first test run of the maglev vehicle of Germany was conducted in 1971, and a passenger-carrying test succeeded in 1976. As a result of practical development of maglev in Germany, the maglev train ran along a half-mile demonstration route at the International Transportation Exhibition in Hamburg.

In 1984, the first commercial maglev operated in Birmingham, England, as a connector between an airport and railway station, offering a max speed of 26 miles per hour. In 1980, Transrapid, a German company developing maglev, began operating 13 miles of high-speed maglev test guideway in Emsland, Germany, which led to the plan of a new commercial high-speed maglev between Berlin and Hamburg. The plan was canceled in 2000, however, due to forecasted financial losses.

The Chinese government also wanted to connect major cities with a high-speed network. Maglev was one of the attractive modes for high-speed mass transit systems. After two years of trial running, in May 2004 the Shanghai Maglev became the first commercial high-speed maglev and set a record speed of 311 mph. The Shanghai Maglev Train (SMT) runs between Shanghai Pudong International Airport and Longyang Road Station. The SMT has a maximum operation speed of 267 miles per hour and can run 19 miles of an elevated guideway in 7 minutes and 20 seconds—which overwhelms the approximately 40-minute average taxi trip on

Super-Maglevs

Magnetic levitation (Maglev) technology has existed for half a century. It is a technological solution for building a transport system in which trains smoothly glide above a
track. Although criticized for being outrageously expensive technology, China's maglev
train is the fastest passenger-carrying train, reaching a speed of up to 260 miles per
hour (mph). Scientists presently are discussing "super-maglev" technology for trains that
could increase the speed up to 2,500 mph. The concept of super-maglev is a mixture
of the Maglev concept and the vacuum train concept—in which a tunnel is built by
evacuating the all air. Japan recently tested super-maglev technology with 100 passengers along a 27-mile route at speeds of up to 311 mph.

Selima Sultana

the same route. The SMT runs a maximum speed of 186 miles per hour in the
morning and the evening for safety and energy efficiency. The SMT carries 574
passengers at a time, with a 15- to 20-minute interval between trains.

The first commercial high-speed maglev shows satisfactory stability of the system and steady ridership. One million passengers used the Shanghai maglev between 2002 and 2004, and the accumulated ridership from the initial operation
passed 7 million in 2006. Daily ridership of the Shanghai maglev exceeded 20,000
passengers in 2007; however, expanding of the high-speed maglev has encountered some obstacles. The major issue for maglev is the uncertain cost of the system. In general, the construction cost of the high-speed maglev is estimated to be
greater than the current high-speed railway (HSR) system. The practical operation
speed of maglev shown in the SMT, however, is not superior to the HSR. The maglev system also is not compatible with current railway systems, thus the efficiency
and flexibility of maglev is not favorable for providing direct-connection service.
Therefore, in a situation where even the current HSR is considered an expensive
transportation system, promoting the competitiveness of the maglev system in
high-speed transportation is not easy. Indeed, a high-speed link between Beijing
and Shanghai was constructed with conventional HSR for a similar reason. Although the maintenance cost of maglev is less than that of HSR, the construction
cost of maglev is significantly greater. The construction cost of SMT was $1.2 bil-
lion for 19 miles, and the construction of HSR from Beijing to Shanghai, which
opened in 2011, cost around $32 billion for 819 miles.

Maglevs, however, still are the fastest trains in the world. Although a current
four-wheeled HSR recorded a maximum speed of 310 miles per hour, it was derived from special preparations for increasing the speed. Additionally, increasing
its operation speed requires the HSR system to overcome the limitation of friction.
Conversely, maglev has the possibility to increase its speed based on its systematic
advantages. A Japanese maglev train recorded a maximum speed of 361 miles per

hour on the testing track for the Shinkansen, which has a planned operation speed of 314 miles per hour. Thus, we can anticipate faster travel in the future.

Hyojin Kim

See also: High-Speed Rail; Monorails; Railroads.

Further Reading

Chen, Xiaohong, Feng Tang, Zhaoyi Huang, and Guangtao Wang. 2007. "High-Speed Mag-lev Noise Impacts on Residents: A Case Study in Shanghai." *Transportation Research Part D: Transport and Environment* 12(6): 437–448.

Givoni, Moshe. 2006. "Development and Impact of the Modern High-Speed Train: A Review." *Transport Reviews* 26(5): 593–611.

Goodall, Roger M. 2012. "Maglev: An Unfulfilled Dream?" Presented at 2012 International Symposium on Speed-up, Safety, and Service Technology for Railway and MAGLEV Systems (STECH '12), September 17–19, Seoul, South Korea.

Luguang, Yan. 2004. "Development of the Maglev Transportation in China." Presented at the 18th International Conference on Magnetically Levitated Systems and Linear Drives, October 26–28, Shanghai, China.

Luguang, Yan. 2007. "The Maglev Development and Commercial Application in China." Presented at the International Conference on Electrical Machines and Systems. 1942–1949.

Xiangming, Wu. 2006. "Experience in Operation and Maintenance of Shanghai Maglev Demonstration Line and Further Application of Maglev in China." Presented at the 19th International Conference on Magnetically Levitated Systems and Linear Drives, September 13–15, Dresden, Germany.

MAMMY WAGONS (WEST AFRICA)

Mammy Wagons are passenger and freight vehicles common on West African roads. The vehicles use a truck body and are modified with seats, roofs, and other personal touches. They often are lavishly decorated with religious or traditional imagery and the text of proverbs. Because of the customization no two Mammy Wagons look alike. These vehicles provide much of the intercity traffic in West Africa, carrying a mix of people, animals, and freight. They take their name from a term that refers to wealthy women who control much of the market trade in West Africa.

Joe Weber

See also: Chicken Buses (Central America); Double-Decker Buses; Jeepneys (Philippines); Ox/Bullock Carts.

Further Reading

Bennett, Martin. Eye on Africa. 2014. "Travels in Africa: A Nigerian Travelogue." http://www.arthurbrooks-images.co.uk/info.html. Accessed July 17, 2015.

MASS RAPID TRANSIT (SINGAPORE)

The Mass Rapid Transit (MRT) in Singapore—often called the Singapore Rapid Transit System (RTS)—is a key metropolitan public transit system that covers the whole city-state. Singapore is the third-ranked most densely populated country in the world, with 5.4 million people and a population density of more than 21,000 people per square mile. The MRT started its operation in November 1987, and the geographic coverage of the network is not as large as other MRT systems in Asia, such as those in Beijing (China), Seoul (South Korea), and Tokyo (Japan). The MRT, however, is known as one of the efficient MRT systems in the world in terms of average daily ridership and rail density.

The plan for the MRT was derived from the idea of integrating overall land-use planning with multimodal transport systems to establish sustainable land use and to accommodate the increase of commuting passengers. According to the Land Transport Master Plan 2013, the land space for roads amounts to more than 12 percent, which influences tighter constraints on the land use planning of Singapore. A public transit system is considered the best alternative way to handle both land use and increasing travel demand in Singapore effectively.

The Singapore MRT began operations in 1987 with two lines that were designed to cover Singapore's heavily populated corridors first, by connecting east to west and north to south from the central business district (CBD). The first two lines were established with route length of 4.97 miles with 14 stations. Early daily

Singapore Light Rail Transit, 2015, operated by Singapore Mass Rapid Transit. (Regina Lui/ Dreamstime.com)

ridership was only 300,000 in 1987, but rapidly increased to 1 million in 1997 and to more than 2.7 million by 2014. To respond to the increased number of passengers, the MRT expanded the network to 106 stations with five lines, and the total network length has grown to 93.7 miles. As a strategy for providing comprehensive and seamless transport service, the MRT network was connected with the Light Rail Transit (LRT), which serves local and regional linkages. Consequently, the entire transit system in Singapore provides passengers with a smoother ride to local housing communities and residential areas. Additionally, a bus system provides last-mile service for areas that are not covered by either MRT or LRT stations. In terms of operation, the two initial lines were constructed and managed by the Land Transport Authority (LTA) of the government of Singapore. Later, in early 2000, the system was handed over to two private transport operators, SBS Transit and SMRT Corporation.

The first line of the MRT is the North South Line (NSL), which opened in November 1987. The line connects the central area of Singapore with northern and southern areas. The East West Line (EWL), currently the longest line of the MRT, commenced operation in December 1987. The EWL encompasses the eastern and western areas, and plans are in place to extend it with additional stations and an extension line named "Tuas West Extension" (TWE). The third line, the North East MRT Line (NEL), commenced operation in 2003 between the southern and northeast areas in Singapore. The NEL is one of the longest driverless lines, stretching 12.4 miles with 16 stations. Eight stations are transfer stations that give passengers alternative route choices to destinations. The fourth line, a circular line named "Circle Line" (CCL), was completed in 2009. This line improves accessibility, mainly covering the central area with the existing lines. The line also is automatically operated for 22 miles with 30 stations. The fifth line, the Downtown Line (DTL), opened in 2013. The DTL currently is only 2.7 miles with 6 stations, but expansion is planned to both the NEL and EWL to make downtown more accessible from suburban areas. Thanks to the completion of these lines, people can access most of the CBD area within five minutes from a MRT station. The MRT will span 224 miles by 2030 with two additional new lines—the Cross Island Line (CRL) and Jurong Region Line (JRL)—and three extension lines adding to the existing CCL, DTL, and NEL Lines.

In addition to supporting increasing transport demand with limited land use, the MRT also is designed to enhance the quality of environment and energy efficiency in Singapore. The strategy to improve energy efficiency includes reducing train weight and utilizing the hump-profile alignment. The hump-profile rail alignment is based on the fact that propulsion is maximally required when accelerating and braking to and from stations. The stations are sited higher than interstation alignment so that the trains naturally slow down and accelerate when entering and leaving stations.

Hyun Kim

See also: Seoul Metropolitan Subway (South Korea).

Further Reading

Department of Statistics Singapore. 2014. "Statistics. Population & Land Area." http://www.singstat.gov.sg/statistics/latest_data.html#14. Accessed December 1, 2014.

Land Transport Authority. 2013. "Land Transport Master Plan 2013." http://www.lta.gov.sg/content/ltaweb/en/about-lta/what-we-do/ltmp2013.html. Accessed December 1, 2014.

Thong, Melvyn and Adrian Cheong. 2012 (May). "Energy Efficiency in Singapore's Rapid Transit System." *Journeys* 8: 38–47.

MASS RAPID TRANSIT, BANGKOK (THAILAND)

An urban transit system is a transportation mode that serves the transportation needs of a large urban or densely populated metropolitan area. The Mass Rapid Transit (MRT) in Bangkok is one of the key public transit systems along with the city's other transit system, the Bangkok Mass Transit System (BTS-Skytrain). Both systems serve the Bangkok Metropolitan Region (BMR) in Thailand, geographically connecting two central business districts (CBD) of Bangkok with nearby provinces, including Samut Prakan, Nonthaburi, Pathum Thani, Nakhon Pathom, and Samut Sakhon, which include more than 6.3 million people. The daily ridership by both mass transit systems is nearly 1 million, and annual ridership can reach 283 million. The metro system in Bangkok is not as big as other transit systems in other metropolitan cities in Asia (i.e., Singapore, Seoul, Beijing, Tokyo) in terms of the number of lines, stations, and size, but Bangkok's mass transit systems have become an important transport mode since they began operating in 1999.

The initial master plan for the MRT was prepared in the early 1970s by a group of experts dispatched from the German government. These experts helped survey Thailand's transportation needs and recommended the ideal transportation mode to serve the growing population of Bangkok. In the early 1980s, Bangkok faced a challenge as the city's road network became severely congested. The dramatic increase in the number of private vehicles and road-based transportation modes was the main problem for traffic management and the cause of environmental degradations such as air pollution and noise. This was particularly true in the CBD areas of Bangkok. Accordingly, the initial plan for the MRT construction project aimed to resolve heavy traffic congestion in areas where population density had significantly surged, and where several informal paratransit services such as motorcycle-taxis, microbuses, and pickup trucks were operated on underdeveloped road systems.

As a first step, the BTS-Skytrain commenced operation in 1999 with two transit lines extending 23 miles. The BTS-Skytrain system's operation was granted by the Bangkok Metropolitan Authority (BMA). This system is operated electrically and without drivers. It runs both north-south and east-west to relieve the highly congested roads—Silom and Sukhumvit. In contrast, MRT—the second metropolitan rapid transit—is a subway transit line system. The stations and lines were constructed underground rather than being elevated. The line is horseshoe-shaped, encircling the area of downtown Bangkok with several stations at which passengers can transfer to the two BTS-Skytrain lines. When it was launched in 2004, the MRT consisted of 18 stations and 17 miles as a single line. The Bangkok Metro Company Limited (BMCL) has served the MRT since 2004. The concession was granted by the Mass Rapid Transit Authority of Thailand (MRTA). If the SRT Red line—which currently operates as a suburban commuter rail line—is included then, as of 2013, the total combined route length served by three public transit systems is more than 62 miles long with 64 stations.

Current mass transit systems in Bangkok have expanded with a series of projects not only to meet the increased transportation demand but also to increase number of trips by passengers via public transit systems. The benefits are twofold, resulting in sustainable land use and reducing environmental impact by private vehicles, but increasing the number of trips made by passengers using mass transit systems. As indicators, from 1993 to 2013 the number of registered private vehicles in Bangkok—both cars and motorcycles—rose significantly, from 1 million to 3.5 million cars, and from 1.1 million to 3 million motorcycles. In contrast, during 2004–2013 the number of passengers using both the BTS-Skytrain and the MRT has increased gradually from 0.36 to 0.56 million, and 0.15 to 0.23 million, respectively. This implies that the number of trips made by passengers on the mass transit systems is declining, which raises questions about the need to expand the networks.

The expansion projects aim to create a greater transit network to cover most of Bangkok and its vicinities by constructing the lines in all geographic directions. These projects mainly are driven by the MRTA. According to the MRTA's 2013 Annual Report, the existing Green Line will be extended and the new Purple Line will open for service in 2017. Other lines—the Blue, Orange, Pink, and Yellow—also are under construction.

Hyun Kim

See also: Mass Rapid Transit (Singapore); Seoul Metropolitan Subway (South Korea).

Further Reading

MRTA Annual Report 2013. "Connecting Today for the Future." http://www.mrta.co.th/en /aboutMRTA/annualReport/All2556eng.pdf. Accessed December 1, 2014.
The National Statistical Office Kingdom of Thailand. 2015. "2010 Population Census." http://popcensus.nso.go.th/en/index.php. Accessed December 19, 2015.
UITP Asia-Pacific. "Bangkok Urban Rail at a Glance." http://www.asiapacific.uitp.org/. Accessed December 1, 2014.

MASS TRANSIT RAILWAY, HONG KONG (CHINA)

Mass Transit Railway (MTR) is the metro railway system in Hong Kong, operated by MTR Corporation Limited (MTRCL). As of 2014, the MTR system has 135 miles of underground and ground-level railways, and 154 stations. The achievement of MTR has been noticeable. In 2001, MTR's total route of 55 miles already covered 41 percent of the total population within one-third of a mile of the MTR station catchment area. According to the 2013 patronage data, MTR comprised 47 percent of public transportation in Hong Kong.

Hong Kong is a world-class city and one of the leading global financial centers and traffic hubs. More than 7.2 million people live in the small area of the hilly inland and large number of islands. Residence is concentrated in the 860-square-mile urbanized area of its 3,600-square-mile territory, thus the population density is extremely high. Additionally, the highly developed central business district (CBD) with its skyscrapers concentrated around Victoria Harbor has created tremendous traffic. To handle the traffic, the transportation policy in Hong Kong has focused on developing a mass-transit system and restricting private car use.

To improve the mobility of people and goods in Hong Kong, construction of the MTR began in 1970—a decision based on several reports suggesting the construction of an urban rapid-transit railway network. After modification of the initial plan of MTR, the Kwun Tong Line in 1979 began service between Kwun Tong and Shek Kip Mei. In 1982, the Kowloon Canton Railway (KCR)—another government-owned railway company serving the commuting railway connection to the New Territories and Mainland China—was electrified to offer heavy rail commuting service to the suburbanized New Territories. In 2000, the MTRCL was privatized and merged with KCR in 2007. The single-fare system arising from the merger reduced charges for passengers.

The MTR lines connect the CBD along Victoria Harbor with residential districts and new towns in Hong Kong. The Kwun Tong Line connects east and west Kowloon. The Tsuen Wan Line runs through Victoria Harbor, Nathan Road, and the densely populated Sham Shui Po area. The Island Line runs through the business center along the northern Hong Kong Island. The Tung Chung Line runs between Tung Chung on the northwestern coast of Lantau Island and the CBD, which offers railway service for Tung Chung, a town recently built near the new Hong Kong International Airport. The Airport Express Line provides express connection between Hong Kong International Airport and the CBD. The Tseung Kwan O Line assists connection between the Kwun Tong Line and the Island Line by passing the Eastern Harbor Crossing tunnel. The East Rail provides connection service with Shenzhen in Mainland China. The Ma On Shan Line supports mass transit for the new town of Ma On Shan. The West Rail connects the northwestern New Territories with Kowloon. Light Rail also offers 12 different tram service routes in the new towns of Tuen Mun and Yuen Long, which are connected with the West Rail.

The MTR network has been extending to areas not served by the regular railway system. The Kwun Tong Line is intended to extend to Whampoa in the Kowloon city district from Yao Ma Tei. In 2014, the Island Line was extended from Sheung

Wan to Kennedy Town. The South Island Line (east section) is under construction from Admiralty to South Horizons. Additionally, the new section of railways will rearrange the current network. Ma On Shan and the new link between Tai Wai to Hong Ham will integrate with the West Rail Line to become the East West Corridor. The East Rail and the new link between Hong Ham and Admiralty will comprise the North South Corridor.

To support these substantial expansions of the MTR network, the Hong Kong government has tried to attract passengers to MTR stations by regulating competition between transit modes. At first, the government restricted competition between MTR and franchised buses after the opening of the MTR network. Then franchised buses were protected regarding competition with nonfranchised buses used in special areas. The policy improved the financial profits of MTR, but the lack of motivation for service improvements became an issue. For that reason the regulation of competition was relaxed in the 1990s and competitive circumstances between the railway and franchised buses resulted in improved service quality. This policy, however, resulted in an oversupply of public transportation in competing sections, potentially undermining the viability of the rail-based transportation policy.

Another strategy for sustaining MTR is inducing investments for development in MTR station areas. The MTR has played a leading role in implementing transit-oriented development. Densely developed high-rise apartments and multipurpose buildings—including shopping centers—have been constructed near railway stations such as Tsing Yi and Tseun Wan. Due to the developed real estate market in Hong Kong, revenue generated from the investment in these properties contributes to covering MTR construction costs. This is referred to as the "Rail + Property model."

Hyojin Kim

See also: Light Rail Transit Systems; Mass Rapid Transit (Singapore); Mass Rapid Transit, Bangkok (Thailand).

Further Reading

Cervero, R. 2010. "Transit Transformations: Private Financing and Sustainable Urbanism in Hong Kong and Tokyo." *Physical Infrastructure Development: Balancing the Growth, Equity, and Environmental Imperatives*, 165.

Cervero, R., and J. Murakami. 2009. "Rail and Property Development in Hong Kong: Experiences and Extensions." *Urban Studies* 46(10): 2019–2043.

Loo, B. P., C. Chen, and E. T. Chan. 2010. "Rail-Based Transit-Oriented Development: Lessons from New York City and Hong Kong." *Landscape and Urban Planning* 97(3), 202–212.

Tang, S., and H. K. Lo. 2008. "The Impact of Public Transport Policy on the Viability and Sustainability of Mass Railway Transit—The Hong Kong Experience." *Transportation Research Part A: Policy and Practice* 42(4): 563–576.

MASSACHUSETTS BAY TRANSPORTATION AUTHORITY (MBTA) (UNITED STATES)

The Massachusetts Bay Transportation Authority (MBTA) manages and operates the majority of bus, subway, commuter rail, and ferry routes in the greater Boston area. The MBTA was "hereby made a body politic and corporate, and a political subdivision of the commonwealth" in 1964, following its predecessor, the Massachusetts Transportation Authority (MTA). The Boston subway is the fourth busiest in the United States, and the Green and High Speed light rail lines comprise the busiest light rail system in the United States. The MBTA also operates an independent law-enforcement agency, the MBTA police.

The MBTA rail system is a direct conversion from the original steam locomotive lines that ran from Boston to Lowell in the 1830s. Mass transportation largely began with investments from private companies, creating minor monopolies in Massachusetts with the power of eminent domain granted by the state legislature. Although monopolies were created, these established necessary right-of-way acquisitions that led to the expansive rail network that later would serve the Greater Boston Area (Eastern Railroad, 1839–1840, 1874–1877). Eventually, the Boston and Lowell line grew to become the Boston and Maine line, when the railway extended into Portland, Maine, with the purchase of the Eastern Railroad in 1900. Additional railway lines throughout New England and the Northeast include Penn Central, Old Colony, and New York Central companies.

The first streetcar companies began to emerge in 1853, starting with Cambridge Railroad. Several competing companies built surface rail lines wherever possible, with most of the companies being conglomerated into just two major companies by the early 20th century—Middlesex and Boston Street Railway controlling the western suburbs, and Eastern Massachusetts Street Railway controlling the northern and southern railway systems.

To alleviate the extreme street congestion created by the amalgamation of streetcars throughout Boston—without adding intersections—the subway and elevated railway were established. The subway system installed in Boston was the first in the United States and was dubbed "Americas First Subway" by the MBTA and other organizations. The Green Line opened during 1897–1898 and the Main Line Elevated (Orange Line) opened in 1901. By 1912, three more major subway lines had opened—the Cambridge Tunnel, Washington Street Tunnel, and East Boston Tunnel. Fares for the subway and elevated rail services were controlled by requiring passengers to prepay before entering the service areas.

By 1962, all of the streetcars and elevated services were replaced with electric trackless trolleys and buses, as resident and commuter desire to use the services decreased with increased personal vehicle usage. The MTA took over the remaining services from bankrupting streetcar companies to ensure that services continued to run. The MBTA succeeded MTA in 1964, and enlarged the service area. The MBTA soon began providing subsidies to the major commuter rail services until it purchased the Boston and Maine line and Penn Central line by 1973. The year

1987 saw the end of rapid transit service to the Roxbury neighborhood; however, the subway system had been vastly improved, reaching north and south and providing a large range of coverage in Massachusetts.

Between 1980 and 1981, Massachusetts secretary of transportation and MBTA chairman Barry Locke, with the assistant director of MBTA's Real Estate Department Frank J. Waters Jr., ran kickback schemes in the organization that were discovered accidentally. MBTA general manager James O'Leary opened an envelope that was meant for Locke, and it contained proceeds from the schemes. Seventeen people were indicted, and Locke was convicted of five counts of bribery and sentenced to 7 to 10 years in prison.

Plans by MBTA are under way to extend the commuter rail lines, including the Blue and Green lines. Additionally, the MBTA and MassDOT are working together to build a new Silver route that intersects with the Blue line at the Boston Airport, and runs from west to east. The MBTA and the Commonwealth of Massachusetts are working together to bring an "Urban Ring" to the state. It would provide commuters a means to reach downtown Boston by using radial corridors that connect to many of the adjoining Massachusetts cities. It is important to note that the Urban Ring primarily would be made of dedicated bus lanes, busways, or tunnels, with a collective ridership greater than the total ridership of the Orange and Blue lines and the entire commuter rail system.

It definitely can be said that the Greater Boston Area can look forward to increased public transit ridership thanks to the overall success of the MBTA and its cooperation with the state government and department of transportation. Transit usage is predicted to increase with each improvement MBTA makes, and it is largely due to a long history of public transit access and availability.

Dylan Coolbaugh

See also: Mass Rapid Transit, Bangkok (Thailand); TransMilenio, Bogotá (Colombia); TriMet MAX, Portland (United States).

Further Reading

Belcher, Jonathan. 2014 (March 22). "Changes to Transit Service in the MBTA District." NETransit. http://www.transithistory.org/roster/MBTARouteHistory.pdf. Accessed April 7, 2014.

Harvard Business School. 2015. "Eastern Railroad Company of Massachusetts Records, 1839–1840, 1874–1877." HBS Baker Library. http://www.library.hbs.edu/hc/sfa/easternrr.htm. Accessed December 14, 2015.

Helman, Scott. 2008. "Lawmakers Seek $700m for Projects, Job Stimulus." *Boston Globe.* http://www.boston.com/news/local/articles/2006/06/15/lawmakers_seek_700m_for_projects_job_stimulus/. Accessed August 12, 2014.

Wall, Lucas. 2005 (August 1). "T Ridership Reaches Low Point of Decade." Boston.com. http://www.boston.com/news/local/articles/2005/08/01/t_ridership_reaches_low_point_of_decade/. Accessed December 14, 2015.

MBTA
See Massachusetts Bay Transportation Authority (MBTA) (United States).

METRO DE SANTIAGO
See Santiago Metro (Chile).

METRO RAILWAY, KOLKATA (INDIA)
The Metro Railway in Kolkata is the first of its kind in India. It serves the city of Kolkata along with the districts of North and South 24 Parganas in the Indian state of Bengal. It covers the busy north-south axis of Kolkata and extends from Dum-Dum, near Netaji Subhas Chandra Bose Airport, to Kavi Subhash station near Patuli. There were two major phases in constructing the Metro railway. Phase I—which covers services from DumDum to Mahanayak Uttam Kumar (Tollygunge) station—includes 10 miles of rail line and was completed in 1995. Phase II, completed in 2009, covers the network from Mahanayak Uttam Kumar (Tollygunge) station to Kavi Nazrul station, stretching a length of nearly 4 miles. A 2-mile network was constructed and opened for commercial operation in 2010. Metro Railway Kolkata further extended its service from DumDum to Noapara in 2013 with a rail length of slightly more than 1 mile.

Commuters on a busy train seek the breeze by an open doorway on a train in Kolkata, India, February 2014. (Zatletic/Dreamstime.com)

A total of 24 stations are served by the Kolkata Metro system. Of these, 15 are underground, 7 are elevated, and 2 are at grade. The average distance between any two stations is 0.71 miles.

Kolkata Metro uses automatic signal technology to operate its trains. It also utilizes a Train Protection and Warning System, which is designed to prevent accidents caused by human error. The interval between any two trains is 10 minutes during off-peak hours and 5 minutes during peak hours on weekdays. Fewer trains run on the weekends as compared to weekdays—270 on a typical weekday versus 92 on a typical weekend. The Metro carries more than 500,000 passengers per day.

The basic features of Kolkata Metro are comparable to metros in other cities around the globe. It is equipped with air-conditioning and ventilation systems and has an automatic ticket vending and checking system. The Metro offers reserved seats for women and for senior citizens. Other facilities include mobile signals in underground tunnels, radio-frequency identification tokens, and smartcard system. The Kolkata Metro has applied strong security measures as well, including closed-circuit cameras, metal detectors, and x-ray baggage scanners. It has become an integral part of the transportation network system in the city of Kolkata.

Taslima Akter and Bhuiyan Monwar Alam

See also: Mass Rapid Transit, Hong Kong (China); Mexico City Metro (Mexico); Taipei Metro (Taiwan).

Further Reading

Indian Railways. 2010. "Metro Railway, Kolkata: India's First Metro." http://www.mtp .indianrailways.gov.in/. Accessed June 3, 2015.

METROLINK, LOS ANGELES (UNITED STATES)

The 20th-century growth of Los Angeles is closely linked to a vast streetcar system—the nation's largest. The system was dismantled in the 1950s, leaving the region dependent on a growing freeway network. In 1992, however, local trains reappeared and are called the Metrolink. These trains were built and are operated by the Los Angeles County Metropolitan Transit Authority (Metro). Metrolink is a commuter rail system that runs passenger trains between Los Angeles Union Station and many outlying areas within the metropolitan area. It has seven lines—the most recent added in 2002—and has a total route length of 388 miles. Several expansions are planned for existing lines as well as a possible new line to LAX airport. There are 55 stations on the system—the farthest being 100 miles from Union Station. The suburban communities of Ventura, Lancaster, San Bernardino, Riverside, and Oceanside are among the endpoints of the lines. At Union Station the system connects with Amtrak intercity trains and the Los Angeles subway and light rail system.

Joe Weber

See also: Coast Starlight, The (United States); Commuter Trains; Los Angeles Freeways; Streetcars (United States).

Further Reading

Crump, Spencer. 1977. *Ride the Big Red Cars: How Trolleys Helped Build Southern California.* Corona Del Mar, CA: Trans-Anglo Books.
Los Angeles County Metropolitan Transit Authority. 2015. LA Metro Home. http://www .metro.net/. Accessed August 31, 2015.

MEXICO CITY METRO (MEXICO)

The Mexico City Metro (Spanish: *Metro de la Ciudad de México*), commonly known as "the metro" (*el metro*) and formally known as the *Sistema de Transporte Colectivo* (STC) is a metro rapid transit system in the metropolitan area of Mexico City. It has been publicly owned and operated by the government of the federal district through the STC since its inauguration in 1969. The system has a network of 12 lines with 175 stations that are served by 302 trains. On average it serves 7.6 million passengers per day and more than 1.6 billion passengers per year. Based on ridership, when compared to other metro systems in the world the Mexico City Metro ranks as number 8 in the top 10—just behind New York City and Tokyo and above London and Paris. Like transport systems and public spaces elsewhere, it reflects historical, social, economic, and cultural aspects of Mexican society.

By the mid-1960s Mexico City had reached a population of 5 million people. Because of inadequate mass transit the city faced crippling congestion of its streets and highways, especially in the city center. An early proposal for public transit included a monorail system, but this was deemed too expensive. During the presidency of Gustavo Díaz Ordaz, Mexican politicians and private business approached French president Charles de Gaulle to seek financial and technological help from France. The French government agreed to help and on April 29, 1967, the STC was created. The groundbreaking ceremony took place in June 19, 1967. Throughout construction, mammoth fossils were uncovered as well as a host of other archeological treasures. In addition to pre-Columbian pottery and statues, the Pyramid of Ehécatl also was uncovered in 1967. The pyramid was excavated and now can be seen on site at the Pino Suarez station.

Each of the 12 lines in the system has a specific color and number. For example, Line 1 is Pink and Line 2 is Blue. Each station has a name as well as a symbol to identify the station. For example, the symbol for the Pino Suarez station is the Pyramid of Ehécatl and the symbol for Terminal Aerea (Air Terminal) station is a plane. The designer Lance Wyman—who also designed the logo for the 1968

Olympics in Mexico City—implemented the design principles of "wayfinding" when designing the symbols for each station. The symbols are derived from local landmarks, local names, or historical events that took place in the vicinity of the station. The wayfinding system is user friendly both for those who speak Spanish and for those who don't—especially tourists. This signage system is implemented throughout the Metro network and it also has been adopted by other mass transit systems in Mexico City, such as the MetroBus Bus Rapid Transit (BRT) system.

The metro network's estimated maximum carrying capacity is 10 million users—or double the current usage. Not all stations are used equally, however. Line 2's Cuatro Caminos station serves 126,000 daily users, for example, and Line 12's Tláhuac station serves 2,000 daily users. Usage will increase when the Metro's Line 12 is repaired. Since its construction, Line 12 has been riddled with flaws that have led to power outages and flooding from Atlalilco station to Tláhuac station.

The metro system has extensive connections to other mass transit systems. In addition to being a metro terminal, for example, Buenavista station also is the terminal of the commuter train to the city of Cuautitlán, in Mexico State. At the southern end, Taxqueña station also is a terminal to an extensive light rail train to the suburb of Xochimilco. The Metro also has extensive connections with the city's MetroBus BRT system.

Future construction projects include the expansion of Line 9 and Line 12 to meet Line 1 at Observatorio station. Observatorio station will be expanded because it will become the terminal of the Mexico City–Toluca intercity express train. Other projects also include the expansion of Lines A and 4 further into Mexico State. Buenavista station also will expand once it becomes the terminal to the Mexico City–Querétaro bullet train.

Along with transporting passengers, the Metro is also a space of contestation between different social groups and governance structures. In an attempt to curtail sexual harassment of women, for example, the first two cars are reserved for women and children. Trans-women (male-to-female transgender persons), however, have had to protest to be able to ride in the "women and children" cars. The Metro's authorities also have tried to expel street vendors—known as *metreros*—that sell items on the trains, with only mixed results. Mexico's uneven economic development forces many people to work in the informal sector. Like other mass transit systems, Mexico City's Metro reflects historical, social, economic, and cultural aspects of its society.

Hector Agredano Rivera

See also: New York City Subway (United States); Santiago Metro (Chile); Tokyo Subway System (Japan).

Further Reading

Castañeda, Luis. 2012. "Choreographing the Metropolis: Networks of Circulation and Power in Olympic Mexico." *Journal of Design History* 25(5): 285–303.

Holman, John. 2014 (January 13). "Mexico: Reports from the Underground" *CCTV America: Biz Asia America*. http://www.cctv-america.tv/2014/01/13/mexico-reports-from-the-underground/. Accessed June 5, 2015.

MINICARS

The minicar is the smallest classification of a car, usually with a 1,000 cubic centimeters (cc) (or less) engine that is powered by battery or gas. The minicar industry began at the end of World War II, when airplane manufactures had to find something else to manufacture instead of war planes. The birth of the minicar also is closely tied to economics. After World War II, most people couldn't afford to buy a "real" car such as the popular Volkswagen Beetle (the "People's Car"), so more affordable minicars became popular for transportation needs.

Messerschmitt AG, the German aircraft manufacturing corporation, produced two of the most iconic minicars, the Messerschmitt KR200 and the BMW Isetta designed by aircraft engineer Fritz Fend. The Messerschmitt actually first was designed for amputees to use as transportation and looked like a plane with its wings clipped. The vehicle basically was a scooter with a cabin built around it and accessed by a hatch that opened upward. Both the KR175 and KR200 had tandem seating, with the person seated in back having to place his legs alongside of the person in front. BMW's

Italian-designed Isetta on display at the BMW Museum, Munchen, 2013. In 1955 the BMW Isetta was the world's first mass-production 3-L/100 km car. (Boggy/Dreamstime.com)

entry into the minicar industry came when it licensed the design of the Isetta from an Italian car-design company. Economically, BMW was suffering after World War II. Sports cars weren't selling well and the company had nothing to build, therefore it turned to Iso SPA for the design. This became a very successful endeavor for BMW, as it sold 168,000 Isettas and it was the top-selling single-cylinder car worldwide. The car was bubble shaped and opened via a refrigerator-type door on the front of the vehicle. The steering wheel was attached to the door.

Other noteworthy entries into the minicar market include King Midget (produced between 1947 and 1970), and the Peel (produced from 1962 to 1965). King Midget started out by buying surplus airplane and World War II equipment parts and offering its customers the parts as a kit including instructions to put the parts together. Economically the kit was very affordable, costing as little as $50. It arrived at the purchaser's doorstep in a crate. Later, the King Midget was sold fully assembled. The Peel was engineered on the Isle of Mann and has the distinction of being the smallest car on record according to Guinness Book of World Records.

The minicar surged in popularity from the early 1950s through the 1960s for many reasons, one being that the very affordable price enabled economically disadvantaged people to have transportation. Those people living in rainier cities particularly enjoyed the covered-scooter aspect of the minicars. In some countries—such as Germany and France—a driver's license was not required for a person to operate a minicar. Further benefits to the consumer included tax and insurance benefits, as the cars were taxed and insured as motorcycles and did not cost much to maintain. In cities, the minicars were easy to park and maneuver around traffic due to their small size. Some minicars even could be parked perpendicular to the curb and could be lifted off the ground by hand.

Despite their practicality, the minicar faded from popularity during the late 1960s to early 1980s. Vintage minicars were stashed in people's garages and barns. Because of their small size, many were forgotten and were allowed to languish for many years. The brief death of the minicar happened for several reasons. One reason was that more people could afford bigger cars. Bigger cars became status symbols and the minicar took on the onus of being a clown car, only fit for the circus. Safety concerns also abounded as the smaller, more fragile car led to more injuries. The insurance companies took note of injuries and consequently increased the consumer cost of insuring minicars.

Minicars have received more attention recently and major car manufacturers are embracing the minicar. From Tango T600, to BMW's Mini Cooper; Toyota's Scion, Honda's N-One, Ford's Fiesta, Chevrolet Spark, and the Smart Fortwo, major manufacturers have been enthusiastically producing minicars for the last decade. With their lower prices, better gas mileage, added safety features, and environmental friendliness, many consumers are buying minicars. In fact, Daimler's production of the Smart Fortwo has surpassed 1.7 million units and is sold around the world.

Selima Sultana

See also: Electric Vehicles; Hydrogen Vehicles; Smart Cars; Sport Utility Vehicles (SUVs).

Further Reading

Alexander, James. 2015. *British Car Manufacture 1946–2013: Part 1—Micro-Cars and 3-Wheeled Economy Transport*. Amazon Digital Services, Kindle Edition.

MONORAILS

Monorail refers to a railway system that operates a train on a single-rail track as compared to conventional dual rails. A monorail's single rail track has the advantage of occupying less land and requiring simplified facilities. After the initial straddle monorail was built in 1825, considering saving construction and maintenance costs, monorails have successfully developed as a medium-capacity transit system. The monorail system is broadly classified as a light rail transit (LRT) system serving 5,000 to 28,000 passengers per hour in many cities. The size of monorail vehicles is relatively small compared to heavy rail trains, and they are mostly elevated, as are many other LRTs. Some monorails, however, are designed for high-capacity transport during peak hours of commuting. Monorails are popular as short-distance people movers in airports, international fairs, and amusement parks. They also assist heavy rails in the metropolis and provide railway service to a trunk line in mid- or small-size cities. Today, the monorail system is a popular railway transportation system, especially in Japan, Germany, and the United States, with 49 lines operating in 2015.

The first modern-style monorail was the Wuppertal Schwebebahn Railway in Germany, which became operational in 1901 and still is serving passengers. The monorail vehicle was suspended on an elevated single rail by electrical power. After the first half of the 1900s, monorails began to be noticed as a new transportation system to solve urban congestion issues. The first commercial monorail in the United States opened in Disneyland Resort in Anaheim, California, in 1959. This influenced the perception of the monorail as being an advanced method of transportation. After the first monorail at Tokyo's Ueno Zoo opened in 1957, Tokyo Monorail Haneda Line began to serve passengers between Tokyo International Airport and central Tokyo during the Tokyo Olympics in 1964. Today, Tokyo Monorail carries 127,000 passengers per day, which has led to continuous introduction of monorails as a mass transit system.

Monorails largely are categorized as straddle-type and suspended-type. Straddle-type monorails run above a steel- or concrete-beam guideway. ALWEG's straddle design mostly is used for monorails. Rubber tires commonly are used for running monorail cars along the beam guideway, which contributes to reduced cost for monorail construction and maintenance. Conversely, suspended-type monorails move below the tracks and the wheels run inside the beam. This type also is called the SAFEGE-type suspended monorail. They operate in Chiba and Shonan in Japan. The SAFEGE type occupies less land (as compared to the straddle type) by using a steel structure for a guideway. Maglev sometimes is referred to as a type of suspended-type monorail because the maglev train runs using magnetic force instead of electric motors to propel the train, and is levitated above a guideway. Maglev is faster than other monorails due to the levitation minimizing friction between the train and tracks.

The monorail has various advantages for urban transportation. A monorail's narrow track makes the viaduct slim, which is a great advantage of this system. The structures of monorails occupy a relatively small area as compared to other elevated railway systems, they do not disturb ground traffic, and they reduce construction cost due to the money saved on land and the reduced construction time. When a city plans to build an additional railway system in an urbanized area, monorails can be built on a more limited budget. Additionally, monorail vehicles have better performance for running sharp curves and slopes. Using rubber tires helps to reduce the noise of trains and helps them to run on steep tracks. Chongquing Metro Line 3 in China, for example, is the world's longest and most crowded monorail system, considering Chongquing's hilly and steep urban area. Moreover, monorails have the environmental benefits of less air pollution, reduced noise, and less vibration, which make them more pleasant for passengers.

Conversely, the cost for monorail tracks and trains can be significant because monorail cars have complicated components and limited flexibility in operation. Also, monorails are not compatible with conventional railways due to the different rail standards, thus operating convenient direct through trains is not an option. Moreover, general straddle-type and suspended-type monorails are low-speed units except for maglev trains. Thus, the role of monorails in urban mass transportation is limited to local service.

The safety of monorails in emergencies should be improved. Monorails operate mostly on an elevated beam track with little available space, thus evacuation in an emergency is difficult as compared to that for other LRTs. Some monorails, however, are prepared for improved emergency measures. The Las Vegas monorail, for example, includes emergency walkways built between the two monorail guideways. Nevertheless, monorails clearly are one of the most attractive urban mass transits. The monorail design—with its aesthetically pleasing appearance and the advantages of technical flexibility—matches modernized urban landscape. That is the reason why monorails continuously receive attention in many urban transportation plans.

Hyojin Kim

See also: Light Rail Transit Systems; Maglevs; Taipei Metro (Taiwan); TriMet MAX, Portland (United States).

Further Reading

Grava, Sigurd. 2003. *Urban Transportation Systems.* New York: McGraw-Hill.

Kennedy, Ryan R. 2007. "Considering Monorail Rapid Transit for North American Cities." *The Monorail Society* 41.

Sekitani, Taketoshi, Motomi Hiraishi, Soichiro Yamasaki, and Takayuki Tamotsu. 2005. "China's First Urban Monorail System in Chongqing." *Hitachi Review* 54(4): 193–197.

MONTSERRAT RACK RAILWAY (SPAIN)

The Montserrat Rack Railway (*Cremallera de Montserrat* in Catalan and Spanish) is part of a larger train system operated by the Ferrocarrils de la Generalitat de Catalunya company. The system links the Spanish port of Barcelona to Santa Maria de Montserrat, a monastery about 35 miles to the northwest. Perched some 2,395 feet above sea level on the slopes of Montserrat Mountain, the monastery dates from the early 11th century and long has been a place of pilgrimage.

Traditionally, Montserrat could be reached only by foot or on horseback. During the second half of the 19th century, however, the Ferrocarrils del Nord Company laid a wide gauge friction (adhesion) railway line from Barcelona to what became the Monistrol de Montserrat station. From 1858 to 1859, the company added a 9-mile road running from the station to the monastery. These improvements made it possible to travel from the port to the monastery by steam train and horse-drawn coach in about three and a half hours.

In 1892, the Ferrocarriles de Montaña a Grandes Pendientes company completed a rack railway that ran from Monistrol to Montserrat, shortening the trip from the station to the monastery to little more than one hour. In 1953, however, an accident on the railway killed 8 passengers and injured more than 100 people. Declining revenues (due in part to the opening of the Aeri de Montserrat cable car in 1930) led to the closure of the aging line in 1957.

Work on a new rack railway operating on the Abt system began in 2001 and was completed in 2003. The route from Barcelona to the monastery now consists of a friction section running to Monistrol de Montserrat, at which point the three-mile electrically powered Montserrat Rack Railway itself officially begins. Confusingly, the line actually operates by friction over the first 0.6 miles of its length. At Monistrol Vila station, however, the line converts to rack traction, running another 2.4 miles and rising approximately 1,804 feet before terminating at the monastery. Traveling the entire route takes about an hour and a half, and the trip is one of the most popular attractions in the region.

From the monastery grounds two funicular railways (twinned cable cars running in tandem on steep slopes) carry passengers to points even higher up the mountainside. Some 1,650 feet long, the Sant Joan funicular was built in 1918 and rises over a maximum grade of slightly more than 65 percent. The Santa Cova, constructed in 1929, is 925 feet long and rises over a maximum grade of 56.5 percent. Both funicular lines were upgraded in 1997 and, like the rack railway, feature cars with panoramic roofs.

Grove Koger

See also: Narrow Gauge Railroads; Rack Railways; Railroads.

Further Reading

Generalitat de Catalunya. 2014. "Cremallera i funiculars de Montserrat." http://cremallerad
    emontserrat.cat/website_cremallera/eng/index.asp. Accessed December 14, 2015.
Steves, Rick. 2013. *Rick Steves' Spain 2014.* Berkeley: Avalon Travel.
Tourist Guide Montserrat. 2014. "Rack Railway to Montserrat: Cremallera Funicular to
    Montserrat in Catalonia." http://www.montserrat-tourist-guide.com/en/transport
    /rack-railway-montserrat.html. Accessed December 14, 2015.

MOPEDS

A moped is a low-powered lightweight hybrid bicycle that is partly motorized and partly human powered. The word "moped" was coined by the Swedish journalist Harald Nielsen in 1952, who described it as a "pedal cycle with engine and pedals." It therefore requires energy from both the rider and a motor for sufficient functionality. The term "scooter" and "moped" are used interchangeably, but they are fundamentally different. Mopeds are considered powered bicycles, or peddle-assisted motorcycles (with 2-stroke engines), or low-speed motorcycles. Scooters are completely motorized and require a mixture of oil and gasoline to run. Mopeds typically travel at the speed of a bicycle and have a maximum speed of roughly 30 miles per hour. Today, mopeds are known as fuel-efficient vehicles that earn praise for saving money (in the United Kingdom mopeds are excluded from road prices), but their environmental friendliness still is subject to multiple studies.

The invention of mopeds can be traced back many years. The first moped basically was a bicycle fitted with an internal combustion engine. It originated in Germany during the late 19th century when bicycles were immensely popular. This model had many technical difficulties. "The Douglas," produced in 1912, was greatly improved from previous versions, especially for climbing hills. This ultimately led to a boom in engine-powered motorcycles. The economic struggles people faced after World War II, however, led to the demand for affordable transportation alternatives to motorcycles and automobiles, culminating in the moped's rebirth.

Bicycle makers such as Peugot and Ducati began offering small auxiliary engines for their bicycles, and various types of mopeds were designed. The design of "Velosolex"—which appeared in France in 1947—is an example of the first "giant leap forward" for mopeds. It was constructed with a 33cc engine and a friction roller over the front wheel. It also came with a lever that could be used to engage or disengage the power. Another very innovative design was known as the "cyclemaster," which came with an 8cc engine and a completely powered rear wheel. This was produced by the German company DKW around 1947.

The introduction of the MS-50 in 1950s by the bicycle maker company Steyr Puch of Austria, however, dramatically changed the design of mopeds. The

New moped on the shore at Varna Bay, Bulgaria, 2013. (Anna Hristova/Dreamstime.com)

MS-50's design deviated from the principles of bicycle design. Although the moped look was preserved, it had more in common with "real" motorcycles than with bicycles. The MS-50 was equipped with a steel pressed frame, fan-boosted engine cooling, two-speed handlebar-shifted gearbox, and an elaborate electrical system. This new design changed the market for mopeds worldwide.

Mopeds were first sold in the U.S. market though large department store catalogues such as Sears and Montgomery Ward during the 1950s and 1960s. By the 1970s, the U.S. market was flooded with at least 125 models of mopeds derived from the MS-50 design, most likely due to high gas prices at the pump. Drivers sought a vehicle that would do the job without spending much on gas and maintenance. The sale of mopeds in the U.S. market grew so fast—especially for the rural population—that many moped manufacturers set up small showrooms for their machines within auto dealerships. The Japanese motorcycle company Honda began producing new models of mopeds for the exploding U.S. market. Even the U.S. AMF Company started producing mopeds equipped with a rear wheel engine.

The moped's popularity also flourished in many countries in Asia and Africa, where personal transportation is a major challenge for many middle-income people. Because of its low price (as compared to motorcycles), low maintenance, and high gas mileage (it can run 100 miles on a gallon of fuel), plus the few existing regulations on operation, mopeds are a very attractive means of personal transportation, particularly for young people, and especially in congested urban settings. Mopeds' status in Europe improved as the economy got better to the extent that there were moped races, clubs, meets, and trips. It has been estimated that millions of mopeds were sold worldwide in the 1970s.

Despite a brief sales slump due to the worldwide recession in the early 1980s, moped sales bounced back again in the 1990s. Today, there are many types of mopeds that evolved from the original version, as many companies started to produce their own models—including adding advanced technology. There are so many types available that it now is hard to distinguish differences between mopeds with pedals, mopeds without pedals (no peds), motorized cycles, and even scooters.

There are different rules and regulations for mopeds worldwide. Denmark classifies mopeds as motorized personal transport, and requires mechanical brakes, lights, and a license plate. License and insurance is mandatory in the United States, although licensing requirements are less rigorous than those for motorcycles and automobiles. There is no doubt that the original moped concept reappears with the goal of keeping consumers' costs at a minimum.

Selima Sultana

See also: Bicycles and Tricycles; Moto-Taxis; Velomobiles.

Further Reading

Horton, Dave, Paul Rosen, and Peter Cox (eds.). 2007. *Cycling and Society*. Burlington, VT: Ashgate.

Van De Walle, Frederik. 2004. "The Velomobile As a Vehicle for More Sustainable Transportation: Reshaping the Social Construction of Cycling Technology." Master's Thesis. Department of Infrastructure, Royal Institute of Technology, Stockholm. http://users. telenet.be/fietser/fotos/VM4SD-FVDWsm.pdf. Accessed July 20, 2015.

MOSCOW METRO (RUSSIA)

First opened in 1935 and expanded over the course several tumultuous decades, the Moscow Metro's elaborate architecture and works of art serve as a museum for the masses, telling the history of Moscow, Russia, and the former Soviet Union. Today it is among the world's busiest, most extensive, and most beautiful underground railway networks, with more than 200 stations and more than 200 miles of track used by about 9 million passengers per day.

A mass transit system for Moscow first was proposed at the turn of the 20th century, when the city's population was nearing 1 million and continued increasing due to the growth of industrial jobs. At the time, Moscow's primary form of transport was a contingency of about 16,000 horse-drawn carriages. Initial plans called for about 40 miles of elevated railway line and 10 miles of tunnels. Construction was delayed repeatedly—first by an uprising in 1905, then by

World War I and the Bolshevik Revolution of 1917—and eventually was scrapped due to high costs.

In 1931, the idea was revived and approved on a larger scale by the government of the Soviet Union. Initial plans called for 10 underground rail lines covering a total of 50 miles. Soviet leader Joseph Stalin ordered some of the Soviet Union's most skilled artists and architects to create an underground transit system that would celebrate the vast communist nation and provide a public art space for the city's working class. Due to Stalin's influence, many of the Moscow Metro's early art displays reflected communist realist themes, including workers' rights.

When the first experimental tunnel was dug in 1931, engineers discovered some of the most adverse soil conditions ever encountered on such a project—including unstable clays, quicksand, and several underground rivers. Although the Moscow Metro's initial engineers came from London's Underground subway system, Soviet workers handled the actual construction, and many died as a result of the dangerous conditions.

Many engineers were arrested and later deported on charges of espionage, having apparently learned too much about the city's aboveground and underground layout. Some historians believe that a separate, secret transit system was built below the Moscow Metro for use by the KGB and for the evacuation of top Soviet officials in the event of a nuclear attack, but its existence never has been confirmed.

When the Moscow Metro opened on May 15, 1935, it comprised just one underground rail line and 11 stations stretched across 7 miles. Yet, on its first day of operation, an estimated 285,000 people used the system. Expansion continued through the early 1950s, with ever-more elaborate art installations in the various stations. Among the most celebrated for their art and architecture are:

- Mayakovskaya Station, which boasts stainless-steel art deco columns and 34 different ceiling mosaics created by Russian artist Alexander Deyneka;
- Ploshchad Revolutsii Station, which features 76 bronze sculptures depicting working-class citizens created by Russian sculptor Matvey Manizer;
- Novoslobodskaya Station, decorated with 32 stained-glass panels;
- Kiyevskaya Station, which has a wide array of murals and mosaics celebrating Russo-Ukrainian unity; and
- Komsomolskaya Station, which was designed in the baroque style and is adorned with murals by Russian painter Pavel Korin that invoke historic Soviet heroes of World War II.

Construction of the Moscow Metro was delayed briefly during World War II, when some of the deeper underground stations were used as bomb shelters. The most profound impacts on the Metro's architecture came after Stalin's death in 1953, however, when new Soviet leader Nikita Khrushchev began to condemn what he called the unnecessary excesses and luxury spending of the Stalin era. Under Khrushchev, Metro construction continued under the slogan: "Kilometers at the expense of architecture."

Thus, stations built between the late 1950s and early 1990s were constructed with budget—rather than art—in mind. VDNKh Station, begun in the early 1950s under Stalin, saw its Vladimir Favorsky mosaics covered over with thick coats of paint under Khrushchev's austerity campaign.

Moscow Metro's style changed again in the 1990s after the collapse of the Soviet Union, with stations again featuring art—although on a more limited scale. Among the most well-known stations of this era is Rimskaya, opened in 1995, which features sculptures of Remus and Romulus, the mythical founders of ancient Rome. Beginning in the early 2000s, transit system architects returned to more elaborate station artwork celebrating both Soviet and Russian history. The Dostoyevskaya Station opened in May 2010 to celebrate the Metro's 75th anniversary, and features murals of Russian writer Fyodor Dostoyevsky and various characters from some of his most famous novels.

Today the Moscow Metro strains under the demands of 9 million passengers a day, and the Russian government is working to further expand it. New stations are being built outside the city to provide better connections for those traveling from surrounding suburbs. For the Metro's 80th anniversary, its 200th station, Kotelniki, was opened. Current plans call for an additional 35 new stations and 75 miles of rail line to be opened by 2020. If construction is successful, then this would make the Moscow Metro the world's fourth-largest mass transit system, after those in Beijing, Seoul, and Shanghai.

Terri Nichols

See also: Madrid Metro (Spain); Mexico City Metro (Mexico); Shanghai Metro (China).

Further Reading

Guardian, The (U.K.). 2015 (May 15). "Celebrate the Moscow Metro's 80th Birthday with a Journey through the City's History—in Pictures." http://www.theguardian.com/cities/gallery/2015/may/14/moscow-metro-80-anniversary-city-history-in-pictures. Accessed September 2, 2015.

Moscow Metro official website. http://engl.mosmetro.ru/. Accessed September 2, 2015.

Railway-Technology.com. http://www.railway-technology.com/projects/moscow-metro/. Accessed September 2, 2015.

Telegraph, The (U.K.). 2015 (May 15). Travel Section. "Moscow Metro: 80 Years of the World's Most Beautiful Underground." http://www.telegraph.co.uk/travel/journeysby rail/11590175/Moscow-Metro-80-years-of-the-worlds-most-beautiful-underground .html. Accessed September 2, 2015.

MOTORCYCLES

A motorcycle is a self-propelled two- or three-wheeled vehicle that travels on roads or on dirt and is suitable for one or two riders at a time. Sometimes called a "bike," motorcycles generally have a 250- to 2,000-cubic-centimeter (cc) gasoline-fueled internal-combustion engine, or they run on electricity. Variations of the motorcycle include motor scooters, which have medium-size engines; minibikes or minicycles, which have small frames and wheels, and 85cc engines; and mopeds—motorized bicycles—which have engines that are 50cc or smaller. Dirt bikes are fast and nimble off-road motorcycles designed for use on rough terrain. They include motocross bikes and dual-sport bikes (for use both on- and off-road). Motorcycles are a means of transportation, a form of recreation, and a symbol of personal identity.

Motorcycle engines can be either two-stroke or four-stroke and have up to four cylinders. Although some are water cooled, most are cooled by air. The motorcycle's steel frame is made up of both tubes and sheets, with aluminum or steel rims with spokes. Strong, lightweight materials such as graphite, composite, and magnesium increasingly are being used for frames. Tires are rubber and inflated but are more rounded than automobile tires to enable the rider to lean into a turn. Standard motorcycle transmissions contain four to six speeds. Riders operate the clutch and throttle by twisting the handgrips. Front brakes are controlled by a lever near the handgrip, and a foot pedal controls the rear brake, which can be either a disc or drum. Electric push-button starters have replaced traditional kick-starters on most vehicles. Steel chains—substituted for leather belts in 1900 because of the unreliability of leather—transmit power from the engine to rear-wheel sprockets. The sidecar—a separate wheeled passenger-carrying vehicle—first was attached to a motorcycle sometime around 1893.

The origins of the motorcycle can be traced to inventors trying to outfit a standard "safety" bicycle with a steam-powered engine. The first person to be successful at this was American Sylvester H. Roper—whose coal-powered, two-cylinder Steam Velocipede was conceived in 1867. Roper's final revision included inflated tires and a smaller 160cc engine that could reach speeds of 30 to 40 miles per hour. The first commercially available self-propelled bicycle was the Butler Petrol Cycle, invented in 1884 by Edward Butler of England. The Butler had three wheels, a 600cc twin four-stroke engine that was started by compressed air, a radiator for cooling, and a throttle-valve lever used to control the speed. The driver sat between the two front wheels and used rocking handles to steer. To brake, the driver raised and lowered the rear driving wheel.

The German Daimler Reitwagen (riding car) is credited as being the first true motorcycle because of its four-stroke, gasoline-fueled internal combustion engine. It was designed and built by Gottlieb Daimler and Wilhelm Maybach in 1885. The first two-wheeled vehicle actually dubbed a "motorcycle" (the translation of the German *Motorrad*) was an 1894 invention of Hildebrand and Wolfmüller of Munich and contained a water-cooled, 1,488cc, four-stroke, two-cylinder engine that could achieve a speed of 40 miles per hour.

The first production motorcycle—those made to be sold to the public—was the Excelsior, built in 1896 in Coventry, England, and the Orient-Aster, built by Charles Metz in 1898 in Massachusetts. Mass production of motorcycles did not begin until 1901, when Royal Enfield produced its first motorcycle and English

Daimler Reitwagen, Mercedes-Benz Museum, 2013. (MF1 collection/Alamy Stock Photo)

bicycle maker Triumph introduced a line of motorcycles. By 1903, Triumph was the largest manufacturer, producing more than 500 units per year. Its capacity soon was overtaken by Indian Motorcycle Manufacturing in Illinois. In 1913, Indian reached record sales of 32,000 units. Harley-Davidson—the creation of Bill Harley and Arthur Davidson—opened its first factory in Milwaukee, Wisconsin, in 1903, and built a one-cylinder cycle for racing. Harley-Davidson soon became one of the most prolific motorcycle brands and ushered in the chopper style of motorcycle—bikes with sleeker, lighter frames and fewer unnecessary parts, such as front fenders, windshields, and crash bars.

Conducting motorcycle trial tests as a way to make reliable machines led to competitions between manufacturers and sparked motorcycle racing. One of the first such races—the Tourist Trophy—was staged on the Isle of Man in Great Britain in 1907. Competitions called "scrambles" were on street bikes until the 1930s, when technical changes made motorcycles appropriate for riding on rough terrain.

At least 80,000 motorcycles outfitted with holsters were used by the U.S. Army in World War I and played an integral part in the war effort. Motorcycles could quickly carry messages to the front line, and were used in dispatch and reconnaissance efforts and for military police operations. By 1918, close to half of Harley-Davidson's output was for the U.S. military. Many other American, British, and

French companies also produced motorcycles for the Allies, including Indian, Excelsior, Henderson, Triumph, and Douglas. Triumph's Model H, with a rear-wheel belt and no pedals, is considered the first modern motorcycle. Motorcycles played a prominent role in World War II as well, although to a lesser extent than in World War I because the production of mobile armored vehicles was stepped up. Bavarian Motor Works (BMW) and Zündapp made large bikes for the Germans. Zündapp's Kettenkrad ("track bike") looked like a tank attached to a motorcycle's front end and could tow small cannons through rough terrain. The U.S. Army learned from the German engineering by dismantling captured BMW bikes and then "borrowed" the technology.

After the war, American veterans who rode motorcycles for the army established motorcycle clubs for camaraderie. The image of bikers was altered after the 1954 film *The Wild One* portrayed bikers as gang members, although only a small percentage actually became outlaw gang members. In Europe, postwar motorcycle production skyrocketed as bikes were recognized as an economical means of transportation. To appeal to the masses, Italy in 1946 introduced the Vespa motor scooter, which gained immediate popularity. Able to get 100 miles per gallon (mpg), the moped also gained acceptance postwar, and sales in the United States then increased dramatically during the energy crisis of the 1970s. Japan's Honda became the dominant player internationally in the late 1960s, as the Japanese—including Suzuki, Kawasaki, and Yamaha—introduced Japanese motorcycles to the United States in 1959 and continued producing stylish, reliable, and inexpensive vehicles.

Motorcycles tend to be a luxury item in the industrial world but are an important and popular means of transportation in the developing world. They are relatively inexpensive to own, are less regulated, and can be faster than traveling by car or bus on congested streets. Motorcycles are the primary mode of travel in large Asian cities and also are becoming popular in South America and Africa. To deter reckless driving some cities, such as São Paulo, Brazil, are testing motorcycle lanes and restricting travel.

Motorcycle safety is a grave concern worldwide and motorcycle helmet laws are controversial. The Insurance Institute for Highway Safety (IIHS) in 2013 reported that deaths by motorcycle were 26 times greater than death by car, and accounted for 13 percent of vehicle fatalities. The IIHS reported that helmet use reduced the number of deaths by 37 percent and reduced the number of brain injuries by 67 percent. Only 19 states, plus Washington, DC, however, mandate helmet use for all riders. Other states have partial-use laws, and in three states—Illinois, Iowa, and New Hampshire—no laws exist. Worldwide, 29 countries had mandatory helmet laws in place by 2003. Great Britain's regulations go back as far as World War II. Opponents of helmet laws—including the American Motorcyclist Association—believe that universal laws restrict personal liberties, and have lobbied for repeal of such laws.

The typical motorcycle gets 43 mpg as compared to an average 23 mpg for cars. Emission standards continue to be strengthened, with the strictest limits coming from the European Union and the state of California, both of which restrict nitric oxides and carbon dioxide in addition to standard hydrocarbon emissions.

Rosemarie Boucher Leenerts

See also: Mopeds; Moto-Taxis; Segway; Velomobiles.

Further Reading

Page, Victor W. 2004. *Early Motorcycles*. New York: Dover.
Siegal, Margie. 2014. *Harley-Davidson: A History of the World's Most Famous Motorcycle.* Oxford: Shire.

MOTO-TAXIS

A paid-for ride on the back of a motorcycle is a common sight in many large cities worldwide. This form of transportation meets the needs of a niche population—people who can pay a bit more to avoid the cramped conditions and inflexible routes of buses but cannot afford a car taxi on a regular basis. As a low-cost transport option, moto-taxies are common throughout the developing world. Names for the service include Moto-Táxi (Brazil), Okada (Nigeria), Boda Boda (East Africa), Motocy (Thailand), and Ojek (Indonesia).

It is difficult to determine exactly when this type of transport developed. It likely has been around for as long as motorcycles have and began as an informal transport service in rural areas where typical transport was limited to bicycles. As access to motorcycles increased, they replaced bicycles in rural-to-urban transport, and then became more prevalent in urban areas. In many cities motorcycle taxi services developed informally and remained difficult to regulate: Drivers operated without specific licenses or training; riders were not necessarily given helmets; engine sizes, maintenance, and even passenger limits (numbers or age) were not mandated. Riders were willing to take risks in exchange for convenience and fast, efficient, affordable transport.

Since the mid-1990s, attempts to regulate and formalize motorcycle taxi systems have met with some success. In some cities, associations have been formed and motorcycle drivers pay an annual membership fee that provides them with rights of operation and certain other benefits. Associations have been known to represent members in legal cases, provide credit and banking services, and enforce fines related to hygiene. State regulation of motorcycle taxis also influences the two-wheeled transport practice. Fee structures and safety standards are in place in many cities.

Today, all within the same city, motorcycle taxis can be seen dashing by carrying a passenger and a pile of packages, goods, or even small livestock; transporting a handful of young children; or carrying a professional hurrying to get to an executive meeting. Drivers in urban areas might be equipped with cell phones to provide quick pick-up and door-to-door service.

In the developing world, reasons for using motorcycle taxis include passengers being in a hurry, embarking on a trip that requires door-to-door service, safety in the twilight and night hours, or a lack of available standard taxis. Studies have indicated that the time efficiency of motorcycle taxi travel has given riders greater opportunity throughout the day and even has increased incomes. Passengers of motorcycle taxis are more reliably on time for work and are able to engage in more activities during the day because motorcycle trips are faster than other available transport options. These factors are particularly important for women, who juggle home and income responsibilities, and who feel safer on a motorcycle taxi if the daylight hours are waning.

In contrast to developing-world contexts in which the motorcycle taxi industry is fluid and at times precarious, motorcycle taxis are now for hire in California and New York City. In the United States, a sleek fleet of upscale vehicles is available through motorcycle clubs. To target business professionals, the taxis provide Bluetooth-equipped helmets and offer quick transport through congested cities. Costs are substantially higher and can involve paying membership fees of motorcycle clubs, but with "lane splitting" being legal in some cities, motorcycle taxis might be a faster option for America's affluent business executives.

Cynthia Sorrensen

See also: Motorcycles; Rickshaws (Bangladesh); Rickshaws (Japan); Yellow Cab, New York City (United States).

Further Reading

Siler, W. 2011. "Motorcycle Taxis Come to America." RideApart. http://rideapart.com/articles/motorcycle-taxis-come-to-america. Accessed August 8, 2015.

Virtual Tourist. "Getting around Hanoi." http://www.virtualtourist.com/travel/Asia/Vietnam/Thu_Do_Ha_Noi/Hanoi-1481679/Transportation-Hanoi-TG-C-1.html. Accessed August 8, 2015.

MUSICAL ROADS

A musical road is a road or a portion of one on which a tune is played when a vehicle travels over it. The sound is caused by tactile vibrations from a vehicle's tires producing an audible rumbling inside the vehicle in a series of notes forming a musical tune.

There currently are eight known musical roads in the world—in Denmark, Japan, South Korea, and the United States. The first musical road ever made is the Asphaltophone, created in 1995 in Gylling in East Jutland (Østjylland), Denmark. Two Danish artists, Steen Krarup Jensen and Jakob Freud-Magnus, designed and installed the Asphaltophone, which is made by a series of raised pavement markers, similar to the "Botts' dots" that separate freeway lanes or "rumble strips" used to alert weary drivers. The markers are strategically spaced out in intervals that produce a short excerpt from Gioachino Rossini's "William Tell Overture" that lasts just a few seconds when driven over.

A team of researchers at the Hokkaido Industrial Research Institute in Japan created several "melody roads" that each play a tune when they are driven on. The tunes are created by a series of grooves gouged into the roadway. The surface of the road plays notes that vary in tone depending on how far apart the grooves are spaced—typically from 2 to 4 inches apart. The roads were conceptualized by engineer Shizuo Shinoda, who had accidentally scraped markings into a road with a bulldozer's teeth and then noticed that a variety of tones were emitted when he drove over the gouges.

There are four musical road sections in Japan. The first of the musical roads was built in the countryside along Highway 272 between Kushi and Shibetsu in the Hokkaido Prefecture of northern Japan. The melody, cut into the highway in 2007, plays for about 34 seconds as drivers pass at 28 miles per hour. The musical road in Gunma Prefecture plays "Memories of Summer," a traditional Japanese tune. The ridges in this 574-foot stretch of musical road 100 miles north of Tokyo are 2,559 in number and cost $12,500 to create. The perfect pitch and rhythm of the song can be heard when the driver travels at 31 miles per hour. Another musical road is found along the highway on the drive to Mount Fuji in Wakayama Prefecture. This stretch plays a 1963 pop song called "Miagete Goran Yoru No Hoshi Wo" ("Look Up at the Stars at Night") by Kyu Sakamoto. It, too, was formed by strategically placing grooves in the roadway. The song plays for about 25 seconds. A fourth musical road was created along Route 331 in Okinawa in November 2012. It was installed by road crews along 1,100 feet of the road and plays a famous Okinawan folk song, "Futami J wa," when vehicles drive on it at 25 miles per hour.

The city of Anyang, South Korea, has a musical road made in the same way as those in Japan, by grooves gouged into the roadway. It plays the tune "Mary Had a Little Lamb," which lasts about 15 seconds when driven over at a speed of 62 miles per hour. Unlike Japan's musical roads, created as a tourist attraction, the musical road in Korea is on a busy highway and was installed to slow down drivers along a fast downhill S-curve. About 68 percent of all traffic accidents in South Korea are caused by inattentive or weary drivers and the highway commission thought drivers would slow down in this dangerous section so they could hear the tune.

The United States has two musical roads, the Civic Musical Road, built in Lancaster, California, north of Los Angeles, installed on September 5, 2008, and the other along historic Route 66 near the town of Tijeras, New Mexico. Originally created for a Honda Civic automobile commercial, the Civic Musical Road was built by the groove method. It first was installed on West Avenue K in Lancaster, but city officials paved over the section just 18 days later after receiving noise and behavior complaints from residents claiming the two-lane road drew tourists because of its publicity. One month later, the musical road was rebuilt on West

Avenue G, miles from residential neighborhoods and on a busy highway. Drivers must travel in the far-left lane at a speed of 55 miles per hour to produce the tune. As with the Asphaltophone in Denmark, the Civic road also produces a clip from Gioachino Rossini's "William Tell Overture," although it plays for about 13 seconds as opposed to just two or three seconds for the Asphaltophone.

The other musical road in the United States is known as the Singing Road. This portion of Route 66 in New Mexico was built by National Geographic for its TV show *Crowd Control* and debuted in October 2014. The show uses experiments to change social behavior and the Singing Road was the show's attempt to keep drivers focused by driving the posted speed of 45 miles per hour to hear the song—an excerpt of "America, the Beautiful." To create the tune, metal plates were placed along the road and were covered in asphalt; then rumble strips were cut into the asphalt.

Rosemarie Boucher Leenerts

See also: Blue Ridge Parkway (United States); Dirt Roads; Parkways.

Further Reading

Johnson, Bobbie. 2007 (November 12). "Japan's Melody Roads Play Music As You Drive." *The Guardian.* http://www.theguardian.com/world/2007/nov/13/japan.gadgets. Accessed June 5, 2015.

Payne, Oliver. 2012. *Inspiring Sustainable Behaviour: 19 Ways to Ask for Change.* New York: Earthscan.

Younger, Emily. 2014 (October 1). "Route 66 'Singing Road' Debuts in New Mexico." *KRQE News 13.* http://krqe.com/2014/10/01/route-66-singing-road-debuts-in-new-mexico/. Accessed June 5, 2015.

NARROW GAUGE RAILROADS

The gauge of railway is defined by the width between the rails. A narrow gauge is any railway that has a between-rail width of 4.5 feet or less. This contrasts with standard gauge (4.7 feet) and broad gauge (5 to 7 feet) primarily because of the costliness of transferring from narrow gauge tracks to either of the other types. More than 60 percent of the world uses standard gauge as its country's main gauge type.

Narrow gauge railroads date back to the 1500s when used in mines in Bohemia, located within modern-day Czech Republic. The smaller width of the railroad line allowed for smaller carts and entryways to be constructed. This required less intrusion into the mountains and provided safer passage into and out of the mines. Although originally designed to operate within mines, the railroads eventually were extended to the destinations to which the minerals needed to be transported. With advances in technology, railways within the United Kingdom used for industrial purposes began to adopt steam locomotives in the mid-1800s, with the concepts slowly moving to countries such as Belgium.

In the United States, narrow gauge railways often were used as spurs for sections of the transcontinental railroad. They also traveled through mountainous terrain, as it was easier for narrow gauge trains to maintain tighter turns. Narrow gauge railways also are cheaper to construct than the lines and bridges for wider gauges, making them ideal choices for short or temporary lines. Born out of a need to continue building railroads without the expenditure of standard gauge lines, narrow gauge sprang forward during the 1870s when mineral deposits were found and cars were needed quickly and at the lowest cost. The savings quickly ran out when the deposits were depleted and in many cases the lines had to be removed less than 20 years later. This was a significant drawback of using narrow gauge rail during one of its biggest construction and use periods.

There are several advantages to choosing narrow gauge railway over standard gauge, including the ability to make tighter turns—which is ideal for mountainous terrain—and the reduction in costs as compared to standard gauge railways. Narrow gauge lines also were used in areas where the terrain was unsuitable for large gauges and where short additions were needed. In the case of the line in North Queensland, Australia, both conditions were true, as new plantations for hemp and the surrounding environment prohibited rail beyond 3 feet and 6 inches (1,067 mm) wide and, in the future, would require small extensions to meet the demand for products and access to the ports. Additionally, narrow gauge rail is

Durango & Silverton Narrow Gauge Railroad, Colorado. (Americanspirit/Dreamstime.com)

useful for temporary projects, in which the rails are removed postcompletion, again due to its cost. This use, however, has been replaced by using trucks. The most common way that narrow gauge has been consistently utilized is for feeder lines that branch off of main railway lines. Cost is the common factor between all of these advantages, leading to the somewhat marginalized nature of narrow gauge rail in industrial settings.

There are some significant issues with transferring freight and passengers between narrow and standard gauges. Also, the low cost of narrow gauge railroad means that it fails to stand up to demand greater than what was originally conceived. For narrow gauge to be made capable of handling higher capacities and speeds, increased costs are accrued, thereby eliminating the edge over standard and broad gauge railroad. Another disadvantage of narrow gauge over wider sizes is the cramped spaces within the cars that ride on the lines.

Narrow gauge railways exist around the world, primarily in South American countries, parts of western and eastern Africa, Australia, Southeast Asia, and parts of Western Europe. Their purposes mostly are industrial, as passenger rail and even light rail transit have switched almost exclusively toward standard gauge railways. This development leaves the future of narrow gauge rails unclear, as their original uses for branch lines—their advantages for tight turning radii—have been replaced by articulated cars and other new technologies that provide these advantages to standard gauge railway.

There are some heritage railways still in operation today that utilize narrow gauge rail lines, including the Cumbres and Toltec Scenic Railroad that runs through Colorado and New Mexico during the summer months. This particular railway service operates an authentic steam-powered locomotive through portions of the line that were saved by their respective states after the Denver & Rio Grande Railroad filed for abandonment in 1969. In 2013, this historic railroad was granted National Historic Landmark designation, ensuring the railroad's survival for years to come. Other examples of preserved narrow gauge railways include the Durango & Silverton Narrow Gauge Railroad and the Albuquerque BioPark.

Dylan Coolbaugh

See also: Alaska Railroad (United States); Darjeeling Himalayan Railway (DHR) (India); Railroads.

Further Reading

Cumbres and Toltec Scenic Railroad. 2014. http://cumbrestoltec.com/. Accessed December 19, 2015.

NATCHEZ TRACE PARKWAY (UNITED STATES)

The scenic Natchez Trace Parkway runs 444 miles from near Nashville, Tennessee, to Natchez, Mississippi, and passes through northwest Alabama. It is a part of America's national park system and is administered by the National Park Service rather than a highway department. The road was built beginning in the 1930s to parallel the old Natchez Trace, a foot trail that ran from the Mississippi River near modern Natchez to the Cumberland Plateau near Nashville.

The trail began as a Native American path before being used by explorers and settlers. Use of the trail declined in the early 19th century when steamboats enabled easier travel on the Mississippi River and other rivers. In the early 20th century, interest developed in Mississippi to preserve the old path. This ultimately took the form of a scenic drive along the path that would commemorate its history. This Natchez Trace Parkway was to be built by the National Park Service, which started construction in 1937. It was designed to allow a constant 50-mile-per-hour speed with no stops required. Views of surrounding lands are carefully screened by trees to avoid any glimpses of modern buildings and signs. The road finally was completed in 2005.

The parkway is reached by a number of access points where other highways cross it. Compared to most highways, the parkway has little traffic, but where it passes the cities of Tupelo and Jackson use is much greater. These cities grew up along the original Trace, and although it was not designed as such, drivers in these cities have found it to be a useful bypass of city streets. The sections within and between these cities carry more vehicles than the northern and southern ends of the parkway. Traffic volumes on the Mississippi section tend to be level throughout the year. In Tennessee, traffic increases in the summer months and during the fall-color season.

A spectacular feature along the road is the 1994 twin arch bridge that carries the parkway over a highway and creek in Tennessee. This has become one of the most famous sites along the road. Experiencing history is a major part of driving the parkway. Tupelo and Brices Crossroads National Battlefields and Natchez National Historical Park are a short distance off the parkway, and Shiloh National Military Park and Stones River National Battlefield are a bit farther away. Tupelo—located near the midpoint of the Parkway—is the site of a decisive defeat of the French by Native Americans at the Battle of Ackia in 1736; the home of Tupelo Automobile Museum, one of the largest collections of antique motor vehicles in the United States; and, of course, the birthplace of Elvis Presley, who provided the soundtrack for many road trips. The old Trace still exists and can be seen and walked upon in many locations along the parkway.

Joe Weber

See also: Blue Ridge Parkway (United States); National Scenic Byways (United States); Parkways.

Further Reading

Bachleda, F. Lynne. 2011. *Guide to the Natchez Trace Parkway.* 2nd ed. Birmingham: Menasha Ridge Press.

Davis, William C. 1995. *A Way Through the Wilderness: The Natchez Trace and the Civilization of the Southern Frontier.* New York: Harper Collins.

National Park Service. U.S. Department of the Interior. 2015. "Natchez Trace Parkway." http://www.nps.gov/natr/index.htm. Accessed December 14, 2015.

NATIONAL HIGHWAYS DEVELOPMENT PROJECT (INDIA)

India has an extensive road network of 2,050,525 miles—the second largest in the world. For the purpose of management and administration, Indian roads are divided into national highways, state highways, major district roads, and rural roads. The national highways, with a length of 57,695 miles, comprise 1.7 percent of this road network and carry 40 percent of the total traffic volume. The state highways and district roads together constitute the secondary system of road transportation in India. This secondary system of roads connects rural areas to the urban areas of the country. This network is essential for the transportation and supply of goods and services to the rural areas and also for interconnecting the industrial areas with other parts of the country. The third category of roads is district roads that comprise village roads. Due to lack of administration these roads are not maintained properly.

Indian roads are of great significance as they have become a necessity for the economic growth of the country. At the time of independence in 1947, India had a poor road network infrastructure and between 1947 and 1988 there were no new major projects. The roads were poorly maintained and all roads were single lane and unpaved. All national highways and state highways were two-lane, undivided roads with uneven surfaces that caused traffic congestion and longer travel times. For the development and management of the highways in India, the National Highway Authority of India (NHAI) was established in 1988. The National Highway Development Project (NHDP) started in 2001.

The National Highway Development Project's prime focus is to better manage India's roads. The NHDP ensures that engineering, construction, design, and maintenance of the highways are done in a comprehensive and integrated manner. These aspects focus on modern facilities available along the roads, such as gas stations, hotels, restaurants, and toilet facilities. The engineering, construction, and design of roads should match international standards for uninterrupted traffic flows, reduce time for travelling, ease traffic congestion, and create well-surfaced roads. The project has been launched through seven phases.

Phase I and II of the NHDP focused on the construction of six-lane highways: The "golden quadrilateral," connects the four metro cities, namely Delhi, Mumbai, Chennai, and Kolkata; the "north-south corridor" connects Srinagar to Kanyakumari; and the "east-west corridor" connects Silcher to Porbander. Phase III focused on widening the highways to ease the pressure on high-density traffic corridors and to connect places of economic importance, such as state capitals, industrial centers, and tourist sites. Phase IV, V, VI, and VII of the NHDP focused on upgrading national highways with paved shoulders, six lanes, construction of stand-alone ring roads, bypasses, grade separators, flyovers, elevated roads, tunnels, roads over bridges, underpasses, and service roads. Implementation of the national highways has taken place in all phases except for phase VI. Phase VI includes the building of "expressways." The construction of "expressways" has been slow in progress due to lack of suitable funding and land. At present, a master plan for expressways has been prepared based on public and private participation. The plan aims to construct 9,693 miles of expressways by the end of 2022.

It is envisioned that the National Highway Development Project will be implemented completely by the end of the Twelfth National Five-Year Plan (2012–2017). Sources of finance for the project include gross budgetary support; central road funding; external assistance through the World Bank, the Japan International Cooperation Agency, and the Asian Development Bank; toll revenue; and private-sector investment.

Despite tremendous development under the NHDP, the national highways in India are facing a number of challenges—mainly environmental, mechanical, and financial obstacles. In the case of large-scale road projects, environmental clearances are required, such as acquisition of land and forest clearances. Attaining the permissions and clearances from the environmental and land revenue department can cause enormous delays. Although the latest technologies are available for road construction and maintenance, there still is the issue of adequate and technical manpower. Lastly, financing road projects is greatly dependent upon federal fund allocation under the annual plans. One of the major obstacles of road development in India is lack of funds from both central and state governments.

As a way forward, the government of India is looking for the most suitable mode of implementation for road projects to avoid long delays in construction. There is focus on advanced technologies for real-time monitoring of projects. The state governments are extending their active cooperation by signing state-support agreements. There has also been a proposal to exempt national highway widening from environmental approval if the project is less than 62 miles in distance, because the approvals generally take more than three years to obtain.

At present, however, there is no independent regulatory authority for national highways. This results in conflict as the central and state agencies are also the implementing bodies, and there is no independent assessment of the project taking place.

Purva Sharma

See also: Interstate Highway System (United States); National Trunk Highway System (China).

Further Reading

Government of India. Ministry of Roads, Transport and Highways. 2011 (April). "12th Five-Year Plan (2012–17). Report of Working Group on Central Roads Sector." http://planningcommission.gov.in/aboutus/committee/wrkgrp12/transport/report/wg_cen _roads.pdf. Accessed December 19, 2015.

Lok Sabha Secretariat. Members Reference Service. 2013 (August). Reference Note No. 23. "National Highways Development Project—An Overview." Parliament Library and Reference, Research, Documentation, and Information Service. India. http://164.100.47.134 /intranet/NHDP.pdf.

PricewaterhouseCoopers. 2012. "The Road Ahead—Highways PPP in India." https://www .pwc.in/en_IN/in/assets/pdfs/publications-2012/the-road-ahead-highways-ppp.pdf. Accessed December 14, 2015.

NATIONAL SCENIC BYWAYS (UNITED STATES)

There are many scenic highways in the United States, and it is perhaps inevitable that a way to recognize the most outstanding roads in the country was created. The National Scenic Byways are those highways recognized by the U.S. Federal Highway Administration as being of outstanding national beauty or historical significance. This program started in 1991 and as of 2015 there are 120 National Scenic Byways across the United States.

There are six ways a highway can qualify for recognition as a Scenic Byway. It could be a scenic road with a spectacular view; it might pass through an undisturbed natural area; it could pass through an area with interesting historical features; it can connect an area with a unique culture, provide a corridor through

National Scenic Byway 12, highway tunnel arch, near Bryce, Utah, 2009. (James Phelps Jr/ Dreamstime.com)

historic or prehistoric archaeological sites, or pass through areas of recreational significance. The road itself also can possess these qualities. Byways can be urban or rural, although the vast majority is rural. California Highway 190 through Death Valley National Park, the Alaska Marine Highway (composed of ferry connections), the Florida Keys Scenic Highway between the mainland and Key West, the spectacular Columbia River Highway in Oregon, the Las Vegas Strip, the Blue Ridge and Natchez Trace Parkways, Route 66, Outer Banks Scenic Byway in North Carolina, Alabama's Selma to Montgomery March Byway, and Woodward Avenue in Detroit all are examples. Freeways can be byways. The Arroyo Seco Parkway in Los Angeles, Merritt Parkway in Connecticut, and I-90 through the Cascade Mountains in Washington are examples.

The most scenic or historically significant of all the byways are All-American Scenic Roads. There are only 31 of these, including the Alaska Marine Highway, Blue Ridge and Natchez Trace parkways, and the Las Vegas Strip. These are considered to be the best of the best among American roads.

The designation of a road as a byway does not change the legal status of the road or require any rebuilding. Rather, it is a way of promoting tourism and economic development. Each byway should have a corridor management plan that describes what makes the highway unique, how the road will be marketed, and how its status will be preserved and also serve the transportation needs of the community.

The U.S. Department of Transportation makes the final decisions about the status of the roads.

Joe Weber

See also: Blue Ridge Parkway (United States); Las Vegas Strip (United States); Natchez Trace Parkway (United States); Parkways.

Further Reading

U.S. Department of Transportation. Federal Highway Administration. "America's Byways." http://www.fhwa.dot.gov/byways/byways. Accessed December 14, 2015.

NATIONAL TRUNK HIGHWAY SYSTEM (CHINA)

The National Trunk Highway System is a nationwide freeway network currently being built in China. The first freeway in China was built in 1988, but the system grew to about 70,000 miles in 2015 and will continue to grow in coming years. The system also is known as the 7918 network, referring to seven routes radiating from Beijing, 9 north-south (or vertical) routes, and 18 east-west (or horizontal) routes crossing the country. The G30, or Lianyungang–Khorgas Expressway, is the longest route on the system, running 2630 miles from the Yellow Sea in the east to the Kazakhstan border in the far west of China. There also are many urban belt-ways in the system.

Although similar in many ways to the U.S. Interstate Highway System in design, it is much larger. Nearly the entire Chinese network is composed of toll roads. For this reason, the highways officially are called "expressways" rather than "freeways." The speed limit throughout the system is 75 miles per hour, and road signs are written in both Chinese and English. The system has many spectacular bridges—particularly bridges using cable-stayed designs.

The former British territory of Hong Kong also has several freeways. People in Hong Kong drive on the left side of the road, however, but in the rest of China they drive on the right side. Highways crossing between Hong Kong and the rest of China must have crossovers to allow traffic to switch sides. Private vehicles, however, rarely are allowed to cross the border, and it is mainly trucks and buses that do so.

Joe Weber

See also: Cable-Stayed Bridges; Freeways; Interstate Highway System (United States).

Further Reading

Trubetskoy, Sasha. 2015 (May 2). "It Took China's Engineers Only 10 Years to Completely Leapfrog the US Interstate System." *Freestyle Geographic* (blog). http://freegeo

.org/2015/05/02/it-took-chinas-engineers-only-10-years-to-completely-leapfrog-the
-us-interstate-system/. Accessed July 17, 2015.

NEW YORK CITY SUBWAY (UNITED STATES)

The New York subway system is among the oldest, largest, and busiest in the world.
The opening date of the underground portion of the system occurred in 1904, but
the city had elevated railroads dating back to 1870. The system has 232 route miles
and has 656 miles of revenue track—the result of utilizing four tracks on many
underground sections of the system. This unique structure allows for the operation
of both express and local trains on the same line. There are 468 stations, more than
any other urban rail operation in the world. The subway delivers more than 1.7 billion rides each year, which translates into approximately 5.5 million riders on an
average weekday, making it the busiest in the United States and seventh busiest in
the world. This entire system operates 24 hours per day, 365 days a year. The subway system operates in four of New York City's five boroughs (Manhattan, The
Bronx, Brooklyn, and Queens); none of the subway system exists beyond the city
limits, and a separate line operates on Staten Island (the fifth borough).

New Yorkers refer to it as the "subway," regardless of whether a train is aboveground or below ground (60 percent of the network is located underground). It is

New York City's Brooklyn Bridge subway station in Manhattan, 2015. (Demerzel21/Dreams
time.com)

a system that residents love to hate, but could not live without. The subway often has been used as a metaphor for the city itself. This can be seen in many different cultural activities, including art, photography, written works, in music, in television, and in motion pictures, and often the metaphor aligns with the status of the city at that moment.

The City of New York has been closely involved with the subway's construction and operation throughout its history, even when privately held. The two original private companies were the Interborough Rapid Transit Corporation (IRT) and the Brooklyn Rapid Transit Corporation BRT); the latter changed to the Brooklyn-Manhattan Transit (BMT) Corporation in 1918. The IRT operated the first subway line that opened in 1904; a year later, the IRT expanded into Brooklyn to compete directly with the BRT. By 1913, however, both companies were part of a dual contract to expand the system. Still, the IRT and BRT could not share tracks because the two companies built their lines to different specifications (BRT trains were wider and longer than IRT trains). Further competition in the system came by 1932, when the city-owned and operated Independent System (IND) opened its first line. The new network's construction costs and a New York state requirement that the system be able to cover its operating costs meant that fares throughout the system doubled from $0.05 to $0.10, a considerable increase during the Great Depression.

By 1934, all three companies were unionized, but only three labor strikes have ever shut down the system. In 1940, New York City bought the IRT and BMT and the three separate operations became one. Because the IND was built to the same train width as the BMT, a gradual amalgamation of the two operations occurred. In 1953, the New York City Transit Authority (NYCTA)—a public authority presided by New York City—took over subway, bus, and streetcar operations from the city. This developed into a state-level Metropolitan Transportation Authority (MTA) in 1968. The MTA created two divisions: The "A" Division (the numbered lines and former IRT) and the "B" Division, which includes all the lettered lines (and was the former BMT and IND). Despite the lack of official use of the monikers, many New Yorkers still use the abbreviations IRT, BMT, and IND.

The subway often reflects the city's economic standing. During the 1970s and 1980s, New York City endured a downturn in its economy that led to budget cuts; crime increased and many middle-class residents moved out of the city to the suburbs. As city services deteriorated, so did the subway. Most evident were increases in crime and the presence of graffiti, which often covered trains. The economic turnaround in the city during the 1990s and the 21st century also has been reflected in the subway; crime sharply decreased, new graffiti-free trains were put in service, and billions of dollars were spent to revitalize the system. More recently, the subway system has endured major events and disasters that affect the city. The terrorist attacks on the World Trade Center in 2001 caused damage to stations in the immediate area (one of which has never reopened). In 2012, Hurricane Sandy's impact inundated the subway's river tunnels and flooded several stations. Both events caused significant modifications to daily operations and affected the commute for millions of daily riders.

Jason Greenberg

See also: London Underground/The "Tube" (United Kingdom); Paris Metro (France); Seoul Metropolitan Subway (South Korea).

Further Reading

Cudahy, Brian. 2004. *A Century of Subways: Celebrating 100 Years of New York's Underground Railways.* New York: Fordham University Press.
Hood, Clifton. 2004. *722 Miles: The Building of the Subways and How They Transformed New York.* Baltimore: Johns Hopkins University Press.
Nycsubway.org. http://www.nycsubway.org. Accessed June 5, 2015.

NORTHEAST CORRIDOR (UNITED STATES)

The most heavily traveled passenger train route in the United States runs from Washington, DC, through Baltimore, Maryland; Philadelphia, Pennsylvania; and New York City, to Boston, Massachusetts. This area sometimes is known as "Megalopolis," "BosWash," or the Northeast Seaboard, and is the largest urban region in the country.

The corridor is unusual among American railroads for several reasons: it is owned and operated by Amtrak, operates almost exclusively as a passenger railroad, is the longest stretch of electrified railroad in the country, has the fastest passenger trains in the country, and is one of the few places where trains compete against air travel. The corridor is 453 miles long, or 2.1 percent of Amtrak's route miles, and carries more than 11 million passengers a year, about one-third of all Amtrak passengers. The corridor carries 75 percent of those traveling between New York and Washington, DC, and 54 percent of those traveling between New York and Boston.

Despite its status as the busiest and fastest passenger train in America, the corridor actually is quite old. It was privately built by two railroads beginning in the mid-19th century and was complete in 1917. Different sections of what is now the corridor were built by many railroads, which were consolidated into fewer and fewer over time. In 1969, the entire line was under the control of the Penn Central Railroad, which went bankrupt. Amtrak was created in 1971 and purchased the line to continue operating it as a passenger railroad. In addition to the Acela Express, 10 other trains operate over part of or the entire corridor, although at slower speeds. Other passenger trains connect into the Northeast Corridor at Boston, Massachusetts; New Haven, Connecticut; New York City; Philadelphia, Pennsylvania; and Washington, DC.

The construction of the corridor involved many spectacular engineering feats. Access to Manhattan required a tunnel built under the Hudson River, Manhattan

Tracked Electric Vehicle Project

Project designer Will Jones—an inventor and a well-known battery designer—created the Tracked Electric Vehicle (TEV) Project. This project promises a new technology that will radically change the capabilities of current highway and motorway infrastructure. Combining the latest technology—such as autonomous vehicles—the TEV project will redesign current highways. Using software to control operations, this electrified infrastructure can continuously charge the rubber tires of electric vehicles, enabling the vehicles to run safely at high speeds (120 miles per hour) using cruise control. This advanced highway and motorway technology also would enable vehicles to be grouped closely together, thus handling greater volumes of vehicles more efficiently. The TEV Project plans to design this transportation solution for long-distance travel only, not for travel within cities.

Selima Sultana

Island, the East River, and the Hell Gate Bridge. The route through Baltimore includes the Baltimore and Potomac Tunnel, built in 1873. The railroads took tremendous pride in their projects and identity and created a number of spectacular train stations, most of which still exist. Washington, DC's Union Station opened in 1907, a few blocks from the U.S. Capitol. New York's Pennsylvania Station was completed in 1910, but was demolished in 1963 and replaced by a smaller station beneath Madison Square Garden (Grand Central Terminal is impressive but technically is not part of the corridor). The northern end of the corridor is Boston's South Station, which opened in 1899.

Electrification of the corridor began in the first years of the 20th century and continued in stages. The entire Washington, DC, to New York City section was electrified in 1935. The line from New York to Boston was electrified as far north as New Haven by 1914. This remaining section was not electrified until 2000. Electrification enabled powerful engines to pull passenger trains at high speeds. The Pennsylvania Railroad introduced the famed streamlined GG-1 electric locomotive on the route in 1935. This engine was kept in service by later railroads until 1983 and pulled trains at up to 100 miles per hour. The Metroliner trains were introduced in 1969 and pushed speeds up to 125 miles per hour. The Acela Express trains began in 2000 and run up to 150 miles per hour. With these engines—the fastest in the United States—trains can run from D.C. to New York in two hours, forty-five minutes, and travel three and one-half hours from there to Boston. The train provides considerable competition to both highway and air travel between these cities.

Unlike high-speed lines in Europe and Japan, the Northeast Corridor was not built as a high-speed railroad but rather slowly improved over time. Railroad crossings remain, as do drawbridges, and many sections are more than 100 years old.

The route also is quite curvy for a high-speed line. The Acela trains are able to reach the speeds they do only by tilting into the many curves on the line. Trains will not be able to go faster without substantial upgrades to the line.

The Northeast Corridor faces many challenges in the coming decades. Many bridges are quite old and in poor condition, and the corridor is vulnerable to disrupted service in the event of a bridge problem. Enormous investment is required to allow the corridor to continue operating.

Joe Weber

See also: Amtrak (United States); High-Speed Rail; High-Speed Rail in the United States; Railroads.

Further Reading

Amtrak. 2012. "The Amtrak Vision for the Northeast Corridor: 2012 Update Report." http://www.amtrak.com/ccurl/453/325/Amtrak-Vision-for-the-Northeast-Corridor.pdf. Accessed August 25, 2015.

Middleton, William D. 1974. *When the Steam Railroads Electrified*. Milwaukee, WI: Kalmbach Publishing Co.

0

OLEV

See Online Electric Vehicle (OLEV) (South Korea).

ONLINE ELECTRIC VEHICLE (OLEV) (SOUTH KOREA)

Electric vehicles (EVs) are considered promising as an alternative mode of transportation, with the potential of reducing pollution and helping to secure a long-term energy supply. Technical and economical limitations such as the bulky size, weight, cost of batteries, the relatively long charging time, and the scarcity of charging stations have hindered the wide adoption of EVs. Research efforts to overcome these limitations focus on reducing the size of onboard batteries utilizing an innovative wireless-charging technology. The basic idea is to provide electricity to EVs "wirelessly" from power transmitters installed on the road. As the motors can be operated on less than the full amount of electricity transmitted to a vehicle, the excess electricity can charge the onboard batteries while the vehicle is on the road or, of course, when it is stationary.

The Online Electric Vehicle (OLEV) is an integrated transportation system that consists of vehicles and power transmitters that are buried underground to supply electricity. The Korea Advanced Institute of Science and Technology (KAIST) in South Korea developed the OLEV by applying "Shaped Magnetic Field in Resonance (SMFIR)" technology that safely transfers large amounts of energy to EVs. Underground power transmitters consist of an inverter and inductive cables. Each transmitter creates strong magnetic fields. A pickup device installed on the underbody of the vehicle collects the magnetic field efficiently and converts it into electricity. A regulator distributes the electricity to an onboard motor or battery, depending on the power requirement of the motor and the battery's charging level. The technology can be applied to land transportation systems such as electric buses, shuttles in airports and parks, seaport cargo trucks, and railroads.

Although wireless power transfer has attracted considerable interest, the low efficiency of power transmission has been a critical barrier—until the development of SMFIR technology. The OLEV system provides power with as much as 83 percent efficiency over an 8-inch gap between the ground and vehicle, with the potential for even greater efficiency (>90 percent) when using a smaller gap. With regard to safety concerns, the magnetic fields meet international standards for electromagnetic field (EMF) radiation and electromagnetic interference (EMI). Additionally, with its "smart function" that recognizes an OLEV vehicle over the power track, the

OLEV system does not waste power or inappropriately generate magnetic fields for other vehicles or pedestrians. Given that the cost of constructing the infrastructure is critical in commercialization, the proportion of the length of the segments with power tracks compared to the total length of road is an important consideration. Typically, a small portion (5 to 15 percent) of the predetermined routes of transits or airport shuttles is required.

The OLEV has received worldwide recognition. The World Economic Forum named it 1 of the top 10 emerging technologies for 2013, and *TIME Magazine* selected it as 1 of the 50 best innovations of 2010. There are successful OLEV pilot projects. The first OLEV tram bus launched its service in July 2011 at the Seoul Grand Park in South Korea. This bus circulates the 1.4-mile road in the amusement park powered by 0.3-mile-long transmitter segments. A second system was deployed in Yeosu Expo 2012 briefly between May and August 2012 to serve visitors on a 0.4-mile route. In KAIST, on-campus shuttle buses went into service in October 2012 and carry up to 1,300 passengers per day. The 65-yard transmitters supply the power for the buses on a 2.3-mile route. Gumi, a southern city in South Korea, introduced OLEV buses to operate on a 15-mile cross-city route in 2013. The buses of Gumi have batteries that are about one-third the size of a standard electric car battery. The length of power tracks in Gumi is 0.1 miles. In 2014, after the pilot project, Gumi deployed two buses to officially operate on the route.

Although OLEV has the virtue of convenience, it still faces barriers. Above all, the cost of building the initial infrastructure is burdensome. The fixed cost for installing the Gumi system is approximately $4.4 million, which includes two $600,000 buses and other costs. Some might argue that it is not economically viable. Conversely, others expect that mass production of power-components will reduce the unit production costs to an acceptable level.

Jong-Geun Kim

See also: Battery Swapping; Electric Vehicles; Plug-In Hybrid Vehicles.

Further Reading

Ko, Young Dae, and Young Jae Jang. 2013. "The Optimal System Design of the Online Electric Vehicle Utilizing Wireless Power Transmission Technology." *IEEE Transactions on Intelligent Transportation System* 14(3): 1255–1265.

Suh, N. P., D. H. Cho, and C. T. Rim. 2011. "Design of On-Line Electric Vehicle (OLEV)." In *Global Product Development* 3–8. Proceedings from 20th CIRP Design Conference, Ecole de Nantes, France: Springer.

ORIENT EXPRESS (EUROPE)

The Orient Express is a name given over the years to several passenger rail services linking cities in Western, Central, and Southeastern Europe. Enjoying a reputation

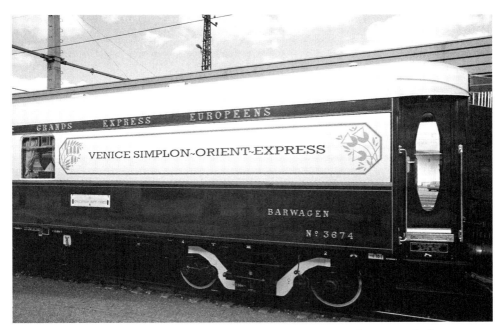

The legendary Orient Express arrives at the station in Venice, Italy, 2011. The luxury train travels between Venice and London. (Peter Lovás/Dreamstime.com)

as one of the most glamorous and exotic trains in the world, the Orient Express has been featured in a number of novels and movies. The first Orient Express was the creation of Belgian engineer and entrepreneur Georges Nagelmackers, who inaugurated a train service in 1872 between the ports of Ostend, Belgium, and Brindisi, Italy. The venture was so successful that two years later the Belgian founded the Compagnie Internationale des Wagons-Lits (CIWL) to provide passenger transportation over several European routes.

The service destined to assure Nagelmackers' fame ran from Paris, France, through the Balkans to the Ottoman capital of Constantinople (now Istanbul). This "Grand Express d'Orient," as it was called, made its initial trip in October 1883, running through Munich, Germany, and Vienna, Austria. Rail service was interrupted at the Danube River, where passengers were ferried from Romania into Bulgaria, and actually ended at the Bulgarian Black Sea seaport of Varna, from which passengers proceeded by ship to Constantinople. The venture was a success, however, and generated substantial publicity.

The length of Orient Express's journey—nearly 82 hours—and the presumed exoticness of its destination were newsworthy in themselves, but the luxurious level of service attracted just as much attention. Unlike the plainer Pullman sleeping cars available on American (and eventually British) trains, the Orient Express offered individual compartments with drawing rooms decorated in Louis XIV style and comfortable fold-down beds. Each compartment had its own toilet, and showers with hot and cold water were available in a separate car. Sumptuous meals were served in an ornately decorated dining car.

An uninterrupted rail journey from Paris to Constantinople became possible in 1889 thanks to further rail construction and a change of routes on the eastern section of the route. Two years later, the train itself was renamed the Orient Express. An important branch line between Ostend and Vienna was added in 1894, creating an Ostend-Vienna Orient Express.

As was the case for many European train routes, the Orient Express suspended operations when World War I broke out in 1914. Service resumed after peace was declared in 1918, and a second line was put into operation. This was the Simplon Orient Express, which followed a more southerly route, passing from Italy to Switzerland via the Simplon Tunnel and continuing on through Venice to the newly annexed Italian port of Trieste. Yet another service, the Arlberg Orient Express, was inaugurated in the 1930s. It covered a route from Paris to the Greek capital of Athens by way of Budapest, Hungary, and Belgrade, Yugoslavia. A link between Paris and the French port of Calais on the English Channel also was added during this period.

In the latter half of the 20th century, CIWL's various Orient Express lines began losing passengers to automobiles and to faster, less expensive trains. The year 1962 marked a major turning point in the company's operations, as several routes were abbreviated. The original Orient Express was shortened to run between Paris and Vienna, with some cars continuing on to Budapest and Bucharest. The latter two cities were dropped in 2001, and in 2009 the entire service was suspended.

The Arlberg Orient Express was downgraded in 1962 to the Arlberg Express with a route running between Paris and Vienna, but the operation lost its identity as a named service in the 1990s. The Simplon Orient Express was downgraded in 1962 to the Direct Orient Express, a slower service with daily runs between Paris and Belgrade and twice-weekly connections to Athens and Istanbul. In 1977, however, this service ceased operations. By that time CIWL had begun leasing its rolling stock to other rail firms, although in many cases it continued to provide staff and catering services. In the early 21st century, a Venice Simplon Orient Express operated by Belmond Ltd. and using restored vintage carriages continues to run between London and Venice.

Graham Greene's 1932 novel *Stamboul Train* (published as *Orient Express* in the United States) is set on the Ostend-Vienna Orient Express, and Agatha Christie's popular 1934 mystery *Murder on the Orient Express* is set on the Simplon Orient Express. Both have been filmed, and one version or another of the service remains a staple of movies and television programs.

Grove Koger

See also: Alpine Tunnels (Europe); Railroads.

Further Reading

Behrend, George. 1982. *Luxury Trains from the Orient Express to the TGV.* New York: Vendome.

Burton, Anthony. 2001. *The Orient Express: The History of the Orient Express Service from 1883 to 1950*. Edison, NJ: Chartwell Books.

Cookridge, E. H. 1978. *Orient Express: The Life and Times of the World's Most Famous Train*. New York: Random House.

Sherwood, Shirley. 1996. *Venice Simplon Orient-Express: The World's Most Celebrated Train*, 4th ed. London: Weidenfeld & Nicolson.

OX/BULLOCK CARTS

An ox cart or bullock cart is a two- or four-wheeled vehicle pulled by an ox or oxen—also known as "bullocks"—which are young bulls or draught cattle. Ox carts are an ancient means of transportation still used in areas today where motorized vehicles are scarce or too expensive. These carts, also called "jinkers," primarily are used for transporting goods. Oxen or bullocks are the animal of choice because of their strong necks that make them ideal for wearing heavy wooden yokes to which a chain or pole is attached to a cart carrying people or other load.

Bullock carts are especially common in countries such as India, the Southeast Asian city-state of Singapore, and the Republic of Costa Rica in Central America (where the cart is known as "*la carreta*"). In Costa Rica it was a popular mode of delivery during the 19th century, when a sturdy vehicle was necessary to transport coffee from plantations through muddy regions, sandy beaches, up hills, and across mountains and rivers. The journey would take from 10 to 15 days. Ox carts, in fact, played a huge role in Costa Rica's becoming a major worldwide coffee exporter by helping transport coffee to the ports for delivery to Europe. The first coffee shipment from Costa Rica to London took place in 1843. The first carts were pulled by men but were replaced by beasts of burden when loads became too heavy and shipments too numerous.

Ox carts also were used as family transportation, especially when large groups needed to travel. Before the railroad between San José and Puntarenas was built in the late 1800s, many families used ox carts to make the several-day journey to the coast. Ox carts also played a role in the building of the railroads by being the number-one mode of transportation to carry construction materials to the railroad sites.

Costa Rican ox carts are known for their elaborate decorations—a custom that started in the early 20th century. Cowherds first chose to make the carts colorful and then individual families added their personal touches to the carts. Each region had a unique design that identified the owner's origin. Bells were added to the wheels to create a unique song. The heyday of the Costa Rican ox carts was from 1850 to 1935. They became obsolete when other modes of transportation—such as trains, tractors, and trucks—were introduced. They now are used as art and decoration and often still are part of celebrations and parades.

Ox carts in Singapore have existed since the 19th century and were used for travel and the transportation of goods. They often were hired out to haul freight including water, hay, coconuts, pineapples, and furniture. As traffic levels increased, ox carts were phased out of production. Inhabitants of southern India and Sri Lanka known as the Tamil owned the largest number of ox carts, which were a major mode of transportation in the 19th century.

Ox carts played an important role in the economic development of Singapore by being a common method for delivering goods at Boat Quay, the busiest part of the Port of Singapore. They also transported water drawn from wells in Ang Siang Hill to inhabitants in other areas. As the carts traveled along the streets, they spilled water that helped keep down the dust of the dirt roads. Ox carts also were used to transport construction materials, and with the carts removed and replaced with attachments, they could mow lawns and level grass.

The large ox carts essential to Australia are referred to as "bullock drays." They were introduced in the nation by early explorers at the end of the 18th century. Drays are large wagons pulled by 20 or more animals and are driven by a person referred to as a "bullocky." The animals of choice to pull the drays were Devon cattle. In the mid-19th century, bullock drays transported essential supplies and food from large towns to isolated areas and would return from the rural areas with produce and wool, wheat, sugar cane, and timber to be hauled to seaports for shipment to other nations. Trips could take weeks, with the animals traveling at a speed of about 3 miles per hour in areas where no roads existed and before the railroads were built. Bullock drays also hauled heavy building equipment to construction sites as the nation developed, and are considered the backbone of the early transport system of South Australia.

Rosemarie Boucher Leenerts

See also: Camels; Llamas; Rickshaws (Bangladesh); Rickshaws (Japan).

Further Reading

Flinders Ranges Research. "William Chace, Bullocky Extraordinary." http://www.southaust ralianhistory.com.au/chace.htm. Accessed August 24, 2015.
UNESCO. 2005. "Oxherding and Ox Cart Tradition in Costa Rica." http://www.unesco.org /culture/intangible-heritage/12lac_uk.htm. Accessed June 5, 2015.

P

PACIFIC COAST HIGHWAY, CALIFORNIA (UNITED STATES)

Pacific Coast Highway (PCH) is the common name for a segment of California State Route 1 (SR 1) or Highway 1. The highway runs along more than 600 miles of the California coast, from Leggett in Mendocino County to the north to Dana Point in Orange County to the south. The PCH is one of the most well-known scenic routes in the United States and is famed for its cliff-hugging turns and magnificent beauty. It is one of the U.S. Department of Transportation's "All-American Roads." These roads are chosen by the agency as having two of six intrinsic qualities designated to National Scenic Byways: being archaeological, cultural, historic, natural, recreational, and/or scenic; and having features that do not exist anywhere else in the United States. The PCH courses through some of California's most-visited landmarks, including Monterey, Big Sur State Park, Hearst Castle in San Simeon, and Mission Santa Barbara.

Construction on the Pacific Coast Highway began in 1919. At that time, it was part of the Roosevelt Highway, named after Theodore Roosevelt, the 26th president of the United States and a renowned naturalist. The 1,400-mile Roosevelt Highway ran along the entire West Coast, from the Canadian to the Mexican borders. Much of the labor used in the construction of Highway 1 (the precursor to the PCH) was done by prisoners from San Quentin Prison in Marin County, who were paid $0.35 a day. The cost to build the highway initially was estimated at $1.5 million, although the actual figure ultimately was more than $10 million.

Seventy thousand pounds of dynamite were needed to blast through the granite, marble, and sandstone in the rocky portions along the coast. The most difficult section was the 65-mile stretch of road from Spruce Creek, near Big Sur, to San Simeon, where 30 million cubic feet of rock was removed.

Along the PCH route is the famous picturesque Bixby Creek Bridge, a reinforced-concrete open-spandrel arch bridge located in Monterey County. It is one of the tallest single-span bridges worldwide and also one of the most photographed. Before the bridge was opened in 1932, transportation to and from Big Sur was nearly impossible, except by an often-impassable single-lane road that led 11 miles to the area's east and often was closed in the winter. Today, the Bixby Creek Bridge stands more than 260 feet high and is 700 feet long.

Until the Roosevelt Highway was built, there was no direct link between the Southern California beach towns of Newport Beach and Laguna Beach in Orange County, nor between Ventura and Santa Monica in Los Angeles County. The

An undated photograph of the historic Bixby Bridge on the Pacific Coast Highway, Big Sur, California. (Michael Flippo/Dreamstime.com)

Roosevelt Highway therefore was much celebrated by locals at the time of its construction. The building of the PCH was not lauded by all, however. A decades-long fight started in the 1890s between a landowner, May Rindge, and the State of California—and almost prevented the Roosevelt Highway from being built across Rindge's property, which hugged the coastline between Santa Monica and Malibu. Because of Rindge's efforts, the Rancho Topanga Malibu Sequit was untouched, even by land-hungry interests, including the Southern Pacific Railroad and homesteaders.

In 1907, when the county considered building a coastal road through Rindge's property, county representatives were forced away by armed guards Rindge had hired. Rindge then challenged the county's power of eminent domain—a right of government to expropriate private property from owners for the benefit of the public—and a stalemate ensued until 1923. In that year a U.S. Supreme Court decision upheld the county's right to take the land for the proposed Roosevelt Highway. The county paid Rindge $107,289 for the property (about $1.46 million in present-day monetary values).

Construction in another section of land, this one along Malibu, also was delayed by litigation, but the way was cleared for the highway to continue through this section in June 1929. For the remaining two decades, the Pacific Coast Highway (renamed in 1941) was adequate to meet the needs of travelers. By the 1950s, however, traffic was too great to carry all north- and southbound automobiles. A new 100-mile freeway that would parallel the PCH—but in some communities also would bisect the coast and consequently restrict public beach access—was proposed. Some communities fought the plan, including the Orange County cities of Laguna Beach and Costa Mesa; and others sought to modify it to continue access to coastal business districts. In the 1960s, strong opposition also was voiced by landowners, civic leaders, and environmentalists when the state wanted to buy right-of-way land that ran through expensive communities, including Malibu. The plan for the proposed freeway was tabled in 1972.

Rosemarie Boucher Leenerts

See also: Alaska Highway (Canada and United States); Blue Ridge Parkway (United States); Natchez Trace Parkway (United States); National Scenic Byways (United States).

Further Reading

Longfellow, Rickie. 2013 (October 17). "Back in Time: California's Pacific Coast Highway–Highway One." U.S. Department of Transportation, Federal Highway Administration. http://www.fhwa.dot.gov/infrastructure/back0403.cfm. Accessed June 5, 2015.

Masters, Nathan. 2012 (May 2). "From Roosevelt Highway to the 1: A Brief History of Pacific Coast Highway." KCET Los Angeles. http://www.kcet.org/updaily/socal_focus/history/la-as-subject/from-the-roosevelt-highway-to-the-one-a-brief-history-of-pacific-coast-highway.html. Accessed June 5, 2015.

PARIS METRO (FRANCE)

The Paris Metro (*Metropolitain*) is an urban rapid transit system that operates primarily within the city limits of Paris, France. Inaugurated in 1900, it is one of the world's oldest transit systems; with more than 1.5 billion passengers per year (or 4.2 million riders per day), it also is one of the busiest systems. The Paris Metro also is one of the densest systems, with 303 stations on 133 route miles—an average of 1 station every 1,600 feet. There are 16 lines (although 2 of them are branches of main lines) that are identified by number and color. Most of the system operates within the city of Paris. Only 4 lines operate beyond the city limits. Most of the Metro is located underground and was constructed using the "cut-and-cover" method; thus, most stations are close to the street surface. Six lines have aboveground sections. Five lines use rubber tires instead of steel wheels—a relatively unique method that is copied in a few other rapid transit operations. The public transport authority, *Régie autonome des transports Parisiens* (RATP), operates the Metro. It also operates part of the suburban *réseau express régional* (RER) network, light rail lines, and bus lines. With its uniform Art Nouveau architecture, the Metro is a symbol of Paris.

The governments of the City of Paris and of France had a historically poor relationship. Originally conceived in the early 1870s, it took 25 years for Paris and the French government to agree to a plan. Construction of the Metro began in the late

1890s. The Metro began operation in 1900 during the World's Fair. The system expanded very quickly and 10 lines were operating by 1920. The size of the metro tunnels deliberately was chosen by the City of Paris to prevent the operation of main-line trains through the city. The stations are close together so that Parisians never are far from a metro station. Station length averages approximately 250 feet. The result of the construction method meant that, as compared to other major rapid transit networks, metro trains are short (the average train has five cars) and have low passenger capacity (550 to 700 passenger capacity per train).

Initially, the Metro did not extend beyond the city limits, but by 1934 the first extension to the inner suburbs occurred. Further expansion came in the 1930s, including the construction of an entirely new line. During World War II, however, all progress halted as Paris dealt with Nazi occupation. World War II had a significant impact on the Metro, and some stations closed permanently. The current operator, RATP, took charge in 1948 when the metro and bus network merged.

The Metro reached a saturation point during the 1950s. Outdated technology limited the maximum number of trains; thus, RATP concentrated on modernization. As Paris's population boomed after World War II, the suburbs further from the city grew. As the number of suburban commuters increased, the main railway stations faced overcrowding during rush hour. In the 1960s, the solution to the overcrowding issue was to join suburban lines to new underground portions in the city. New underground station construction occurred within the city from 1969 to 1977. The new network officially opened with two lines in 1977; a third line opened two years later. Today, the five lines of the RER contain 257 stations (33 of which are within the city limits of Paris) and operate over 365 miles of track, including 47.5 miles underground. The RER is operated partly by RATP and partly by the national rail operator, *Société nationale des chemins de fer Français* (SNCF) (National Society of French Railways or French National Railway Company).

In October 1998, Line 14 became the first new metro line in 63 years. It is one of the two fully automatic lines within the network (the other is Line 1). It became the first line constructed with station platform screens to prevent accidents and suicide attempts. Unlike the traditional metro station proximity, many of Line 14 stations are at least a half a mile apart. As with the urban RER lines, nearly all of Line 14 stations offer connections with multiple metro lines. Although Line 14 represents modern Paris, the Metro remains an endearing symbol of the city for its Art Nouveau design that originated more than 100 years ago. The Metro continues to preserve this iconic style, with many station entrances exhibiting Hector Guimard's design. Stations can be attractions all by themselves. The Louvre station on Line 1, for example, actually includes exhibits from the Louvre art museum.

Jason Greenberg

See also: London Underground/The "Tube" (United Kingdom); New York City Subway (United States).

Further Reading

France official website. 2014. "Brief History of the Paris Metro." http://www.france.fr/en
/paris-and-its-surroundings/brief-history-paris-metro.html. Accessed June 5, 2015.

Ovendon, Mark. 2009. *Paris Underground: The Maps, Stations, and Design of the Metro.*
London: Penguin Books.

Régie Autonome des Transports Parisiens (RATP). 2014. "The Metro: A Parisian Institution."
http://www.ratp.fr/en/ratp/r_108503/the-metro-a-parisian-institution/. Accessed June
5, 2015.

PARKWAYS

A parkway is a landscaped roadway set within a linear park, often with restrictions banning commercial truck traffic and buses. The parkway was a 19th-century innovation that played a key role in helping Americans take pleasure in their country's landscapes and remake them through suburbanization. The design of parkways was inspired by the grand boulevards of European capital cities and, in turn, parkways influenced the development of both freeways and greenways.

The term "parkway" was coined by famed landscape architects Frederick Law Olmsted (1822–1903) and Calvert Vaux (1824–1895), whose achievements included New York's Central Park. They proposed Eastern Parkway as a pleasure road serving as the approach to their crowning jewel, Prospect Park, in Brooklyn. Olmsted and Vaux drew inspiration from Unter den Linden in Berlin and from Baron Haussmann's Avenue de l'Impératrice (now Avenue Foch) in Paris; both are broad, tree-shaded boulevards serving as gateways to urban parks.

Built between 1870 and 1874, Eastern Parkway included a central roadway for "pleasure-riding and driving" set between two grassy malls, each planted with three parallel rows of trees. Eastern Parkway extended to the outer limits of the city and was envisioned as part of an interconnected network of regional parks and parkways that would give city dwellers access to the tranquil beauty of the countryside.

Olmsted and Vaux later designed park and parkway systems for Boston, Massachusetts; Buffalo, New York; Chicago, Illinois; and Louisville, Kentucky. For Minneapolis, Minnesota, Horace Cleveland (1814–1900) proposed a system of interconnected parks and parkways that would encompass most of the city's lakes, creeks, and Mississippi River shoreline. Today, Minneapolis's Grand Rounds includes approximately 50 miles of scenic parkways, bicycle trails, and walking paths that encircle the city.

During the era of the recreational automobile, roughly 1920–1945, cars primarily were used by city dwellers for weekend outings. Parkways built during this period often followed scenic waterways, such as Chicago's Lakeshore Drive, New York's Bronx River Parkway, and Philadelphia's scenic drives along the Schuylkill River. This generation of parkways harmonized the road and the landscape, meandering around natural features, and rising and falling with the topography.

Recreational automobile era parkways featured picturesque lighting and bridges, picnic grounds, and other leisure facilities along the route. Parkways provided motorists safe respite from congested city streets by limiting access points and providing grade-separated crossings.

An ambitious program of rural parkway construction took place during the Depression as part of the U.S. government's National Industry Recovery Act. The Blue Ridge Parkway and Natchez Trace Parkway both were constructed under the act. The Blue Ridge Parkway was built between 1935 and 1987 to link Shenandoah National Park in Virginia with Great Smoky National Park in North Carolina. The 469-mile roadway at the time was the longest road planned as a single unit. The road functions as an elongated national park, immersing travelers in the beautiful natural and cultural landscapes of the Southern Appalachians.

Beginning in the 1920s, the New York region's master builder Robert Moses (1888–1981) presided over the construction of a vast network of parkways connecting New York City with Westchester County and Long Island. Unlike Minneapolis's Grand Rounds, New York's radial system emphasized access to the countryside. After World War II, as automobiles became more popular, the New York region's parkways became heavily used commuter routes. To add capacity, they were widened and redesigned for greater speeds, and the once pastoral parkways now facilitate widespread suburban decentralization—ironically altering the very countryside they were designed to conserve.

Blue Ridge Parkway, Linn Cove Viaduct, North Carolina, October 2012. (Daveallenphoto/ Dreamstime.com)

In Los Angeles, the Arroyo Seco Parkway, which opened in 1940, marked a transition between the recreational parkway and the contemporary freeway. The Arroyo Seco Parkway connects Los Angeles and Pasadena in a 10-mile sunken corridor along a seasonal riverbed. When it opened, the Arroyo Seco Parkway was noted for its lush plantings and scenic views. It later was renamed the Pasadena Freeway and became the first section of the region's vast system of high-speed freeways. Within that system, it developed a reputation as being neglected and unsafe. Recent efforts to restore the roadway have included returning to its original name, adding safety features, and enhancing views of mountains and the Los Angeles skyline. Like other parkways, it has been designated a civil engineering landmark and a National Scenic Byway for its historic, natural, cultural, and recreational features. In returning the Arroyo Seco Parkway to its original vision, there is recognition that—compared to the freeways they inspired—parkways offer a more pleasurable experience of scenic travel through a picturesque parkscape. Designed to savor rather than conquer the landscape, parkways suggest the possibilities for harmonizing the built and natural environments.

Mark Bjelland

See also: Blue Ridge Parkway (United States); Boulevards; Freeways; Greenways.

Further Reading

Caro, Robert. 1975. *The Power Broker: Robert Moses and the Fall of New York.* New York: Knopf Doubleday.

Rybczynski, Witold. 1999. *A Clearing in the Distance: Frederick Law Olmsted and America in the Nineteenth Century.* New York: Scribner.

PERSONAL RAPID TRANSIT, WEST VIRGINIA (UNITED STATES)

The Personal Rapid Transit (PRT) system in Morgantown, West Virginia, has been operating for 40 years and is the oldest and most extensive PRT system worldwide. It is an efficient transportation option for both West Virginia University (WVU) and the community at large, by connecting the three university campuses and downtown Morgantown. It is funded by the Federal Transit Administration. Unlike other mass-transit systems, the PRT carries passengers continuously to their destination points without stopping at other stations. This feature is meant to minimize trip time and maximize convenience. Due to the nature of this feature, however, the PRT is a more effective service for a smaller, compact area as opposed to a larger and more densely populated metropolitan region.

The Morgantown personal rapid transit system originally was implemented in response to traffic congestion that increased during the early 1970s as a result of the growing WVU population and subsequent university expansion into multiple campuses, coupled with limited existing roads. The benefit of PRT is that it not only alleviates road traffic, by transporting riders who would otherwise contribute to automobile or bus traffic, it also is entirely separated from the road network.

The Morgantown PRT is composed of 71 automated van-size electric cars that travel along 8.7 miles of guided track (guideway) and reach a maximum speed of 30 miles per hour. The majority (65 percent) of the guideway is elevated, and the remaining 35 percent is located at or below grade. Although the route is programmed to a specific schedule based on the WVU academic calendar peaks, the system operates on demand during periods when reduced ridership is predicted. The Morgantown PRT was the first fixed-guideway transit service to implement on-demand mode. Cars are programmed to arrive at a station within five minutes of a passenger either swiping a university ID or depositing the fare at a station turnstile. Ridership is free for West Virginia University students, faculty, and staff. Riders who are not affiliated with WVU pay $0.50 per ride. After either swiping the card or paying the fare, riders press a button to indicate their desired destination before entering the station platform. Boarding instructions are displayed above the loading gate to direct passengers to the specific car that will travel to the desired destination. Each car contains eight seats and four poles for standing passengers, holding up to 21 riders.

Since its inception, the Morgantown PRT has been an invaluable transportation alternative for a population of more than 50,000 by alleviating automobile traffic congestion and by providing sheltered transportation for those who would otherwise have difficulty biking or walking across the hilly terrain, especially during inclement weather and in winter. Operational hours are limited, however, to accommodate the WVU academic year. During fall and spring semesters, service is available from 6:30 a.m. through 10:15 p.m. on weekdays and from 9:30 a.m. through 5:00 p.m. on Saturdays. During the summer semester, weekday operation only extends through 5:15 p.m. The PRT is closed during university holidays and academic breaks. Ridership is estimated at 15,000 people per day during the school year.

Rachel Rupe

See also: Buses; Light Rail Transit Systems; Rapid Transit.

Further Reading

Raney, Steve, and Stan Young. 2005. "Morgantown People Mover—Updated Description." Paper presented at the Transportation Research Board Annual Meeting, Washington, DC. January 2005. http://www.cities21.org/morgantown_TRB_111504.pdf. Accessed December 15, 2015.

Sproule, William J., and Edward S. Neumann. 1991. "The Morgantown PRT: It Is Still Running at West Virginia University." *Journal of Advanced Transportation* 25(3): 269–280.

Sulkin, Maurice. 1999. "Personal Rapid Transit Déjà Vu." *Transportation Research Record* 1677: 58–63.

West Virginia University. 2015. "Personal Rapid Transit (PRT)." http://transportation.wvu.edu/prt. Accessed March 13, 2015.

PLUG-IN BATTERY ELECTRIC VEHICLES (BEVs)

Plug-in battery electric vehicles (BEVs)—sometimes called "all-electric vehicles"—employ electricity to power the vehicle's motor, and this electrical energy is stored in a rechargeable onboard battery. Unlike a hybrid electric or plug-in hybrid electric vehicle, which each have an internal combustion engine that still can operate when the battery's charge is depleted, BEVs have only an electric motor. Depending on the automobile manufacturer and battery type, a fully charged battery can provide a driving range of 100 to 200 miles, which is less than the range of gasoline-powered vehicles. Driving ranges also can vary based on individual drivers' habits and on extreme heat or cold.

Electricity technically is an energy carrier and not a fuel itself, but electricity is considered an alternative fuel under the Energy Policy Act of 1992. Because electrical energy stored inside the battery powers the vehicle, no emissions are produced when driving a BEV, but the energy sources used to produce that electricity could be responsible for carbon dioxide and greenhouse gas emissions. A BEV's operation is much quieter than gasoline- or diesel-powered vehicles, and BEVs could offer a significant decline in air pollution if they became the dominant mode of personal transportation, depending on the source of the electrical energy. BEVs also are two to three times more efficient at converting the energy source into energy used to power the vehicle than are internal combustion engines.

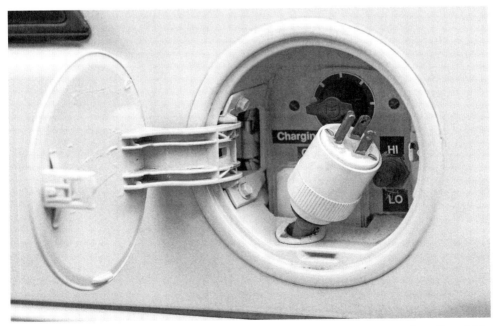

Where the gas tank normally is, there is a plug in this electric car. This eco-friendly vehicle may someday replace gasoline-powered cars. (Karen Keczmerski/Dreamstime.com)

Recharging plug-in battery electric vehicles requires vehicles to be plugged into the electrical grid, and thus they are more reliant on a public refueling infrastructure than are plug-in hybrid electric vehicles. To accelerate BEV adoption and to alleviate "range anxiety," which is a BEV driver's fear of being stranded on the roadway because of a depleted battery, charging stations for BEVs have been installed in recent years by private companies in locations such as workplaces, government buildings, and shopping malls. These stations generally feature both Level 2 and DC fast charging. Charging times vary on the amount of battery depletion and the charging type selected, but times currently range anywhere from 30 minutes to 8 hours. An extensive network of battery switching stations could reduce the wait time necessary to fully charge a battery, but would require an extensive infrastructure build-out. Rates paid by BEV drivers for public recharging of their battery does depend on the geographic area, but generally are cheaper than gasoline per unit of energy. Vehicle purchase prices, however, are higher for BEVs than for conventional vehicles.

Battery electric vehicles are not a new technology but only recently have again received increased consumer attention. The first battery-powered electric vehicle was developed in the 1830s and, by the turn of the century, electric vehicles were common sights on American roadways. The development of the internal combustion engine and gasoline-powered cars, coupled with cheap and abundant domestic petroleum and a lower vehicle production cost, stalled the widespread adoption of BEVs at that time. Now, most major auto manufacturers again feature a BEV model, including the Nissan Leaf, the BMW i3, the Mitsubishi i-MiEV, and the Tesla Model S.

Scott Kelley

See also: Biofuel Vehicles; Electric Vehicles; Hydrogen Vehicles; Plug-In Hybrids.

Further Reading

Grech, Johnaton. 2014 (August 29). "The Top 5 Reasons Why We Chose EV Charging to Battery Swapping." LinkedIn. https://www.linkedin.com/pulse/article/20140829105008-28164290-5-logical-reasons-why-we-chose-ev-charging-to-battery-swapping. Accessed December 15, 2015.

M-Beam. "Modular Battery Exchange and Active Management." http://www.modularexchange.com/. Accessed November 22, 2014.

PLUG-IN HYBRIDS

A plug-in hybrid is a hybrid electric vehicle (PHEV) that combines the use of battery power and an electric motor, as well as a fuel tank and an internal combustion engine (ICE). The purpose of having both systems in one car is to introduce electric grid power into the expense of driving a car but retaining the greater range of a gas-powered internal combustion engine. Although research for PHEVs began in the early 20th century, the first production model did not appear until 2008. Today there are many different styles of plug-in hybrids, including compact city cars such as the BMW i3, sedans such as the Chevy Volt and the Toyota Prius PHEV, and luxury cars such as the Porsche Panamera S and the BMW i8. The cost of electricity is less than the cost of gas making it cheaper per mile to power a car with electricity. Plug-in hybrids therefore offer an environmentally friendly and economically beneficial alternative to standard ICE cars.

Plug-in hybrid electric vehicles are the first step in the process of moving away from gas-powered vehicles. Although PHEVs still contribute to carbon dioxide emissions because of their battery-powered motors, they do not produce carbon dioxide emissions when running in electric mode. Compared to conventional ICEs, PHEVs produce about 9,000 pounds of carbon dioxide emissions annually, versus approximately 13,000 pounds for ICEs. By contributing to the reduction of carbon dioxide emissions in the atmosphere, plug-in hybrid cars help improve air quality by lessening the amount of the most abundant greenhouse gas on earth.

As the environmental advantages of plug-in hybrids become more widely known, state and national governments in the United States are beginning to offer incentives for owning any type of electric or alternative-fuel vehicle. From the beginning of the movement toward plug-in hybrids (around 2002), the U.S. Department of Energy saw the long-term benefits of electric powered vehicles and offered funding to different research projects across the country. By encouraging companies to develop the new technology, production PHEVs became available faster than originally anticipated. The government incentives, however, did not end after production model PHEVs became available. State governments in the United States continue to pass laws that offer a wide variety of incentives for PHEV owners, including income and sales tax exemption, property tax discounts, state inspection exemption, carpool lane access, parking meter exemption, and discounted electricity rates available during off-peak hours. Some states even reimburse PHEV owners for part—or all—of the cost of their in-home charger installation, which can be quite expensive. California offers tax incentives of up to $1,500 for PHEVs. The amount of the tax incentive increases for all-electric vehicles. The amount of incentives that federal and state governments have offered toward the advancement and continued use of plug-in hybrid technology, it is easy to see the political influence of PHEVs. New ordinances promoting the use of PHEVs continually are being passed as the technology continues to grow.

From a cultural perspective, PHEVs have affected the automotive industry in more ways than initially anticipated. The production of plug-in hybrids had the desired effect of promoting environmental consciousness and increasing sales for many automotive companies. As the sales of PHEVs continue to increase, however, the development and improvement of the technology behind PHEVs are making them even more successful. The battery technology of PHEVs that initially took so long to develop, for example, now is being improved by many different companies to make all-electric vehicle production ready. Tesla Motors is a good example.

Offering the first production model EV that could hold an entire family and give the range of a full tank of gas on an ICE vehicle, Tesla Motors utilized the battery technology from plug-in hybrids, improved it to operate as the only means of power in a car, and then streamlined the process to mass-produce its cars. Plug-in hybrids also have created a more widespread acceptance of electric power as an alternative fuel. One other cultural effect of PHEVs is that automotive companies now are seeing the benefits of producing luxury PHEVs and EVs to suit all types of environmentally conscious customers. With the production of the Tesla Model S, the BMW i8, the Fisker Karma, and many more luxury PHEVs and EVs, automobile manufacturers are taking advantage of the success of PHEVs not only economically but culturally as well, to become even more widely used worldwide.

Selima Sultana

See also: Battery Swapping; Electric Vehicles; Hydrogen Vehicles; Online Electric Vehicles (South Korea); Plug-In Battery Electric Vehicles (BEVs).

Further Reading

United States Environmental Protection Agency, Office of Transportation and Air Quality. 2007 (October). Plug-In Hybrid Electric Vehicles. http://permanent.access.gpo.gov/lps90342/420f07048.pdf. Accessed July 8, 2015.

Q

QINGHAI-TIBET RAILWAY (CHINA)

The Qinghai-Tibet Railway—acclaimed as the "Sky Road"—is the world's longest and highest railway. It began full-fledged operations on July 1, 2006. The whole railway runs 1,215 miles, connecting Xining in Qinghai Province to Lhasa in the Tibet Autonomous Region in the People's Republic of China. Most of the line is located more than 13,000 feet above sea level, and the highest point is Tanggula Pass at 16,400 feet above sea level. There are 45 stations along the railway and 11 of them are staffed. The maximum operational speed is 75 miles per hour on ordinary terrain and 62 miles per hour over permafrost.

The plan to build a railway to Tibet was conceived in 1919 by Sun Yat-Sen—the first president of the Republic of China. After the creation of the People's Republic of China in 1949, Chairman Mao Zedong started to make the plan a reality. The actual construction of the Qinghai-Tibet Railway started in 1956 but was delayed by technical difficulties and the subsequent Great Chinese Famine and Cultural Revolution. It was not until 1981 that the first phase of the project was completed. This section connects Xinin to Golmud in Qinghai Province and runs 526 miles. Most of this part of the railway is at relatively low elevation.

Because of the technical difficulties in constructing railroads on permafrost, the second phase of the project—connecting Golmud to Lhasa—was delayed for 20 years. It was not until 2001 that the second phase of the project was formally resumed under the call to develop western China. More than 30,000 workers were stationed on the plateau during the construction. The project was completed in 2005.

Passing through a high-plateau region required the project to conquer tremendous engineering challenges. The air on the Tibetan Plateau is thin and cold. In most parts of the region, the oxygen pressure is only 50 percent to 60 percent of that at sea level. To account for oxygen shortages, seven oxygen generators and 25 hyperbaric oxygen chambers were deployed along the route to supply oxygen to construction workers. During the construction of the Fenghuoshang Tunnel (the highest railway tunnel in the world), an oxygen station was built to provide a continuous supply of oxygen in the tunnel. To tackle the issue of track subsidence caused by permafrost melting during the summer, track was laid on raised bridges with pile-driven foundations and cooling devices—such as liquid ammonia pipes and sun shields—were used to reduce foundation temperatures. Another great challenge of the project is to protect the fragile ecosystem of the Tibetan Plateau. To lessen the impact on wildlife habitats, the route avoided major natural reserves.

Thirty-three wildlife crossings over 36 miles were constructed to ensure the normal migration of wildlife such as Tibetan antelopes. According to a recent monitoring report, the usage of the crossings is steady.

The railway is operated by the Qinghai-Tibet Railway Group Co., Ltd., a state-owned enterprise under the China Railway Corporation. Four pairs of passenger trains run between Golmud and Lhasa, and each can carry 936 passengers. Designed for the high-altitude environment, the trains are equipped with a specially developed oxygen-generating system. It produces enriched airflow with sufficient oxygen content at a comfortable level. Individual oxygen nozzles are provided under passenger seats. The windows are specially designed to reduce the ultraviolet radiation at high elevation. Prior to the construction of the Qinghai-Tibet Railway, Tibet was the only Chinese province that could not be reached by train.

Xuwei Chen

See also: Darjeeling Himalayan Railway (DHR) (India); Guoliang Tunnel (China); Narrow Gauge Railroad.

Further Reading

China Internet Information Center. "Qinghai-Tibet Railway." http://www.china.org.cn /english/features/Tibet/171193.htm. Accessed December 15, 2015.

Kable Intelligence Limited. "Qinghai-Tibet Heavy Rail Line, China." http://www.railway -technology.com/projects/china-tibet/. Accessed December 15, 2015.

State Council of the People's Republic of China. "Special Report: Qinghai-Tibet Railway." http://english1.english.gov.cn/special/qztl_index.htm. Accessed December 15, 2015.

R

RACK RAILWAYS

Rack railways (also known as cog railways or rack-and-pinion railways) operate on slopes that are too steep for the more standard friction or adhesion trains. Conventional trains of normal weight cannot operate on grades that are greater than 2 percent, although lighter ones on narrow gauge tracks can manage grades approaching 4 percent. To overcome this problem, a revolving cogwheel (pinion) beneath the locomotive engages with a toothed track (rack), forcing the locomotive forward and upward. Rack railways can travel on grades of up to 48 percent, and, depending on the layout of the particular line, can run on friction rails where appropriate.

The first rack railway began operation in 1812, and was the work of John Blenkinsop, who worked as a steward at the Middleton Colliery (coal-mining operation) near Leeds, England. Blenkinsop thought that iron locomotive wheels might not have enough friction on iron rails to pull a heavy load. In search of a solution, he substituted a rack for one of the two rails on the railway serving the colliery and arranged with a nearby engineering firm to fit a pinion onto a steam locomotive of his design, dubbed the *Salamanca*. The apparatus did, in fact, help pull the locomotive, although it later was determined that it was not necessary when on level ground.

The oldest rack railway still in operation, and the first built on mountainous terrain, was the three-mile Mount Washington Cog Railway (the "Cog") in New Hampshire, completed in 1869. It was built by Sylvester Marsh, who modified Blenkinsop's design by laying a rack *between* two friction rails—an improvement incorporated into almost all subsequent rack lines. Marsh also utilized an open rack design through which snow would fall rather than accumulate. Within a short time, European engineers began building rack railways in the Alps. The first was Niklaus Riggenbach, who opened the Vitznau-Rigi rack railway in Switzerland in 1871. A second important Swiss rack railway was the Pilatus, designed by engineer Eduard Locher and opened in 1889. The Pilatus Railway utilizes horizontal pinions engaging both sides of the rack, enabling the train to operate on a grade of up to 48 percent. The Pilatus is the steepest rack railway in the world and, like the Vitznau-Rigi, still is in operation. Some years later, one of Riggenbach's former employees, Carl Roman Abt, engineered a design that was more efficient than Riggenbach's. Abt's rack-and-pinion design utilized multiple offset rows of teeth, enabling the parts to be engaged constantly. The first use of this system was in 1885 on a narrow gauge line on Harz Mountain in Germany. In the early 21st

Rack-and-pinion train heading to the Snowdon summit, Snowdonia, Wales, 2011. (Pere Sanz/ Dreamstime.com)

century, rack railways remain in operation in mountainous locations around the world, and most are used in the tourist industry.

Grove Koger

See also: Montserrat Rack Railway (Spain); Railroads.

Further Reading

The International Steam Pages. "Rack (Cog) Railways of the World." http://www.internatio nalsteam.co.uk/hills/rackrailways.htm. Accessed December 15, 2015.
Mount Washington Cog Railway. "The Mount Washington Cog Railway." http://www .thecog.com/. Accessed December 15, 2015.
Wolmar, Christian. 2014. *The Iron Road: An Illustrated History of the Railroad.* New York: Dorling Kindersley.

RAILROADS

Railroads or railways are wheeled vehicles that operate on parallel rails. These vehicles usually are coupled together to operate as a train, pulled (and sometimes pushed) by one or more powered locomotives. Their history can be traced back to Roman times; however, current rail technology originated in Britain during the Industrial Revolution. The ability to mass-produce iron and later steel rails, coupled with the development of steam engines, led to railroads becoming a crucial transport technology. The first railroad in the United States, the Baltimore & Ohio, was built in 1830.

Railroads can be distinguished in many different ways. One of the most basic differences is between freight and passenger operations. Railroads have been powered by a variety of propulsion types—horses originally provided the power, but soon were replaced by steam locomotives that burned wood or coal. In the 20th century, steam locomotives slowly were displaced in most countries by diesel-electric locomotives, with diesel engines powering electric motors. A few lines were powered by electricity provided by overhead wires.

Railroads can be distinguished by the gauge (width) of the rails. Many different gauges are used around the world but several have become standard. In North America, the standard gauge is 4 feet, 8½ inches. As the names indicate, narrow gauges are narrower than standard gauge, and broad gauges are wider than standard. The continent of Africa has at least nine different gauges in use, including meter gauge, Imperial (3 feet, 6 inches), and standard—each introduced by different colonial powers.

Railroads can be classified by purpose or use. For example urban passenger railroads include subways, elevated trains, streetcars, and cable cars. Small trains are used to transport ore in underground mines. Tourist or heritage railroads are maintained for tourist operations.

In the United States, freight operators are divided into common carrier or private railroads. Common carriers haul freight for other parties, and private railroads only carry freight for the owner. Mining companies, for example, often have private railroads to haul ore. Common carrier freight lines have consolidated into a few major systems in the United States: CSX, Norfolk Southern, BNSF, and Union Pacific. These are all that is left of the many large companies that existed over the years, including the New York RR, Southern Pacific, Baltimore and Ohio. Many smaller railroads (shortlines) exist as well. Modern freight railroads are technologically sophisticated operations and have changed tremendously since the 1950s. Cabooses have been replaced by sensors, containers have replaced boxcars, and the major U.S. freight railroads serve as a land bridge between Asia and Europe.

Passenger service generally has declined since the 1930s, except during World War II when wartime conditions caused all-time high passenger volumes. The decline in demand for passenger service resulted in continuing cutbacks until 1971, when Amtrak—a government agency—took over passenger operations. Since that time, Amtrak has operated a limited national network with greater frequencies in the Northeast Corridor.

An important way that passenger trains could compete with air travel or highways is by making them much faster. High-speed trains are those that run significantly faster than regular trains, and 150 miles per hour (mph) often is used as a threshold. In the United States, trains do not operate faster than 79 mph. Although in the 19th century steam-powered trains commonly could run more than 100 mph, the first true high-speed trains are the electrically powered bullet trains of Japan. The Japanese trains began operations in 1964 and could travel 130 mph. These trains have been improved since then, and the current generation can achieves speeds of 200 mph. In Europe, high-speed trains first were developed in France. Subsequently, many other countries built and integrated high-speed trains into a European high-speed network. The Northeast Corridor is the fastest line in the United States but runs well below the 150-mph standard. Even Amtrak's sleek Acela express trains operate at an average speed of 79 mph—slower than the trains of the 1960s. The U.S. Air Force uses rocket sleds to test objects at very high speeds. These sleds run on rails that are similar to railroad lines but are built to much higher tolerances. Some test tracks are almost 10 miles long. The greatest speed ever achieved on a test track was Mach 8.6 (6,416 mph), which was attained in 2003 in New Mexico.

In the United States railroad construction proceeded rapidly throughout the 19th century, peaking in the 1920s. Since that time, nearly 200,000 miles of lightly used tracks have been abandoned. The United States still has the longest railroad network of any country; when ranked by passenger use, however, it is eleventh. Indian railroads have the most passenger use, followed closely by China. Each of these carries 37 times more passengers than do U.S. railroads. Chinese railroads carry more freight (by weight) than any other country; the United States ranks a close second.

Joe Weber

See also: Amtrak (United States); Light Rail Transit Systems; Narrow Gauge Railroads.

Further Reading

Armstrong, John H. 1994. *The Railroad: What It Is, What It Does.* 3rd ed. Omaha, NE: Simmons Boardman.
Vance, James E. 1994. *The North American Railroad: Its Origin, Evolution, and Geography.* Baltimore: Johns Hopkins.

RAILROADS, NARROW GAUGE

See Narrow Gauge Railroads.

RAPID TRANSIT

This term refers to specific urban passenger transit systems, usually which run on rails. This term encompasses subways and elevated railroads, but not streetcars or light rail. The primary distinction between these types of trains is that rapid transit does not run in streets but rather on a physically separate system. Subways and elevated trains today commonly are known as rapid transit. These have been built since the late 19th century, but the term only came into use in the 1960s, when

Tubular Rails

Tubular rail technology retains the same characteristics of conventional rail transport, but reconfigures them in highly innovative ways. Instead of building rail track on the ground, trains themselves carry the tracks as they pass through elevated rings that contain wheels and electric motors. The train can hold nearly 500 passengers and can run at a speed of up to 150 miles per hour. The innovative idea of tubular rails is the brainchild of Robert C. Pulliam, chairman of Tubular Rail Inc., to provide clean, safe, and efficient mass transit for cities. It has been claimed that Tubular rail is a low-cost, low-impact solution for light or commuter rails. It also has been said that because the system can be built very quickly, there is less disruption to existing infrastructure, and that the cost is 60 percent less than that of building conventional urban rail tracks.

Selima Sultana

construction started on postwar subway systems in American cities. At the time, subways had a somewhat negative image, thus a new term—"rapid transit"—was adopted to make them seem new and exciting. Bay Area Rapid Transit (BART) in the San Francisco area was the first system to make use of the term. Today, however, rapid transit also includes earlier subway or elevated systems.

Rapid transit's scope does not include streetcars or similar systems, which are powered by an electrified third rail and run in streets. Rapid transit systems do not run in streets but rather on separate routes, which could be aboveground, underground, or elevated (or a combination). Rapid transit railcars are much larger than streetcars and usually are connected into a train. The floors of the cars can be as much as four feet above the ground, therefore passengers must board from stations with high platforms. Because they are not limited by traffic flow, the trains achieve a considerable speed between stations; this greater speed enables them to carry a much greater capacity of passengers per hour than light rail or streetcar systems can.

In addition to San Francisco, several other cities have built rapid transit systems. Atlanta, Georgia, and Washington, DC, are the two most extensive new U.S. systems. These systems are expensive to build, as they require building tunnels under city streets or building unpopular elevated sections. The difference between rapid transit and streetcars is not always obvious. Some light rail operates trains and could run in streets, central medians, below ground, and elevated above ground. The Portland MAX light rail system has a long subway section and includes the deepest rail station in North America. The Los Angeles rail transit system includes examples of both types.

Bus rapid transit sometimes denotes bus-based transit systems that operate in such a way as to have higher speeds and fewer stops. They have their own roads or lanes, called "busways," and passengers can board only at stations with raised platforms. The city of Curitiba, Brazil, has a bus rapid transit system that is considered a model example of this rapid transit mode.

Joe Weber

See also: Bay Area Rapid Transit (BART) (United States); Buses; Light Rail Transit Systems.

Further Reading
Cervero, R. 1998. *The Transit Metropolis: A Global Inquiry.* Washington DC: Island Press.
Grava, S. 2003. *Urban Transportation Systems: Choice for Communities.* New York: McGraw-Hill.

RICKSHAWS (BANGLADESH)

Rickshaws are a popular mode of transport in Bangladesh. The capital of Bangladesh, Dhaka, also is called the "Capital of Rickshaws." The rickshaw was invented in Japan in the early 19th century and in 1919 came to Chittagong in Bangladesh. The rickshaw was introduced into Dhaka in 1938 when European jute exporters imported the cycle rickshaw from Calcutta for their personal use. Initially, the use of rickshaws was confined to rural areas, but it now is one of the major non-motorized modes of urban transport in Bangladesh.

In Bangladesh, a rickshaw is a three-wheeled nonmotorized vehicle. It generally has a structure made of wood and iron. There is a seat at the back where two passengers can sit comfortably, and one seat in the front where the rickshaw "puller" drives the rickshaw via two peddles. In Bangladeshi streets, rickshaw pullers are predominantly male. Every rickshaw has a hood to protect the passengers from adverse weather conditions, but the rickshaw puller does not have such protection.

There are different types of rickshaws in Bangladesh, such as the cycle rickshaw and auto rickshaw. Cycle rickshaws are used to travel short distances and auto rickshaws are used to reach destinations further away. Bangladeshi people generally only refer to cycle rickshaws as rickshaws; the auto rickshaw is called a "CNG," "baby taxi," or "scooter." These are important modes of transportation in Bangladesh for low- and middle-income people who don't have personal transportation such as a car or motorcycle. Bus service throughout the country is not efficient in terms of numbers and quality of service, thus rickshaws are the primary mode of transportation for most low- and middle-income people in urban areas of Bangladesh.

Rickshaws are a major source of revenue in the transportation sector. Thousands of rickshaws run along the roads of Dhaka city every day because rickshaw pulling is one of the major means of income for more than 3 million people. The maintenance and repair of rickshaw parts and painting rickshaws also are primary sources of income for many people in the country. Rickshaw painting has been

Rickshaws transport passengers in Dhaka, Bangladesh, 2014. About 500,000 rickshaws cycle daily in Dhaka, nicknamed "the rickshaw capital of the world." (Dmitry Chulov/Dreamstime .com)

recognized as one of the integral components of Bangladeshi art and culture. Colorful decoration and painting of Bangladeshi rickshaws began during the 1950s. There have been many national and international exhibitions of rickshaw painting, and it has been recognized worldwide as an art form. The Bangladesh National Museum recently added a collection of rickshaw art and has recognized the art as a part of Bangladeshi culture. In addition to painting them, Bangladeshi rickshaws also are decorated with flowers and star-shaped plastic.

The Bangladeshi government tried to ban rickshaws from city streets several times to reduce traffic congestion. These bans never worked due to huge public transportation demand and the rickshaw's popularity as a low-cost transportation mode. The increasing number of rickshaws in the capital city has been remarkable. In 1941, Dhaka City had only 37 rickshaws, and by 1998 reached 112,572. This increase was due to the high rate of population growth in Dhaka City and associated increases in transportation demand. In 2014, Dhaka City had approximately 6,000 rickshaw garages and 225,000 rickshaws on the street.

Most of the rickshaws on the streets of the capital have no legal documentation. Rickshaw pullers also are not specially trained—many start pulling rickshaws within a few days of arriving in the city, to earn cash for their poverty-stricken families. Most pull rented rickshaws; very few rickshaw pullers are the owners.

Lastly, rickshaws are one of the most environmentally friendly vehicles and have little negative impact on the environment.

Neelopal Adri

See also: Bicycles and Tricycles; Ox/Bullock Carts; Rickshaws (Japan); Segway.

Further Reading

Hoque, Rezaul. 2014 (October 21). "Illegal Rickshaws Dominate Dhaka Streets." bdnews24.com. http://bdnews24.com/bangladesh/2014/10/21/illegal-rickshaws-domi nate-dhaka-streets. Accessed December 15, 2015.

Zaman, Niaz. 2015. Rickshaw. *Banglapedia: National Encyclopedia of Bangladesh*. http://en.banglapedia.org/index.php?title=Rickshaw. Accessed December 15, 2015.

RICKSHAWS (JAPAN)

Invented in Japan in the 19th century as an everyday mode of transportation, the rickshaw remains an essential part of everyday mobility to ordinary people throughout Asia and Africa. The word rickshaw came from the Japanese word "jin-rikisha," which means "human-powered vehicle." These two- or three-wheeled carts began operation in Japan around 1869, when missionaries and rich members of the Japanese society were pulled along as they travelled across the country. This form of transportation became very popular in Japan in the 19th and 20th centuries during a time when there were many new possibilities for travel, including the railroad locomotive. It would not be exaggerating to claim that rickshaws altered the way the majority of Japanese experienced travel in urban and rural areas alike. Its popularity perhaps was not only due to its simplicity, low cost, convenience, and speed, but also because it evolved as a source of employment for youths in most Japanese cities and towns, such as Tokyo. The appeal of rickshaws grew so fast in Japan that by 1872—only three years after its first appearance—there were 40,000, and by 1875 there were more than 100,000 rickshaws on the streets of Tokyo. By 1896, rickshaws emerged as the "new mobility machine" for ordinary people all over the country and the number of rickshaws in use peaked at 210,000.

Soon rickshaws got their popular name of "Daisuke's Car" when wealthier households started purchasing them for personal transportation. Owning a rickshaw became a symbol of new social status, as it reflected the modernity of status based on wealth and social standing rather than hereditary status. For ordinary people, riding in rickshaws was an exciting experience of the new modern era of transportation and an opportunity to demonstrate their elevated status over walkers. The speed, comfort, and convenience that rickshaws offered at that time were noted by the Western world as well.

The geography of Japan also positively influenced the use of rickshaws in transport. Most of Japan consists of relatively flat ground, as is the case with most other islands. This makes it much less difficult for rickshaw operators to move the vehicles over long distances. The climate of Japan also is conducive to this mode of transport—the country experiences mild weather conditions under which rickshaw operation is practical.

After the advent of motorized vehicles, however, rickshaws eventually became a less-preferred mode of transportation in Japan. In particular, engine-powered trains started serving the major cities in Japan. Rickshaw operators became jobless. Additionally, rickshaws later were banned by the government as an exploitative transport option. Despite the ban, rickshaws were improved by the addition of an engine to power them.

These engine- and human-powered rickshaws presently have returned to some areas of Tokyo and other major cities. Indeed, there are some established companies that specialize in providing rickshaw transport services. As the urban environment becomes more and more congested, rickshaws are preferred for their ability to easily maneuver through the growing human and vehicle traffic and to deliver goods and passengers to their exact destinations—unlike motor cars. This flexibility coupled with the adventurous experience of riding in these vehicles makes them the preferred mode of transport for locals and tourists.

Selima Sultana

See also: Bicycles and Tricycles; Rickshaws (Bangladesh); Segway; Wheelchairs.

Further Reading

Jun, Jong S., and Deil S. Wright (eds.). 1996. *Globalization and Decentralization: Institutional Contexts, Policy Issues, and Intergovernmental Relations in Japan and the United States.* Washington, DC: Georgetown University Press.
Stille, William M. 2014. "Mobility on the Move: Rickshaws in Asia." *Transfers* 4(4): 88–107.

RUN-COMMUTING

Run-commuting is a transportation practice in which people run to and from work. It is a curious form of transportation emerging in cities across the "developed" world, yet it is by no means new. At its core, running is about the traversal of space and locational displacement. Running was human's second technology for overcoming the distance of time and space and always has been a means of transportation. Although this understanding still prevails in parts of the world, current "western" discourses about running have shifted away from transportation and toward sport and fitness. This does not mean that running for transport is an extinct practice, however. Linger long enough at any train station and you will witness hundreds of people running to catch their trains; running that might be undesired but running that is used to traverse space.

Run-commuting represents intentional instances of running being used for transportation. It is a specific idea that only came to prominence in the 2010s, emerging for a variety of personal, cultural, and geographical reasons. Individuals often take up the practice as a time-saving mechanism, allowing them to harmonize the rhythms of life and work by utilizing incumbent moments of mobility to fit in their running. Culturally, places where run-commuting is pervasive—such as London—hold strong aspirations for people to lead healthier workplace lifestyles, establishing run-commuting as acceptable. Geographically, dense urban cities offer the most attractive spaces for run-commuting as destinations often are not too distant; the street layout allows for many routes to be carved through it; congestion makes road-based traffic slow; and subterranean transport can be hot, hurried, and crowded. There is growing public interest in run-commuting and this enthusiasm perhaps can be best epitomized by the trend of racing the subway—trying to outrun a subway train between stations. The point? It can be just as quick and a lot more fun to get around by running.

The *International Survey of Run Commuters 2014* provides some initial insights the "brute facts" of run-commuting. A run-commuter is most likely to be a white male between the ages of 30 and 44 and most likely runs between three and seven miles on any commute. In cases where the total commute is farther than this, run-commuters use another mode of transport to complete the rest of their journey. The majority of run-commuters only do so two to three times a week. This often relates to the logistical complexities involved with run-commuting and the need for "things" as well as bodies to commute. Backpacks help to overcome this but only can be used for the bare essentials given the difficulty of wearing a backpack while running.

Unlike historical uses, this current trend for using running as a form of transportation is one made out of choice rather than necessity. But is it any good? Run-commuting has many positives for physical and mental well-being, as well as being environmentally friendly, relatively safe, and low cost. Conversely, running can be difficult, resulting in sweaty, tired bodies severely limited in distance and speed. It also is a transportation mode that heavily relies on other modes to enable commuters to complete their journeys if running cannot and for transporting "stuff" that runners cannot. Perhaps most notably, however, many people simply do not possess the ability or proclivity to consider run-commuting. Yet there *is* need to start considering it and other modes—beyond automobiles—and although there are currently no formal policies for run-commuting, a whole range of creative, community, and business-driven solutions have been established around the world to begin to fill the gap. In the United Kingdom a lobbying group has been founded called "*run2work*." The group lobbies workplaces to provide sufficient facilities, petitions government for tax exemptions, and hosts monthly awareness-raising "run2work days." The difficulty of running with a backpack has seen the rise of cycle escorts, such as Home Run London, which will transport a run-commuter's bag as the commuter runs. The loneliness of the run-commuter also is being mitigated through the establishment of run-commuting groups or running buddies as seen in the Brazilian initiative "*Corrida Amiga*." As run-commuting continues grow, however, many runners are calling for more comprehensive strategies to tackle the logistical difficulties, infrastructural necessities, and normalization of the practice to help the transportation mode flourish.

Simon Cook

See also: Bicycles and Tricycles; Sidewalks; Walking School Buses (United States).

Further Reading

Bale, John. 2004. *Running Cultures: Racing in Time and Space*. London: Routledge.

Kemp, Rob. 2014 (November 10). "All Change Please: The Rise of Run-Commuters." *The Telegraph*. http://www.telegraph.co.uk/men/active/11216715/All-change-please-the-rise-of-the-run-commuters.html. Accessed December 15, 2015.

Larouche, Richard, Adewale L. Oyeymi, Antonio Prista, Vincent Onywera, Kinsley K. Akinkroye, and Mark S. Tremblay. 2014. "A Systematic Review of Active Transportation Research in Africa and the Psychometric Properties of Measurement Tools for Children and Youth." *International Journal of Behavioral Nutrition and Physical Activity* 11(1): 129–146.

SAN DIEGO TROLLEYS (UNITED STATES)

The San Diego Trolley is the oldest of the modern—or second-generation—light rail systems in the United States, having begun service in July 1981. It is operated by the Metropolitan Transit System (MTS) in the city and county of San Diego, California, and is known by its red color. The trolley has a weekday ridership of 122,000 people and an annual ridership of 34.5 million.

San Diego's history of public transit dates back to 1886, when the San Diego Street Car Company—founded by Hamilton Story and Elisa Babcock—began running open-air streetcars pulled by either mules or horses in downtown San Diego. In the late 1800s, while Northern California was experiencing a population boom because of the gold discovered there, a new boom in land was developing in San Diego. Several wealthy San Francisco industrialists—including Alonzo Horton, John D. Spreckels, and William Heath Davis—bought land on which to build a city. San Diego's population increased from 8,600 in 1880 to more than 40,000 just seven years later, creating a need for inner-city transportation. In June 1887, the San Diego Land and Town Company began a steam line that shuttled new home-owners to suburban housing divisions, and in November 1887 the first electric streetcars began running.

Sugar heir Spreckels's acquisitions included the San Diego Street Car Company (which he renamed the San Diego Electric Railway Company), and the San Diego Cable Company, Citizens Traction Company, and the O. B. Railroad. New streetcar lines and a power-generating plant were needed to power the Electric Railway Company. Spreckels built two power plants just in time for the tourism industry to take off in connection with the 1915 Panama-California Exposition in San Diego's Balboa Park.

In 1922, the first motorized bus—made with hard rubber tires, mechanical brakes, a four-cylinder engine, and a plywood body—began service. In the 1930s, 222 new buses took to the streets, and bus ridership continued to increase in the 1940s, when World War II turned San Diego—home to a major U.S. Navy base and many defense-industry companies—into a boom town once again. During the war years, public transit use increased 600 percent. Business on the electric car line also surged but, by the mid-1940s, buses began to replace the street car lines. By 1947, only three street-car lines were operating. Spreckels sold the San Diego Electric Railway Company to Jesse L. Haugh in 1948 and the name was changed to the San Diego Transit System. Streetcars were retired in 1949, having been replaced by new General Motors buses. As an upgrade to the system, 58 percent of the bus line was replaced in 1950.

Passengers board the San Diego Blue Line trolley at the southernmost station stop at the U.S.–Mexico border crossing in San Ysidro. The Blue Line runs along Interstate 5, making numerous stops between San Ysidro and downtown San Diego. (Czuber/Dreamstime.com)

The boom in the automobile ownership in the 1950s coupled with the population shift from cities to suburbs caused a 47 percent reduction in public transit ridership nationwide, and San Diego was no different. Although the population numbers soared, ridership in the sprawling city did not improve from the era prior to World War II. To improve ridership, Haugh added charter coaches with bathrooms and refreshment centers to the fleet. In collaboration with San Diego's premier department store, Marston's, Haugh also created the Fashion Bus, with dressing rooms and a runway.

In 1966, voters approved San Diego Transit's purchase by the city, making San Diego Transit a nonprofit corporation operated by the city in 1967. An intergovernmental agency known as the Comprehensive Planning Organization (CPO) was charged with studying plans for the city's long-range transportation needs. Suggestions included restoring the 1949 streetcar system for $1.3 billion; building a rail rapid transit; creating an elevated rail line; adding a short light rail line from the airport to downtown; and adding express buses to the area's busy freeways. The decision came down to building either a light rail system or rail rapid transit. It was determined that a rail rapid transit system would cost $1.5 billion initially and officials were worried about the price tag as well as possible construction delays often associated with such systems. Proponents of rapid transit, however, expressed concerns that a light rail system would be too slow and less efficient than a rapid transit line.

In 1976, the Metropolitan Transit Development Board was created to help make a decision, keeping in mind that the new system should have both high- and low-speed trains, be affordable to build, and be built along existing public corridors to keep costs down. The decision was made to build a light rail line, with funding coming from the MTDB's enabling legislation, which was set aside for guideway construction through 1981. Part of the rail line was to be built along the old San Diego and Arizona Eastern (SD&AE) tracks, after the SD&AE's Desert Line had been wiped out by Tropical Storm Kathleen in 1976 and the Southern Pacific Railroad petitioned to abandon the SD&AE in August 1977. In 1978, the MTDB purchased the SD&AE lines from Southern Pacific.

San Diego Trolley Incorporated (SDTI) was founded in 1980 and construction of the trolley line soon began, with 40 percent of the old SD&AE ties requiring complete replacement. To save funds, the SDTI ordered just 14 German-built Siemens-Duewag vehicles. On July 26, 1981, the trolley went into service, with the first line—the South Line—running from downtown to the U.S.-Mexico border. The East Line, running to El Cajon, opened in June 1989; an extension to it, east to Santee, was added in August 1995. There were several other extensions of the existing lines and, in 1997, the names became color coded: the South Line becoming the Blue Line and the East Line becoming the Orange Line. A third line was added in July 2005 and named the Green Line. This line includes the only underground station, which is located at San Diego State University. A fourth line, the heritage streetcar line, is called the Silver Line and is a limited service run on a circle loop within downtown.

There are 53 stations along the 53.5-mile track running 134 light rail trains made by Siemens. Self-service ticket-vending machines are found at each station and riders must have a ticket or a prepaid electronic ticketing card (Compass Card) before boarding. Several realignment projects are in the works, including one to continue service to the University of California, San Diego, in the community of La Jolla.

Rosemarie Boucher Leenerts

See also: Light Rail Transit Systems; San Francisco Cable Cars (United States).

Further Reading

Holle, Gina. 1990. *San Diego Trolley*. Glendale, CA: Interurban Press.
Metropolitan Transit System. "MTS Historical Timeline." http://www.sdmts.com/MTS /timeline.asp. Accessed June 5, 2015.

SAN FRANCISCO CABLE CARS (UNITED STATES)

The iconic San Francisco cable cars are the last manually operated cable cars in the world. They are a system of transportation served by the San Francisco Municipal Railway (MUNI) in California. Twenty-three lines were built between 1873 and 1890, and three remain running. The cable cars are both a major tourist attraction and a mode of transportation for Bay Area residents, although a great majority of the 7 million annual riders are visitors. The cable cars are listed on the National Register of Historic Places as worthy of preservation.

The idea for the cable cars came from former Englishman Andrew Smith Hallidie, who in 1869 had witnessed a tragic accident involving a horse-drawn carriage sliding backward down one of San Francisco's notoriously steep hills on a rainy day and dragging five horses to their deaths. In England, Hallidie's father, Andrew Smith, had acquired the first patent for wire rope—or cable—in 1835, and had applied his invention to strengthening ships' rigging and as a stronger alternative to hemp haulage rope in London's railroad industry. Smith teamed up with Robert Newall, who had improved the speed of manufacturing the rope by using a machine called a strander. Their rope consisted of six strands of fiber-core wire wound around a central fiber core. Smith and Hallidie—who had taken his new surname from his uncle— left England for California during the Gold Rush. Having been disappointed in the mining prospects in Mariposa County, California, Smith returned to England, but the teenage Hallidie stayed on, trying his hand at mining gold. While in California, Hallidie applied his father's cable invention in bridge building and in the mining industry, hoisting ore cars up steep hillsides. In 1857, Hallidie abandoned mining altogether and moved to San Francisco, opening a cable-manufacturing plant called A. S. Hallidie & Co. and continuing to build bridges.

Bridge building was strenuous work and took Hallidie away from the city, so he decided to devote his efforts to manufacturing wire rope, just in time for the discovery of silver in Nevada in 1859, which put it in high demand. After witnessing the carriage accident in San Francisco, Hallidie set about making plans to use the wire rope in a railway system with cars propelled by underground cables. Public opinion was not high and funds were difficult to obtain, so Hallidie partnered with three friends and business associates to help finance the $85,000 venture. The cable railway was to start at Clay and Kearny streets and travel 2,800 feet uphill. In May 1872, the project commenced. A deadline of August 1, 1873, was given for completion. If the cable cars were not running by that date, the rights to the project would expire. On August 1, a test run was made on Clay Street. A car was pushed forward and inserted into the slot in the road. The onboard grip that Hallidie invented moved up and down and clamped onto an underground cable. When the car reached the bottom of the hill at Kearny Street, the car was rotated on a turntable and sent back up the hill.

The second cable car line, opened in 1877, was run by the Sutter Street Railroad. Hallidie had patented his cable grip and the Sutter Street designer did not want to pay to use Hallidie's invention so he devised a side grip and lever operation. An advantage of the side grip was that it enabled cable cars to cross at intersections. A year later, Leland Stanford began the California Street Cable Railroad, or Cal Cable. Following Cal Cable was the Geary Street, Park and Ocean Railway and the Presidio and Ferries Railway (P&F), opening in 1881. The P&F was noteworthy for being the first cable company to incorporate curves in its routes. The cars negotiated curves when the operator, or gripman, released the cable and allowed the car to coast using its own momentum.

1883 saw the opening of the Market Street Cable Railway, operated by the Southern Pacific Railway—which soon was the largest cable car operator. Other cable car companies included the Ferries and Cliff House Railway, whose Powell-Mason line still runs today along the original route and its Powell-Washington-Jackson line is today's Powell-Hyde line. The railroad also built the car barn and powerhouse at Washington and Mason streets, which still is in operation. The Ferries and Cliff House Railway purchased the Clay Street Hill Railway. In all, 23 lines were built.

Cable cars felt competition when electric streetcars were proposed in the city. Those wanting to preserve the cable cars objected to the more expensive electric cars' need for overhead electrical lines to power the cars—which preservationists thought were unsightly. The debate ended, however, with the San Francisco earthquake of April 18, 1906. The earthquake and resulting fire destroyed several cable car power houses and car barns, as well as 117 cable cars stored in the barns. Rebuilding the city now included adding electric cars to replace most of the destroyed cable cars. Six years after the fire, just 12 lines remained; however, they were being replaced by buses that could better handle San Francisco's hills. By 1944, only five lines remained—the two Powell Street lines, now run by the city, and three independent lines run by Cal Cable. In 1947, Mayor Roger Lapham wanted to shut down the remaining two Powell lines, which angered many groups, including the newly formed Citizens' Committee to Save the Cable Cars. A referendum to an amendment to save the cable lines was put on the ballot and passed.

The high cost of insurance caused Cal Cable to sell its three lines to the city in 1952. The city had plans to discontinue these lines in favor of buses, but a compromise was reached when the city once again felt pressure from sources interested in saving the cable cars. The compromise entailed keeping the California Street line as well as the Powell-Mason line intact, and forming a third hybrid line by combining sections of the Hyde Street line and the Powell-Washington-Jackson line, making the Powell-Hyde line. The three remaining lines stopped running for seven months in 1979, and underwent repairs totaling $60 million. A second overhaul was performed between 1982 and 1984, and included replacement of track along 69 city blocks. The lines reopening coincided with the 1984 Democratic National Convention held in the city.

Today there are 28 Powell Street cars and 12 California Street cars. The single-ended older cars run along Powell and are "combination cars," both an open grip dummy and closed-car train in one. The double-ended California cars first were used by Cal Cable. These cars have two sets of control levers, one at each end, and do not require a turntable to rotate the cars to go back in the direction from which they came. The California cars have covered inner seating as well as outward-facing seats.

Rosemarie Boucher Leenerts

See also: Light Rail Transit Systems; Streetcars (United States).

Further Reading

Cable Car Museum. http://www.cablecarmuseum.org/heritage.html. Accessed June 5, 2015.

Kahn, Edgar Myron. "Andrew Smith Hallidie." Virtual Museum of the City of San Francisco. http://www.sfmuseum.net/bio/hallidie.html. Accessed June 5, 2015.

San Francisco Cable Car website. "Cable Car History." http://www.sfcablecar.com/history.html. Accessed June 5, 2015.

SANTIAGO METRO (CHILE)

Santiago, the capital city of Chile, is located in an enclosed valley at the foothills of the Andes Mountains and has a metropolitan population of 6.1 million people. Santiago has experienced an economic boom since democracy returned in 1990, after almost two decades of Pinochet's dictator regime. During this economic boom, many businesses have relocated their multinational regional headquarters to the Chilean capital of Santiago. As is the case for any city experiencing such economic growth, some problems arise, such as traffic congestion and air pollution—not only from automobile and bus combustion, but also from industry.

The realization that Santiago eventually would need a new form of transportation grew out of the significant traffic problems and emissions from cars and city buses in the 1960s. The construction of the first metro line began in 1968, with a stretch of slightly more than 5 miles and including 12 stations. The line was inaugurated in September 1975. This section paved the way for the development of the most modern metro system in Latin America. During the months following the line inauguration, Santiaguinos (residents of Santiago) were drawn to the novelty of the new system and the construction taking place. By March 1977, the first expansion of Line 1 of the Metro was completed, adding five more stations in two miles. By 2009, Line 1 had 27 stations along 11 miles. Today, the Santiago Metro system contains 108 stations distributed through 5 lines, transporting 2.2 million people daily. The Santiago Metro has seen significant expansions during the past 15 years, including extensions on the east end of Line 1 and the completion of lines 4, 4A, and 5. In 2007, the metro system began integration with the city bus system called "Transantiago."

One of the unique impacts of the Santiago Metro is cultural. A project known as the "MetroArte" was begun in 1992 to showcase the art of the country along the routes of the metro system. Many metro stations of Santiago expose their riders to artwork that was donated by corporations and cultural associations. All of the art projects are created by Chilean artists. In all, there are 30 art projects displayed throughout the system. Riders of the system can experience the rich cultural and artistic heritage of the country through this unique project. Another cultural feature is the "BiblioMetro," established in 1995. This program introduced libraries

Electric trains at Central Railway Station in Santiago, Chile, 2013. (Serjio74/Dreamstime.com)

into three metro stations along Line 1. By 2007, 10 more locations were added across all of the lines. Established by collaboration between the National Library and the Metro, the BiblioMetro encourages riders to read.

Urban development is the second major impact of the Santiago Metro. Not only is the Metro able to connect communities, thus fostering interaction of citizens from different parts of the city, it also influences the development of housing and traffic patterns along the route. As Agostini and Palmucci (2008) noted,

> Investing in public transport infrastructure has a strong influence on urban development patterns and on the spatial distribution of property development. Building or improving highways and mass public transport not only reduces transport costs and traffic congestion in urban areas, therefore facilitating the efficient operation of cities, but also affects land use patterns and property values.

Such has been the effect of Santiago's Metro on both housing and traffic. Housing prices near the Metro have increased more than the prices of housing further away. Also, the construction of the system has been able to improve local traffic patterns, facilitating better and smoother movement of cars and reducing traffic congestion.

The third major impact of the Metro was that envisioned in the initial planning stage of development—environmental impact. The Metro of Santiago has been able to reduce air pollution indices due to the decreased use of cars. The city has enacted a policy of vehicle restriction based on the last number of a driver's license plate. Air pollution is especially problematic during the winter months (between April and September) due to an atmospheric inversion layer, trapping the smog in the valley near the surface. A similar phenomenon occurs in Los Angeles, California, and Mexico City, Mexico.

Today, the Santiago Metro contains 108 stations and extends 64 miles, making it one of the largest systems in Latin America. By 2018, the completion of Lines 3 and 6 will add a total of 136 stations, covering 87 miles. By increasing the capacity of the system, Santiago will be able to compete globally in a more sustainable way, making it a more attractive destination to visit and to establish new businesses.

Alejandro Molina

See also: Delhi Metro (India); Metro Railway, Kolkata (India); Mexico City Metro (Mexico).

Further Reading

Agostini, Claudio A., and Gastón A. Palmucci. 2008. "The Anticipated Capitalisation Effect of a New Metro Line on Housing Prices." *Fiscal Studies* 29(2): 233–256.
Metro de Santiago. 2014. http://www.metrosantiago.cl. Accessed June 10, 2014.
Schwandl, Robert. 2007. "Santiago." UrbanRail.Net. http://www.urbanrail.net/am/sant /santiago.htm. Accessed December 16, 2015.

SCHWEBEBAHN-WUPPERTAL (GERMANY)

The Schwebebahn-Wuppertal is a high-speed suspension monorail that began transporting passengers in 1901 in Wuppertal, Germany. Its full name is Eugen Langen Monorail Suspension Railway (German: *Einschienige Hängebahn Eugen Langen*), but the locals affectionately call it the "Old Lady" or "Iron Wyvern" (Iron Dragon). Meaning "floating railway," the Schwebebahn is the oldest electric suspension railway in the world, and is an important means of transportation for citizens of Wuppertal, with 75,000 to 85,000 passengers riding it daily and approximately 25 million annually. It is also a tourist attraction. The 8.3-mile line travels at 37 miles per hour and stops at 20 stations. The train, which looks like it is upside down, takes 35 minutes from the first stop to the last stop. It is suspended 39 feet above the Wupper River and 26 feet above roadways. The Schwebebahn-Wuppertal is known for its near-perfect safety record, although a fatality did occur in 1999.

The idea for a railway in the region around Wuppertal first was considered in 1824, when Englishman Henry Robinson Palmer proposed his idea for a suspended single-rail system pulled by horses, which would aid in transporting people and cargo through the hilly region and over a river known for its flooding. Famous German industrialist and politician Friedrich Harkort latched on to the idea, hoping the railway would help transport coal in the Wuppertal Valley and aid in his business ventures by connecting his factories. In 1826, Harkort built a prototype segment of the Palmer System railway in his steel mill in Elberfeld and presented it to the town council. Plans fell through, however, when owners of mills not along the proposed route objected. In 1898, a second attempt at a railway was made—this time an electric-powered suspended monorail. Engineer Eugen Langen had built and tested a proposed suspension railway in Cologne, German, in the 1880s, with plans to install it in Berlin. Construction in the Wuppertal region, overseen by the German government's master builder Wilhelm Feldmann, began at the end of the century. More than 19,000 tons of steel were needed to build the supporting frame and transit stops. The project cost 16 million gold marks. Although dates conflict, one record reports that the first test run of the monorail left the station on October 24, 1900, with Emperor Wilhelm II and his consort Auguste Viktoria taking the inaugural ride. The car they rode in was named the Emperor's Carriage and still is in operation. The suspension monorail opened to passengers on March 1, 1901. The supporting frame and tracks of the Schwebebahn-Wuppertal consist of 486 pillars and sections of bridgework. Similar to aboveground metro systems, trains turn around at the end of the route to take passengers in the opposite direction. A single car can hold 200 passengers.

The monorail, which runs between Vohwinkel and Oberbarmen, has come off its track only once, in January 1917, when one train rear-ended another train. Two minor injuries resulted. In 1968, a truck collided with a pillar and collapsed it, but no trains were involved. To prevent this from happening again, concrete walls were built around pillar anchors at all stops. The only fatal accident to date occurred in 1999 near the Robert-Daum-Platz station. Workers failed to remove a metal claw from the track and the first train of the day hit the claw while traveling 31 miles per hour. The train derailed and fell into the Wupper River, killing five passengers and injuring 49 others. The salvage operation took three days and the Schwebebahn was shut down for eight weeks, costing the system DM 8 million (US $5 million). Another crash in 2008 near Hammerstein station resulted in one minor injury to the driver of a crane truck that was making deliveries under the track. The truck crashed into an oncoming train and a tear nearly 33 feet long was made in the floor of the train car.

The most bizarre incident on the Schwebebahn-Wuppertal occurred on July 21, 1950, when an elephant named Tuffi, placed aboard the monorail as a publicity stunt to promote the upcoming Althoff Circus, panicked and burst through the side of the train. The elephant fell about 30 feet into the river below. The elephant as well as one passenger and two journalists covering the story experienced minor injuries and Tuffi lived 39 years more. The short-lived stunt has been commemorated in a 1983 children's book, *Tuffi und die Schwebebahn* (*Tuffi and the Schwebebahn*), by Ernst Ziegler and Marguerita Eckel.

Several modernization and reconstruction efforts have taken place on the system. One of the stops in Elberfeld—the Kluse Theater stop—was destroyed in World War II, and in 1995 the supporting frame underwent extensive repairs.

Rosemarie Boucher Leenerts

See also: Berlin U-Bahn/S-Bahn (Germany); Monorails.

Further Reading

Atlas Obsucra. "Schwebebahn Wuppertal: This 'Floating Railway' Is the World's Oldest Monorail System." http://www.atlasobscura.com/places/schwebebahn-wuppertal. Accessed June 5, 2015.

Zweisystem. "Wuppertal Shwebebahn: The 108-Year-Old Gadgetbahnen." Rail for the Valley (blog). http://railforthevalley.wordpress.com/2009/07/15/schwebebahn-wuppertal-the-108-year-old-gadgetbahnen/. Accessed June 5, 2015.

SEGWAY

The Segway is a self-balancing, two- or three-wheel electric-powered mobility vehicle. The rider stands upright on a platform and maneuvers the device by shifting the balance of his or her body. Often, it is classified as a nonconventional mobility aid, as this device initially was developed for special-purpose situations to enhance the mobility of the disabled on short-range trips. In the United States, Senate Bill 2024 defines the Segway as an electric personal assistive mobility device designed for transporting only one person, and the rider can use it on sidewalks for pedestrians or in bicycle lanes. This distinguishes Segway from similar motorized vehicles, such as motorcycles and scooters.

Dean Kamen (1951–), the inventor of the original Segway, introduced this device in 2001 under the product name of "Segway." The device's earlier veil of secrecy included the project's code name, "Ginger." After the release of advanced models, it has come to the fore as an eco-friendly alternative transportation mode because of its potential benefits for reducing pollution and urban traffic congestion. Currently, due to advancements in technology in electric propulsion, inertial sensing systems, and extended electric battery life, the Segway's capability is widely discussed for use for various purposes, such as commuting in urban settings; leisure; and special occasion use for security agencies (military and police), emergency services (hospitals), and public transportation facilities (transit stations, airports, and terminals). The types of Segway models vary from unicycle to two- , three- , and four-wheeled models. Because of its increasing popularity, different manufacturers have produced similar devices worldwide in Europe, Asia, North America, and South America.

The key technology enabling operation of this device without using any gasoline-based power is called "dynamic stabilization." The rider shifts his or her weight to control motion forward and backward and the speed. The device detects the change in the balance of mass, in particular the rider's center of gravity, using embedded gyroscopic and tilt sensors. Based on the sensor information, the microprocessors compute velocity, tilting, and propulsion to control the drive of the electric-powered motor for a smooth ride. The hand bar is used to control the direction (turn left or right). The maximum distance of operation ranges from 12 to 17 miles, and top speed is 12 miles per hour.

As an eco-friendly potential transport alternative with zero emissions, the Segway doesn't produce pollutants. Its main contribution, however, is reducing traffic

A group of tourists riding Segways past the World War II memorial on the National Mall, Washington, D.C., 2014. (Lee Snider/Dreamstime.com)

congestion—especially in densely populated areas, where urban commuting is made using short trips, and suburban areas with public transportation systems that transport passengers from the station or terminal to near downtown. Reducing congestion results in an environmental benefit—a reduction in total emissions of hazardous materials such as carbon monoxide, hydrocarbons, nitrogen oxides, and other harmful chemical materials. It is expected that the Segway can replace cars for short-range journeys (1 to 2 miles in urban areas).

Although these benefits are well identified, the main issue of the Segway is two-fold: safety and limited mobility. At the early stage, during 2001–2002, 32 states in the United States enacted legislation permitting this vehicle to operate on side-walks and in bike lanes. Many organizations and political groups, however, have raised concerns about safety—despite the fact that the Segway is acknowledged as a revolutionary technology. Additionally, the original Segway model was too heavy to maneuver on sidewalks effectively, as it weighed about 4,000 pounds. Consequently, its mobility was limited when people drove it in rugged terrain, hilly areas, and discontinued bike lanes. The device took up the width of an entire sidewalk, which affected other pedestrians' mobility. Thus, its size brought with it the possibility of injury due to collisions and produced pedestrian discomfort.

In the United States, several projects have been proposed since 2001 to evaluate the Segway's safety, productivity, efficiency, and applicability. Evaluations were conducted by commercial companies (such as GE and Delphi), law enforcement (police departments), public delivery services (such as the U.S. Postal Service), municipal transportation, and leisure agencies (National Park Service). Since 2001, all but five states have passed legislation allowing this device to be used on sidewalks, in bicycle lanes, and in other designated areas. Often its operation might be prohibited in the interest of public safety, however, and require the approval of municipal officials. Regulations differ by country. Some countries in Europe (i.e., France, Germany, Norway, Portugal, Ireland) and Asia (i.e., China, Japan, Singapore), allow the use of the Segway in designated areas but include restrictions on age or require insurance and licenses.

Hyun Kim

See also: Bicycle Libraries; Bicycles and Tricycles; Electric Vehicles; Sidewalks.

Further Reading

European Commission. 2008. "Consultation on New Framework Regulation on European Type Approval of Two and Three Wheels Vehicles." http://ec.europa.eu/enterprise/sect ors/automotive/files/consultation/2_3_wheelers/segway_en.pdf. Accessed December 1, 2014.

Segway website. http://www.segway.com/. Accessed November 14, 2014.

U.S. Department of Transportation. Federal Highway Association. 2002. "Safe Pedestrians and a Walkable America." *Pedestrian Forum* vol. 20.

SEIKAN TUNNEL (JAPAN)

A modern marvel, the 33.5-mile Seikan Tunnel is the world's longest undersea tunnel. It connects the two main islands of Japan—Honshu, the central island, and Hokkaido, the northern island. Approximately 14.5 miles of the tunnel run under the seabed of Tsugaru Strait, and at certain points the water reaches a maximum depth of 787 feet. The Seikan tunnel originally was proposed around 1939, but serious work (e.g., survey and excavation) did not begin until after World War II, when Japan lost many overseas territories and had to accommodate a large influx of people and growing economy on its four major islands, especially in Hokkaido.

Japan has many typhoons and the use of ferryboats between islands frequently is dangerous. During a typhoon in 1954, a ferryboat run by the Japanese National Railway sunk, killing an estimated 1,150 people. The ferryboats even close during certain seasons for safety, making them a somewhat unreliable source of transportation. After World War II, postwar economic growth was so rapid there was an increase in traffic congestion between the two islands. The number of ferry passengers doubled, and cargo levels increased by nearly double between 1955 and 1965. An underground tunnel between Hokkaido and Honshu became essential for providing safe transportation and meeting the needs of growing industry, trade, tourism, agriculture, fishing, and daily life.

The excavation of the Seikan Tunnel began in 1964 and construction of the main tunnel began in 1971. This mega-engineering project was not completed until 1988. Building it took nearly 24 years, the efforts of 16 million construction workers, and 6.9 billion yen. Construction required drilling and detonating explosives through an earthquake-prone zone. When traditional boring technology proved to be inadequate for the tough seabed, a new laser surveying style was used. The unusual terrain of silt, tuff, shale, and other rocks was so challenging that it took construction workers four months to move 40 yards ahead. The construction project not only was time consuming, it also proved to be very dangerous. Approximately 36 workers were killed in accidents during construction. Engineers found several major geological faults, and flooding halted the project numerous times.

There are three tunnels within the Seikan Tunnel. The main tunnel is used for running trains and has a triple-slab track to enable the passage of Shinkansen bullet trains in the future. Two other tunnels are pilot and service tunnels. The pilot tunnel mainly is used for geological surveys, and the service channels connect with the main channel for services. There are two stations inside the tunnel—Tappi Kaitei on the Aomori side, and Yoshioka Kaitei on the Hokkiado side—and both are known as "submarine stations."

By the time the mega project was completed, the tunnel was much less popular than expected. The trains were outdated and could not compete with the quickly growing technology in transportation, such as that of road transportation (auto) and aviation. At the time that the tunnel was finished, planes could travel from Tokyo to Sapporo in just an hour and a half. The growing popularity of air travel between Honshu and Hokkaido caused the tunnel to see much less traffic than was expected. As a result, the tunnel constantly is improved to keep up with transportation technology. Presently, it once again is undergoing construction to enable use of the newer bullet trains. If construction goes well, then the Shinkansen trains should reduce travel time significantly. Although the tunnel perhaps has not had the economic success that Japan hoped for, it created a transportation legacy that has been—and will continue to be—adopted by other countries for years to come. It also is a prime example of how a transportation technology can change faster than construction projects can be completed.

Selima Sultana

See also: Channel Tunnel (France and United Kingdom); Stormwater Management and Road Tunnel (SMART Tunnel) (Malaysia); Tünel (Turkey).

Further Reading

Ikuma, Michitsugu. 2005. "Maintenance of the Undersea Section of the Seikan Tunnel." *Tunneling and Underground Space Technology* 20(2): 143–149.
Kitamura, Akira. 1986. "Technical Development of the Seikan Tunnel." *Tunneling and Underground Space Technology* 1(3): 341–349.

Matsuo, Shogo. 1986. "An Overview of the Seikan Tunnel Project." *Tunneling and Underground Space Technology* 1(3): 323–331.

SEOUL METROPOLITAN SUBWAY (SOUTH KOREA)

The Seoul Metropolitan Subway system is a metro system that serves Seoul and the surrounding cities comprising the Seoul Metropolitan Area (SMA). Multiple operators serve 627 stations on 18 lines and 581 miles of routes consisting of the urban subway network of Seoul and Incheon, commuting rail lines (the Gyeongeui-Jungang, Gyeongchun, Bundang, and Shinbundang lines), and local light rails in Euijeongbu and Yongin. Most of the network is under a single-fare system that charges by distance, with free transfers via buses, thus the entire line comprises a single network system. This system serves 9.8 million passengers per day—which is 36 percent of all trips made per day.

The Seoul Metropolitan Subway network has been extending to cover the heavy commuting demand from new residential developments in the SMA. Seoul has experienced rapid economic growth since the 1960s. Seoul's 1990 population was 10 million—which was double its 1970 population. To solve the congestion problem, Seoul propelled large-scale residential development in the city during the 1980s and in suburban areas during the 1990s. Since the construction of five new suburban towns—Bundang, Ilsan, Jungdong, Pyeongchon, and Gunpo—accommodating 1.17 million people, many new towns have developed. Thus, 25 million people now live in the SMA—which is half of the total population of South Korea in 2015. To cover the commuting demands of the initial new towns, Lines 3 and 4 and 7 extensions, as well as the Bundang Line were constructed.

The subway network of Seoul plays a major role in the Seoul Metropolitan Subway. It consists of Line 1 to Line 9, which together total 206 miles and include 311 stations operated by the two of Seoul's public corporations—Seoul Metro and Seoul Metropolitan Rapid Transit—except for a private operator for Line 9. Line 1 originally referred to the first subway line of South Korea, a 4.8-mile underground subway section in central Seoul opened in 1974. The section directly connects with the Gyeongbu, Gyeongin, and Gyeongwon lines for through service. To cover the major cities of the SMA, such as Incheon, Suwon, Anyang, and Euijeongbu, 124 miles of track were built for this line. Line 2—a long, circular line stretching 30 miles—passes Seoul's major urban centers, including Euljiro, Shinchon, Gangnam, Jamsil, and Dongdaemun, and connects most of the other subway lines. Line 3 connects northwestern and southwestern Seoul via its two central business districts—Jongno and Gangnam. Line 4 serves the northeast and south parts of Seoul, and was extended to the southwestern cities of the SMA.

Construction of subsequent lines was aimed to relieve the congestion of the existing lines during the first phase, and to expand the subway coverage. Line 5 was designed to serve east and west Seoul via central Seoul. Line 6 connects the outskirts of Seoul. Line 7 improves access to the business district of Gangnam from northeastern and southwestern Seoul, and the termini were extended to Incheon via Bucheon to reduce the congestion of the Gyeongin Line. Line 8 mainly serves southeastern Seoul and Seongnam, a city adjacent to Seoul. Line 9 provides a fast connection between west and east Seoul via Gangnam with express service, which attracts numerous commuters. Lines 4, 5, 6, 7, 8, and 9 will be extended to suburban residential areas.

Recent subway network expansions have focused on improving express service and light rail, as well as extending the existing lines. The locations of additional new towns are farther from Seoul, thus the express railways are designed to reduce excessive commuting time. The high-speed commuting railways will connect suburban centers to central Seoul.

Based on financial feasibility, light rail transit (LRT) presently is considered to be an efficient type of transportation to supplement the existing network. Seoul's additional subway lines after Line 9 will be constructed as LRT, and the Ui LRT presently is under construction.

The Seoul Metropolitan Subway provides significant convenience for passengers through developed IT technology. Its single-fare system is perfectly combined with an RFID payment system—passengers pay their fees by tagging a plastic prepaid card, a credit card providing deferred payment, or a capable smartphone. Passengers can check the real-time location of their train on full-color LCD or LED screens located on the platform and at the subway entrance, or via the system's Internet website and mobile applications. Passengers also can access the Internet while on board the running trains as well as in stations; access is supported by Wi-Fi, WiBro, and 4G LTE technologies. A mobile TV service also is provided on the subway trains via supported devices such as MP3 players and smartphones. These IT-based services have improved the subway trains by making them more productive and enjoyable spaces. Additionally, trains provide accommodations for disabled persons, foreign-language signs and voice announcements, and comprehensive signage for transfer and exit information. The installation of platform screen doors for better safety and air quality in all of the stations is scheduled for completion by 2018.

Hyojin Kim

See also: Beijing Subway (China); Light Rail Transit Systems; New York City Subway (United States).

Further Reading

Bae, Chang-Hee Christian, Myung-Jin Jun, and Hyeon Park. 2003. "The Impact of Seoul's Subway Line 5 on Residential Property Values." *Transport Policy* 10(2): 85–94.

Bhang, Youn-Keun. 2003. "Completion of Phase 2 of Seoul Metropolitan Urban Rail." *Japan Railway & Transport Review* 35: 16–20.

Kim, Gyengchul, and Jeewook Rim. 2000. "Seoul's Urban Transportation Policy and Rail Transit Plan: Present and Future." *Japan Railway & Transport Review* 25: 25–31.

Kim, Jeonghyun, and Seungpil Kang. 2005. "Development of Integrated Transit-Fare Card System in the Seoul Metropolitan Area." *Knowledge-Based Intelligent Information and Engineering Systems* 3683: 95–100.

Lee, Jeong-Geun, Sang-Hoon Byeon, and Jai-Hyo Lee. 2009. "The Effect of Platform Screen Door (PSD) for Fine Particles at Subway Train in Seoul, Korea." 2012 3rd International Conference on System Science, Engineering Design and Manufacturing Informatization, October 20–21, 2012. Vol. 1. 154–159.

Lee, Seugjae, Shinhae Lee, and Young-Ihn Lee. 2006. "Innovative Public Transport Oriented Policies in Seoul." *Transportation* 33(2): 189–204.

Lee, Seungil, Changhyo Yi, and Soung-Pyo Hong. 2013. "Urban Structural Hierarchy and the Relationship between the Ridership of the Seoul Metropolitan Subway and the Land-Use Pattern of the Station Areas." *Cities.* Vol. 35: 69–77.

SHANGHAI METRO (CHINA)

Shanghai, one of four central government–controlled municipalities in the People's Republic of China (PRC), is the largest city in the country and one of the largest in the world. In a 20-year period, the city grew from 13 million residents (in 1990) to 24 million (in 2010). The metropolis covers a large area—more than 3,930 square miles. The city's rapid population growth is matched by its economic growth. Both trends necessitated the rapid development of an urban rail system. Today, Shanghai's metro network features 14 lines, 337 stations, and an operating route length of 341 miles, making it the longest in the world (based on route miles). The system ranks third in the world for annual ridership; more than 8 million people use the network daily (or 2.5 billion trips annually).

Compared with other major cities, Shanghai entered the urban transport age relatively late. Spearheaded by the Shanghai Municipal Government, it has been instrumental in the design and construction of this urban rail project. The government, however, enlisted private developers to build and operate the system; for example, the railcars are from several different builders including Alstom, Bombardier, and Siemens. The infusion of private money, coupled with local government oversight, has led to an unprecedented completion rate for such a large-scale project.

Construction on the first line began in January 1990. The first portion of Line 1, a mere 2.7 miles long and serving just 4 stations, opened on May 28, 1993. Extensions to Line 1 opened in 1995 and 1996, expanding the single route network to 15.5 miles and 16 stations. This first line operates in a western portion of the city and in a northeast-southwest direction. In 1999, a 10.1-mile second line operating from east to west opened. Line 3 opened at the end of 2000, adding an additional 15.3 miles to the network, and opening the western section of a route that eventually became an inner loop.

Over the next six years, extensions to existing lines and two brand-new ones began operation. On December 29, 2007, two extensions to existing lines and three new lines began operation on a single day. Growth of the system continued;

Waiting for the metro train in Shanghai, China, November 2010. (Typhoonski/Dreamstime. com)

by 2010, both of Shanghai's airports (Hongqiao and Pudong) were connected to the Metro. The latter already had been part of a transport achievement when, in 2003, an 18-mile maglev line began operations—and became the only maglev line in the world. The maglev line, however, is not part of the Shanghai metro system. Another milestone was achieved in 2010, when the Shanghai metro network became the longest in the world in terms of route miles.

Although the entire network has been built to a standard gauge (4 feet, 8½ inches, or 1,435 mm.), Shanghai's unusual feature is that 13 of 14 lines receive their power supply from overhead wires (most metros are powered using a third rail). Only the newest line (called Line 16, even though there presently is no Line 14 or Line 15) uses a third rail. As with many modern rail systems, many of Shanghai's metro stations feature platform screen doors; the train doors line up with marked sliding platform doors. The advantage of this feature is that it reduces the decibel level of incoming and outgoing trains in subway stations; it also prevents passengers from falling onto the tracks. The disadvantage of this feature—particularly on the Shanghai Metro—is the long wait time for trains in stations, which also increases journey times.

The Shanghai Metro has had a significant impact on the city and its residents. Prior to the Metro, journeys made via other transport means (particularly by bus) could take three times longer than the same journey when taken by rail. The Metro has made commuting more efficient and has eased traffic congestion in some areas.

Unfortunately, Shanghai remains hampered by its growth; traffic remains congested because the population's prosperity continues and the demand for private automobile ownership increases. Further, China's largest city has some of the worst air pollution in the world—a by-product of the exhaust from millions of vehicles.

The goal of becoming a more transit-oriented city remains alive, however. As with many other urban railways, this network has spurred considerable residential, commercial, and economic development along the corridors of the routes. The expansion of the system continues unabated; more lines are under construction and the goal is to have 18 lines spanning a 500-mile network by the year 2020. As far as commuting patterns, Shanghai's Metro could handle 60 percent of the city's public transportation needs. If achieved, these amazing feats will have occurred in the span of less than three decades.

Jason Greenberg

See also: Beijing Subway (China); Madrid Metro (Spain); Paris Metro (France).

Further Reading

Railway Gazette. 2010. "Shanghai Now the World's Longest Metro." http://www.railway
  gazette.com/news/single-view/view/shanghai-now-the-worlds-longest-metro.html.
  Accessed August 8, 2015.
Reidel, H-U. 2014 (November). "Chinese Metro Boom Shows No Sign of Abating." *International Railway Journal*. http://www.railjournal.com/index.php/metros/chinese-metro
  -boom-shows-no-sign-of-abating.html?channel=525. Accessed August 8, 2015.

SHENZHEN METRO (CHINA)

The Shenzhen Metro is the urban subway transit system for the city of Shenzhen in the Guangdong province of the People's Republic of China. Owing to China's open-door policy, Shenzhen was transformed from fishing villages to a modern city with a population of roughly 10 million. Neighboring Hong Kong, Shenzhen is the most successful special economic zone and by 2005 soared up to be the fourth largest urban agglomeration in mainland China. The enormous increase in population and product flow has caused the Shenzhen's urban transportation system to undergo significant changes.

The Shenzhen Metro opened on December 28, 2004, making Shenzhen the fifth city to have a subway system (the first four are Beijing, Shanghai, Tianjin, and Guangzhou on mainland China). The Shenzhen Metro currently covers five administrative districts with 5 lines, 131 stations, and more than 110 route miles. The daily passenger flow on the Metro reached nearly 2.7 million in 2014. That equates to about 27 percent of the total number of passengers using public transit in the city (Shenzhen Government Online 2014).

The planning of the Shenzhen Metro was initiated in the late 1980s. The first phase of the project, including Line 1 (or Luobao Line) and Line 4 (or Longhua

Line), started in 1999. The two lines were completed and went into operation on December 28, 2004. Although the first phase covered a total length of only 13.6 miles with 19 stations, it was the largest municipal project ever completed in Shenzhen (People Daily 2004). The Huanggang Station on Line 4, however, remained inactive until June 28, 2007, to accompany the opening of the Futian Checkpoint. That connection facilitated passengers visiting Hong Kong via the Futian Checkpoint. The Huanggang Station later was renamed to Futian Station. Both Line 1 and Line 4 were designed to facilitate the connection with Hong Kong. Passengers can transfer to Hong Kong's MTR East Rail Line via either LuoHu/Lo Wu on Line 1 or Futian Checkpoint/Lok Ma Chau on Line 4. Line 1 also connects with Shenzhen Bao'an International Airport.

On January 16, 2007, Shenzhen won the bid to host the 2011 Summer Universiade, which prompted the start of the second phase of the subway project. In addition to the expansion of Line 1 and Line 4, construction was started on the second phase of the project: Line 2, Line 3, and Line 5. In June 2011, all 5 lines were completed. Each line is connected with at least 1 other line, and passengers can transfer between 2 lines at interchange stations without having to leave the stations. The second phase significantly expanded the coverage of the Metro and promoted ridership. The third phase of the project was approved in 2011 and started in 2012. Five additional lines (Line 6 through Line 9, and Line 11) presently are under construction. They are expected to be completed by the end of 2016 and include 102 more stations. By then, the total mileage will be nearly doubled, reaching 216 miles. Early in 2014, Line 16 was renamed to Line 10 and incorporated into the third phase plan. In the long run, the Metro aims to reach more than 435 miles, include more than 20 lines, and carry about 38 percent of the public traffic (Shenzhen Metro Group Co., Ltd. 2014).

The Shenzhen Metro greatly enhanced the service level of the city's public transportation. To promote the use of public transit, the government invested 17 billion Chinese yuan to build six transportation hubs in the city. Currently the Shenzhen Metro links five transportation hubs in Shenzhen (Shenzhen North Railway Station, Qianhaiwan Station, Futian Railway Station, Buji Station, Shenzhen Bao'an International Airport) but not Pingshan Railway Station. Among them, Shenzhen North Railway Station presently is the biggest transportation hub in mainland China. It provides connections to the Guangzhou-Shenzhen-Hong Kong Express Rail and the Hangzhou-Fuzhou-Shenzhen High-Speed Rail. On July 30, 2014, Shenzhen Metro introduced free Wi-Fi service on Line 1 and Line 2, making Shenzhen the first city in China to provide Wi-Fi on the subway.

As a convenient and efficient way of transport, the Metro greatly facilitated the connection between the city and the suburban districts. For those who live in the suburbs and work in the city, they no longer need to wait in the long lines at bus stations during peak hours. People also use the subway as a primary mode of transport on weekends and holidays. The changes brought by the Metro not only shape the life of people, they also have important implications to regionalization in the area. The increasing connectivity between Shenzhen, the neighboring towns and cities in the Pearl River delta, and the rest of the country is a new chapter in Shenzhen's development.

Xuwei Chen

See also: Mexico City Metro (Mexico); Paris Metro (France); Shanghai Metro (China).

Further Reading

China's National News Online. 2011 (November 5). "Map of Shenzhen Metro." http://www.china.com.cn/info/2011-12/05/content_24076034.htm (in Chinese). Accessed November 13, 2014.

Kable Intelligence Limited. 2014. "Shenzhen Metro, Rapid Transit System, China." http://www.railway-technology.com/projects/shenzhen-metro-rapid-transitsystem-china/. Accessed December 15, 2014.

People Daily. 2004 (December 28). "Shenzhen Metro Officially Opened Today and Will Connect with Hong Kong MTR in the Future." http://www.people.com.cn/GB/shizheng/14562/3085576.html (in Chinese). Accessed November 13, 2014.

Shenzhen Government Online. 2014. "Checkpoints and Transportation." http://english.sz.gov.cn/economy/201408/t20140814_2545895.htm. Accessed November 13, 2014.

Shenzhen Metro Group Co., Ltd. "Plan of the Shenzhen Rail Routes (2011–2016)." http://www.szmc.net/page/plan/planning.html (in Chinese). Accessed November 14, 2014.

SHINKANSEN (JAPAN)

The Shinkansen is Japan's high-speed, intercity passenger rail network, and the first of its kind in the world. The term has reference to both the rail network and its high-speed trains. English speakers often refer to it as the "bullet train." The first Shinkansen trains traveled between Tokyo and Osaka on the new Tokaido Shinkansen line on October 1, 1964. This revolutionary advancement cut travel time from 6.5 hours to 4 hours, with trains reaching an initial top speed of 150 miles per hour (mph). Today that same trip can be completed in 2.5 hours. In 2014, the Shinkansen network consisted of 1,484 miles of high-speed rail (HSR) spanning the country and using a variety of rolling stock with maximum speeds ranging from 170 mph on the Tokaido line to 200 mph on the Tohoku line.

With a large population and high population density, rail transport is a popular, convenient, and efficient way to move people around modern Japan. The idea for Shinkansen has roots in the Pacific War (World War II), but it was put on hold until high demand for rail travel during the postwar era brought congestion and a string of high-profile, deadly transportation accidents. Initial plans for an HSR on dedicated tracks to alleviate such congestion were met with skepticism from politicians, bureaucrats, and the general population. Following the inaugural run, however, opinions shifted dramatically; jeers became cheers and "Shinkansen fever" took hold (Hood 2006, 30). Subsequently, a grand plan for a nationwide 4,350-mile

Shinkansen network was proposed. This was scaled back during economic uncertainties of the 1970s when a new plan was adopted, and it remains in place today with minor modifications.

Shinkansen lines connect major Japanese cities in broad coastal planes using an elaborate system of tunnels through mountainous areas to reduce both distance and the need to climb steep grades. Four main lines connect Aomori to Sendai and then to Tokyo (Tohoku Shinkansen); Tokyo to Nagoya, Kyoto, and Osaka (Tokaido Shinkansen); Osaka to Kobe, Hiroshima, and Hakata in northern Kyushu (Sanyo Shinkansen); and Hakata to Kumamoto and Kagoshima at Kyushu's southern end (Kyushu Shinkansen). Four additional lines extend from the Tohoku Shinkansen: the Joetsu Shinkansen connects Tokyo to Niigata; the Hokuriku Shinkansen connects Tokyo to Takasaki and Nagano; and Yamagata and Akita are serviced by lines branching off from Fukushima and Morioka, respectively. Additional lines under construction include the Hokkaido Shinkansen to connect Aomori with Hakodate and Sapporo through the Seikan Tunnel; an extension of the Hokuriku line to Kanazawa; a branch line connecting Hakata with Nagasaki; and the Chuo Shinkansen, a super-high-speed maglev train connecting Tokyo to Osaka thus shortening that trip to one hour traveling at a maximum speed of 310 miles per hour.

The Shinkansen plays a major role in local and national economies. Shinkansen stations are regional hubs for feeder rail and road networks. Shopping, dining, accommodations, meeting space, and office space often develop in and around them, yielding many local economic benefits. Train stations often are the social and economic core of Japanese cities. Indeed, the direct translation of Shinkansen as "new trunk line" is an apt moniker considering its place in Japan's broader circulatory passenger rail network. The system also is an affordable and convenient option for business travelers and makes new locations more accessible for both domestic and foreign tourists. An affordable rail pass is available for non-Japanese travelers and includes unlimited use of most Shinkansen trains. This attracts thousands of foreign tourists to the country every year.

The Shinkansen is a pioneer in HSR. The system is admired around the world for its expansive network, passenger volume (more than 350 million riders per year), record of promptness and reliability (an average overall delay of less than one minute), reputation for safety (zero fatalities and a novel earthquake accident-prevention technology), and dedicated rail infrastructure (other HSRs use an integrated system). Shinkansen engineers continually work to improve technology and meet Japan's strict noise pollution and energy-efficiency standards. Other countries have looked to Japan as a model for HSR development, but thus far Taiwan is the only country to directly import Shinkansen technology.

The Shinkansen is central to Japan's national identity. Beginning with its 1964 unveiling, the Shinkansen has been associated with the country's postwar emergence as an economic and technological world power. Further, the Shinkansen is now ingrained in Japanese popular culture. Lastly, the Shinkansen has become a recognizable symbol of modern Japan, especially when paired with the iconic Mt. Fuji. Indeed, the aptness of the name "new trunk line" extends beyond a literal meaning and represents the train system's place at the technological and cultural core of the country.

Rex J. Rowley

See also: High-Speed Rail; Maglevs; Seikan Tunnel (Japan).

Further Reading

Hood, Christopher P. 2006. *Shinkansen: From Bullet Train to Symbol of Modern Japan.* London: Routledge.

Ministry of Land, Infrastructure, Transport and Tourism. 2013. "Shinkansen Japanese High-Speed Rail." http://www.mlit.go.jp/en/tetudo/tetudo_fr2_000000.html. Accessed December 1, 2014.

Okada, Hiroshi. 1994. "Features and Economic and Social Effects of the Shinkansen." *Japan Railway & Transport Review* 3: 9–16.

SIDEWALKS

Every day people use sidewalks—or at least see one—but rarely are their origin and purpose examined. In general, a sidewalk infers a separated raised path designed for pedestrian traffic next to a vehicular road. Although the term "sidewalk" is used in the United States, other parts of the world call it something different. The term "pavement" or "footway" is used in the United Kingdom, and people in other countries call them "footpaths." The materials that are used to build sidewalks also vary. Brick and stone materials commonly were used to design sidewalks for aesthetic purposes in the late 20th century, and these are very common in urban centers of European cities. Concrete slabs most commonly are used in the United States, but wood also is used in coastal areas.

For more than a century, sidewalks have been played an important role in transportation by providing safe routes for pedestrians. Because safety is a big concern for sidewalks, they take different forms based on location, surrounding demographics, and uses. Usually the wealthy part of a city has much wider and nicer sidewalks than do less-fortunate areas. Sidewalks typically were built on each side of the vehicular roads in many cities that were built during the industrial era. In the United States, sidewalks are most common in urban areas, but are less common in suburban areas, especially suburban areas that were built during the 1960s—a time when automobiles became the primary mode of transportation. Decades of research, however, have documented the importance of sidewalks for reducing reliance on automobiles and for their health benefits. Today, building only sidewalks is not enough as it has been recognized that well-connected street networks with sidewalks between residential areas, schools, parks, and small shopping centers are conducive to an increase in walking. Therefore, accessible, safe, and comfortable sidewalk and crosswalk design are essential components of a comprehensive transportation planning agenda today.

Although the construction of sidewalks serves the primary purpose of safe transportation for pedestrians, sidewalks also represent a symbol of public space, where people can stroll, spend time, enjoy life, and build their communities. Greeks were strong believers in public spaces and most likely they were the first to build sidewalks. There also is evidence that Romans were particularly fond of sidewalks. By the late 19th century, spacious sidewalks were considered an important component of urban sophistication and were built in major cities in Europe and elsewhere. The grand boulevards that were built in Paris, Vienna, and Barcelona reserved generous sidewalk spaces for citizens to walk and enjoy urban life.

Concurrently, Americans were ambivalent about the use of sidewalks for urban life. The notion of building sidewalks in the United States therefore was carried out as a means of transportation. Many municipalities built the sidewalks with strict regulations as to how they were to be used. After World War II, to protect their privacy many homeowners were against building sidewalks, and until the 1990s this remained the trend in the United States. As a result, many suburbs built during this period did not include sidewalks.

Selima Sultana

See also: Bicycle Lanes; Complete Streets; Cykelsuperstier (Denmark); Stairways.

Further Reading

Loukaitou-Sideris, Anastasia, and Renia Ehrenfeucht. 2009. *Sidewalks: Conflict and Negotiation over Public Space.* Cambridge, MA: MIT Press.

SKI LIFTS

The term "ski lift" refers to a variety of vertical transportation systems used to carry skiers up a mountainside so that they can ski back downhill. Ski lifts are unique among transport systems in that they function as one-way systems—people ride them in only one direction. They can include technologies such as aerial tramways, chairlifts, or funicular railways. Tramways and chairlifts have compartments or open seating and ride on steel cables suspended between towers. The individual units can be pulled along the cables or might be permanently attached to the cable, which itself is pulled in a loop.

Although early ski lifts were in use hundreds of years ago, it was the development of skiing as a major recreational activity—beginning in the 1920s—that led to the modern ski lift. The first chairlifts were built in the 1930s at ski resorts in Idaho and at many other ski areas. A ski resort could have single or multiple lifts and lifts can be arranged in a network enabling passengers to select the direction and distance they want to travel up the mountain. Summertime use of ski lifts also is common. A ski area on the San Francisco Peaks in Arizona operates in the

A ski lift carrying skiers to the top of the ski trail. (Volodymyrkrasyuk/Dreamstime.com)

summer, allowing visitors to climb to 11,500 feet elevation in a few minutes. From there they can hike down the mountain or even climb higher to the very top.

Joe Weber

See also: Aerial Tramways; Funiculars; Stairways.

Further Reading

Skilifts.org. http://www.skilifts.org/old/. Accessed July 17, 2015.

SLUGGING (UNITED STATES)

"Slugging" is a relatively recent phenomenon occurring in many large metropolitan areas. It is a form of ridesharing in which car drivers on their way to work will pick up commuters standing in lines at particular locations so that they may drive in car-pool or high-occupancy lanes on highways. Both driver and passenger arrive at work sooner by doing so. The practice also has environmental benefits in that there are fewer vehicles on the road and they can use the freer-flowing high-occupancy lanes.

The practice first was recorded in the Washington DC, metro area, which suffers from heavy traffic congestion. The city also has many employment concentrations, which means there are many workers heading to and from the same destination each day. The term is said to derive from the fake coins, called "slugs," used in vending machines. In this case, slugs would be "fake" passengers or those that are not from the household that owns the vehicle.

Slugging is unique among many forms of travel in the United States in that it is not organized by, paid for, or regulated by the government. Passengers and drivers have created this system along with informal policies about how it should work. It differs from hitchhiking in that it is not random or occasional. Instead, passengers form lines at particular locations where drivers will seek them out. It differs from carpooling in that it requires no formal commitments from any of the participants and likely involves different people each day. Informal or casual carpooling sometimes is used to describe it.

Joe Weber

See also: Car Sharing; Hitching; Lyft; Uber.

Further Reading

Sluglines.com. http://www.sluglines.com/. Accessed August 26, 2015.

SMART CARS

A smart car, also called an "intelligent vehicle," contains onboard technology allowing linkage with the vehicle's surrounding environment including the transportation infrastructure. The enabling technology that allows such linkage is called Intelligent Transportation Systems (ITS). An ITS performs sensing, communications, computing, and algorithmic procedures to produce automated reasoning that solves problems that frequently would be solved by a human operator. Antilock braking, traction control, and electronic stability control are examples of present-day safety systems that provide primary control. Smart cars that can assess risk or sense the environment represent an advancement of current automotive safety systems. There is a distinction between smart cars that use automation for specific tasks, such as navigation or distance separation (semiautonomous), and cars that rely solely on automation (autonomous).

In 1925, the Houdina radio-operated automobile was an early example of an autonomously guided vehicle. The vehicle was a production model 1926 Chandler outfitted with an antenna and radio receiver controlled by an operator in a nearby vehicle. The operator used a transmitter to send signals to small motors that controlled different systems in the autonomous vehicle.

At the 1939 New York World's Fair, industrial designer Norman Bel Geddes created "Futurama," a popular exhibit at the General Motors pavilion. The exhibition featured a large-scale model of a futuristic city of 1960 that foretold the morphological landscape that would arrive with the Interstate Highway System. Bel Geddes was an active proponent of removing the human factor from driving and was

---

Vehicle-to-Vehicle (V2V) Communications Technology

Vehicle-to-vehicle (V2V) communications technology is a component of what often are called "intelligent transportation systems" (ITS). The goal of V2V is to use new technologies to improve the capacity and safety of highways. A recent innovation is communication technology to enable cars to automatically communicate with each other in traffic. Radio signals indicate a car's speed, acceleration, deceleration, and direction to all other cars in the area. Information received from other vehicles in turn can be used to warn drivers of hazards, such as speeding vehicles or cars coming to a halt on the route ahead. Although the technology is far from proven, it offers great potential for reducing traffic accidents.

Joe Weber

---

instrumental in sparking the public imagination with the possibilities of intelligent vehicles.

In the 1970s, the U.S. Government Office of Research and Development, Public Roads, sponsored research of onboard vehicle navigation and route-guidance systems. The Electronic Route Guidance System (ERGS) was part of an initial research effort for ITS. The ERGS was a destination-oriented system that required the human driver to enter a destination code into the vehicle system. The vehicle communicated with an instrumented intersection where the target code was decoded and routing information then was sent to the vehicle.

In the 1990s, developments in Autonomous Intelligent Cruise Control (AICC) used a safety distance separation rule. The AICC system indicated performance superior to that of human-driver models. Faster responses and smoother traffic flow were shown in studies using computer simulations.

From 2003 to 2007, the U.S. Defense Advanced Research Project Agency (DARPA) sponsored a series of three "Grand Challenges" with teams representing major research universities. Progress over the three challenges was gradual with significant improvements in the final challenge that demonstrated a 60-mile urban course in which autonomous vehicles operated alongside human-driven vehicles. The DARPA "Grand Challenges" are significant as they show the thrust of research shifting from infrastructure automated systems to semiautonomous and autonomous vehicles that depend little on highway infrastructure. The DARPA Challenges also solidified partnerships between the automobile manufacturers and academia.

Since the time of the DARPA Challenges, Google has begun commercial research of an autonomous vehicle, drawing on the technologies tested in university laboratories. Google's autonomous vehicles use Light Detection and Ranging or LIDAR, a remote sensing method that examines the earth's surface and adapts to "make sense" of the vehicle's surrounding environment. As of April 2014, Google's

Driverless Car program logged approximately 700,000 autonomous miles (Urmson 2014).

Collision avoidance and lane-departure warning systems use radar, lasers, and cameras to detect a collision. Collision avoidance systems can alert the driver to react manually to an imminent collision, or it can operate autonomously by taking control of the vehicle's steering and braking. Collision avoidance systems usually work in concert with adaptive cruise control (ACC), which uses behind-the-grill radar to adapt a vehicle's distance and speed to that of the vehicle ahead of it.

Presently, intelligent safety features such as collision avoidance and ACC are optional equipment available only on high-end vehicles. In 2009, the U.S. National Highway Safety Administration (NHSTA)—a federal agency

The model of the city Futurama depicts a city design for 1960 at the New York World's Fair of 1939. The model was designed by Norman Bel Geddes for General Motors. (Corbis)

charged with enforcing vehicle safety standards—studied whether lane-departure warning, forward-collision warning, and ACC should be mandatory features on all highway vehicles sold. Other intelligent safety features such as electronic stability control, automatic braking, tire pressure monitoring, and speed-limiting technology are available on vehicles in most price ranges, but are not yet mandatory equipment. The National Transportation Safety Board (NTSB) advocates that these options become standard equipment in the near future.

Recently developed technology enables vehicles to communicate with each other and with the infrastructure. The "V2V" is a short-range communication system that enables vehicles to "talk" to each other ("V2I" means vehicle communication with infrastructure). Vital information exchanges as fast as 10 per second would be sufficient for cars to calculate a danger within about 1,000 feet and could alert drivers to take action or engage automated collision avoidance systems.

Joe Di Gianni

See also: Autonomous Vehicles; Flying Cars; Intelligent Transportation Systems.

Further Reading

Bishop, Richard. 2005. *Intelligent Vehicle Technology and Trends.* Boston: Artech House.

Putic, George. 2014 (April 4). "Vehicles may Soon Be Talking to Each Other." *Voice of America.* http://www.voanews.com/media/video/vehicles-may-soon-be-talking-to-each-other/1886889.html. Accessed January 8, 2015.

RAND. 2014. Autonomous Vehicle: A Guide for Policymakers. RAND Corporation.

Urmson, Chris. 2014 (April 28). The Latest Chapter for the Self-Driving Car: Mastering City Street Driving. *Google Official Blog*, Self-Driving Car Project.

Wilson, Christopher. 2012. "Probes and Intelligent Vehicles." *Handbook of Intelligent Vehicles.* Edited by Azim Eskandarian. 1145–1171. New York: Springer.

SMART TUNNEL

See Stormwater Management and Road Tunnel (Malaysia).

SOLAR ROADS

Solar energy technologies convert sunlight into usable energy. Power generation depends on the amount of the sun's energy that reaches the solar collectors. Photovoltaic (PV) is the major technology for generating solar power. Solar (PV) cells that convert sunlight to electricity are used along the roads of different developed countries. Transportation engineers also have considered generating solar power directly from or in the road. Examples of the past attempts include road studs that contain small solar panels and emit LED light to illuminate roadways at night; interseasonal heat-transfer systems incorporating solar energy collectors in the road; and shallow insulated heat stores located in the ground.

The idea of embedding solar cells in the roads has attracted the most attention in recent years. In exploratory research, the Idaho-based company Solar Roadways has been awarded grants from U.S. Department of Transportation (DOT) to develop and build a solar parking lot prototype. The company's technology has been categorized as "Solar Power Applications in the Roadway," as compared to other technologies classified by the DOT as "Solar Applications Along the Roadway." The phase 1 prototype was a 12-foot by 12-foot road panel that was completed in 2010. It was developed using polycarbonate in place of glass due to funding limitations. Additionally, a prototype storm-water redistribution system and a 3-foot by 3-foot crosswalk panel were also built. The crosswalk panel recognizes a weight (such as pedestrian or animal), begins to flash, and sends a signal to the solar road panel instructing incoming vehicles to reduce their speeds. For phase 2, in 2014 a parking lot prototype covered with hexagonal glass on a concrete base was revealed. It is fully functional and includes a hexagon-shaped panel of solar cells for covering curves and hills, LEDs to display messages, and heating elements to melt snow and ice. A 250,000-pound load can be supported, and roughly 7.6 kWh of power can be generated by the panels per day.

Glow Roads

The invention of headlights for cars made night driving possible, but nighttime driving still can be challenging and dangerous. A new innovation is to use glow-in-the-dark paint for lane markings on roads. The paint is charged by sunlight during the day and then glows for up to eight hours at night. A project in the Netherlands has used this successfully on a road, and there is considerable interest in lighting up the night this way elsewhere, as well. There are still many details to be worked out, however, including the cost and whether glow-in-the-dark paint is safer for drivers than the reflective paint presently in use.

Joe Weber

The Solar Roadways vision for the project is to harness solar energy to provide light and heat for all roads in the United States. The project includes a technology combining a transparent driving surface with solar cells, electronics, and sensors beneath it to act as a programmable solar array. Estimates by Solar Roadways show that by covering roads with the solar panels—which actually pay for themselves—the panels can produce more than three times the amount of electricity that currently is used in the United States and can cut greenhouse gas emissions by 75 percent. Other benefits of solar roads have been recognized by the Solar Roadways company founders and include economic recovery, national security, snow/water management, traffic management, wildlife protection, and health benefits. Nevertheless, doubts about the feasibility of the project on a national scale have been expressed ever since the research began.

Unlike the project by Solar Roadways, the Netherlands "SolaRoad" came into reality by following a single objective: To generate electricity from roads. At the invitation of the Netherlands Organization for Applied Scientific Research (TNO) in 2009, the province of Noord-Holland road construction company Ooms Civiel, technical service provider Imtech formed, together with TNO developed the core consortium that is building SolaRoad. It was developed as prefabricated slabs of concrete with a translucent layer of tempered glass on top. Crystalline silicon solar cells were placed beneath the glass. As a result, in 2014 a 230-foot cycle path that absorbs sunlight via solar cells and converts it into electricity was developed in Krommenie, a village northwest of Amsterdam. The bike path generated more than 3,000 kWh of electricity in its first six months—which is enough to power a single-person household for a year. The goal of the program is to cover large parts of the roads in the Netherlands with solar panels to generate electricity that can be used for street lighting, traffic systems, household uses, and electric vehicles.

The Solar Roadways project and SolaRoad are not the only solar-power projects. Onyx Solar developed a solar panel walking path that was installed by students at

Bicyclists are forced to use the sidewalk as they pass a stretch of bicycle path where a solar panel roadway is being constructed in Krommenie, north of Amsterdam, Netherlands, 2014. (AP Photo)

the Solar Institute at George Washington University in 2013. The 100-square-foot walkway generates a peak energy output of 400 watts and is used to power 450 LEDs located under the panels and illuminate the pathway at night. The walkway also is connected to a solar-powered trellis, creating energy that is fed back into the university's Innovation Hall building. Van Gogh-Roosegaarde bike path—which opened November 13, 2014, for the 125th anniversary of van Gogh's death—is another solar-powered path located in Nuenen, Netherlands. Roosegaarde, the designer of the path, was inspired by Vincent van Gogh's painting, *The Starry Night*. The cyclists' path is lit at night by 50,000 stones covered by a phosphorescent paint and with solar-powered LEDs. The aim was to make cycling safer, more beautiful, and energy efficient. As demand for energy increases, using more sustainable resources and improving materials and techniques that are more energy efficient and more cost effective seems necessary to sustain the future.

Nastaran Pour Ebrahim

See also: Electric Vehicles; Musical Roads.

Further Reading

Solar Roadways. http://www.solarroadways.com/intro.shtml. Accessed August 26, 2015.
SolaRoad. http://www.solaroad.nl/. Accessed August 26, 2015.

U.S. Department of Transportation Federal Highway Administration. Office of Planning, Environment, and Realty. 2012. "Alternative Uses of Highway Right-of-Way." http:// www.fhwa.dot.gov/real_estate/publications/alternative_uses_of_highway_right-of -way/rep03.cfm. Accessed August 26, 2015.

Weinberger, Hannah. 2014 (November 19). "Netherlands Unveils 'Starry Night' Solar Bike Path." *Outside.* http://www.outsideonline.com/1927506/netherlands-unveils-starry -night-solar-bike-path. Accessed August 26, 2015.

SPORT UTILITY VEHICLES (SUVs)

A sport utility vehicle (SUV) is a large, modern passenger vehicle similar to a station wagon or van, but built on a truck chassis and often equipped with four-wheel drive for on- and off-road maneuverability. The SUV as a family vehicle gained popularity in the late 1990s and the first decade of the new century, but sales of the SUV declined in the next decade as gasoline prices soared and SUVs were seen as too expensive to fuel. To combat a drop in sales, manufacturers of the SUV began making smaller, fuel-efficient versions such as crossover SUVs.

Sport utility vehicles were first created by Volkswagen in Germany during World War II. Called a "*Kubelwagen*" ("bucket car"), German government officials of the ruling Nazi Party were interested in incorporating into their war effort a small vehicle that could travel over difficult terrain. German chancellor and Nazi Party leader Adolf Hitler initiated production in 1940. The *Kubelwagen*s were similar to a Volkswagen Beetle and had two-wheel drive. They functioned well and were easy to maintain. Nearly 60,000 were produced before the end of the war. Production ended when British troops shut down the Volkswagen factories in Wolfsburg, Germany.

During the war, the U.S. military also sought out a small rugged vehicle. Out of the many prototypes submitted, Ford and Willys beat out the other candidates and the first contract was awarded to Willys to produce 16,000 vehicles for the U.S. Army. A year later, after the United States entered the war following the Japanese bombing of the military installation at Pearl Harbor, Hawaii, on December 7, 1941, both Willys and Ford were contracted to produce vehicles. The vehicles were referred to as General Purpose, or GP; the abbreviation was shortened to the popular name "Jeep." By the end of the war, 600,000 Jeeps were produced. After the war, the vehicle gained credibility with civilians and evolved into the CJ (standing for "Civilian Jeep") and Jeep Wrangler models. American Motors Corporation (AMC) began manufacturing Jeeps in 1970. In 1987, Chrysler bought out AMC and began producing Jeep products.

The first passenger SUVs remained close in functionality and style to the military models. In addition to the Jeep CJ were the Ford Bronco; the Chevy Blazer, produced by the Chevrolet division of General Motors (GM); and the Toyota Land Cruiser, imported from Japan. The first Chevy Suburbans and Jeep Wagoneers—even larger than the CJ and Blazer models—were trucks with a solid shell and remained a niche vehicle for those living in rural areas or those who sought adventure on rugged terrain.

With the introduction of the Jeep Cherokee, SUVs of the 1980s began to appeal to urban dwellers by combining the off-road capabilities of the early SUVs with passenger-vehicle sensibilities, including more seats and four doors. The Ford Explorer was a larger version of the Bronco and had a more powerful engine and comfort, which fueled sales. Paralleling the booming economy of the 1990s, SUV production stepped up to meet the growing demands of the general public. A mid-size line of SUVs entered the market and, in the late 1990s, full-size SUVs were added to conform to the public's desire for bigger and better vehicles. Among this new wave of bigger SUVs that were built on full-size pickup truck chassis were the Ford Expedition, the Chevy Tahoe, and the GMC Yukon. The three-quarter-ton Suburban—produced under the Chevrolet and GMC lines of GM—and the Ford Excursion increased SUV sizes even further. Also in this category were the Lincoln Navigator, a Ford luxury SUV product that was introduced in 1998, and GM's Cadillac Escalade, another luxury SUV that hit the market in 1999. These American models competed with other luxury vehicles such as the British Land Rover and Range Rover, and Japanese models produced by Lexus, Acura, and Infiniti, as well as vehicles from the so-called German Big 3 automakers: BMW, Mercedes Benz, and Audi. What motivated buyers to purchase full-size SUVs and those labeled as "luxury" vehicles was the towing capacity, better visibility, roominess, increased stability in foul weather, and a perceived sense of safety. Image also played a part in sales, as consumers equated the size of the vehicle with status. Automakers were happy to comply with these consumer perceptions as markups on such vehicles were in the thousands of dollars versus a few hundred on the smallest cars.

As the world economy weakened and fuel prices increased to more than $4 a gallon in much of the United States in the first two decades of the 2000s, the demand for expensive-to-own, fuel-thirsty automobiles such as full-size SUVs sharply declined. Production of the Excursion and Suburban was discontinued and automakers turned to manufacturing small, midsize, and compact SUVs and crossover vehicles. An emphasis on fuel economy and concern for the environment also had an impact on sales of larger SUVs as more and more consumers turned to electric, hybrid, alternative-fueled, and smaller vehicles. Crossovers (SUVs built on car chassis) were introduced and are a compromise between a full-size SUV and a sedan.

Rosemarie Boucher Leenerts

See also: Alternative Fuels; Dirt Roads; Electric Vehicles; Plug-In Hybrids.

Further Reading

Wolf, Calvin. 2013 (April 8). "A Short History of the SUV." http://www.autohub360.com /index.php/a-short-history-of-the-suv-243/. Accessed June 5, 2015.

STAIRWAYS

Stairways have existed for thousands of years and enable a person to ascend or descend a set of steps. They are a basic form of vertical movement inside and outside of buildings. Indoors they might be thought of as architectural features, but they also are important for transportation. In some cities stairways are an important part of the city's pedestrian network. Cincinnati, Pittsburgh, San Francisco, Los Angeles, and Seattle are particularly good examples. Smaller towns, such as the former mining town of Bisbee, Arizona, also can have many public stairways due to steep terrain. The world's longest stairway is in Switzerland. It rises 5,476 feet and has 11,674 steps.

The design of stairways now usually is subject to building codes, which can specify minimum or maximum sizes for each step, the height of the rise between steps, and the presence and height of railings. The run (horizontal distance) and rise (vertical distance) will determine the steepness of the stairway, with an angle of about 30 degrees for the slope of the stairway considered standard. Although short stairways can be straight, longer ones likely will have turns to save space. Many different spiral or helical stairways also can be found.

Stairway on a hill in San Francisco, California. (Randy Miramontez/Dreamstime.com)

Stairways are simple and reliable ways of traveling vertically, but there are many competitors. Elevators move vertically in buildings and were first widely manufactured in the mid-19th century, although they are much more expensive. Building codes likely still will require stairways to serve as emergency exits in case of a fire or an elevator failure. Escalators are motorized stairways, first built in 1896. Escalators commonly are used to move between floors in large buildings such as shopping malls or to reach underground train stations, but also can be used outdoors. They have the advantage of serving as stairways in the event of power failure. The Hong Kong Central-Mid Levels escalator system is the largest outdoor covered installation of escalators. It spans 443 feet of elevation and 2,600 feet horizontally and is made up of 20 escalators and various walkways. The system is free to use, takes about 25 minutes (if people are standing and not climbing), and is heavily used by commuters and shoppers.

Ramps or slanted walkways are a possibility for more moderate grades, and have become more common in the United States since legislation for disabled access has moved builders away from staircases. For cities with substantial vertical differences funicular or inclined railroads might be built. These are small streetcars that run diagonally up an inclined track, connecting valley bottom cities to hilltop neighborhoods. They were common in the late 19th and early 20th century when electric streetcars were an important form of transportation. Hilly cities such as Pittsburgh and Cincinnati possessed many of these, but few survive today. The Angels Flight funicular in Los Angeles is one of the few surviving today, and perhaps the best known. Although the cable cars of San Francisco are famous for climbing steep hills they were not designed as a vertical transport system.

Joe Weber

See also: Aerial Tramways; Cable Cars; Elevators; Escalators; Sidewalks.

Further Reading

Loretto Chapel Miraculous Staircase. 2015. http://www.lorettochapel.com/staircase.html. Accessed August 27, 2015.
Stair Maps. 2015. http://publicstairs.com/index_000002.htm. Accessed August 27, 2015.

STORMWATER MANAGEMENT AND ROAD TUNNEL (SMART TUNNEL) (MALAYSIA)

Kuala Lumpur, the capital of Malaysia, lies at the confluence of two major rivers, the Klang and the Gombak. The city faced two major flooding events in 1926 and 1971; the rapid urbanization occurring since 1985 has led to an almost annual occurrence of flooding in the city. The Klang River's average annual flood was at a stable rate of 39,500 gallons per second until 1985. With a sharp increase within 10 years, however, it reached more than 116,000 gallons per second by 1995. This situation has been constantly monitored by Malaysian government to prevent the flooding. As a result, the Klang River Basin Flood Mitigation Project (KRBFMP) has been implemented involving various strategies such as creating holding ponds and increasing river channel capacity. This led to an innovative concept called the Stormwater Management and Road Tunnel (SMART)—the first tunnel of its kind in the world which opened in 2007.

The project was started as a joint venture between Gambuda Berhad and MMC Corporation Berhad under the Malaysian federal government in 2003. The primary aim was to divert flood flows from the Klang River in the north through a bypass tunnel under the city and then to release it into the Kerayong River in the south. The three hydraulic components of tunnel include holding ponds, which draw water from the river and pass it into the tunnel; the diversion tunnel, which is the bypass connecting the holding pond to the storage reservoir at Desa Water Park; and the storage and release system, which provides temporary storage for the floodwater and then conveys it to the Kerayong River. The idea of using a tunnel as an alternative traffic route also was considered to decrease the traffic congestion and make the project more cost effective. The tunnel design includes a double-deck toll roadway, each of which has two traffic lanes for cars and light vans. Ingress and egress connections to the tunnel link the southern gateway to the city center. Commuters can travel on the road decks while water flows below. What makes it truly SMART, however, is the feature that would allow the entire tunnel to be used (including the road decks) as a flood channel to save the city in case of a giant river surge.

The tunnel was built with the help of giant tunnel boring machines (TBM) that drill and also build waterproof concrete rings to form the lining of the tunnel. The flood tunnel is 6 miles long and 1.8 miles of that is occupied by the highway tunnel. To ensure the safety of vehicles, the SMART tunnel has some unique features, including automated flood-control gates, ventilation and escape shafts, and air quality–monitoring equipment. Service and emergency gates were designed to prevent flood water from entering the highway space. Service gates at both ends of the highway tunnel close the upper and lower decks. Emergency gates close both road decks.

A flood warning system with rain gauges spreading through the flood catchment area has been developed. Computers in the Stormwater Control Centre receive rainfall and river level information to forecast probable flood magnitude and determine which one of the three operational modes will be activated for the tunnel. When there is no rain or when the flow is less than 18,500 gallons per second, mode 1 is activated to allow the traffic in the tunnel while preventing water from diverting into the holding pond and tunnel. Mode 2 is started with moderate rainfalls and flow rate of 18,500 to 39,500 gallons per second. Only 13,000 gallons per second are allowed to flow downstream and excess floodwater is diverted into the holding pond by opening radial gates. Once the water in the holding pond reaches the crest of a bellmouth weir, it will spill over into the tunnel. The void under the lower road deck can be used to pass the flows and the highway section still could be open. In case of any major storm event when the flow rate forecast exceeds 39,500 gallons per second, mode 3 is activated. Only 2,600 gallons per second are allowed to flow downstream and one or both road decks are used for flood flow. The radial gates are opened to divert water into the holding pond between 70 and 100 minutes after the start of the storm. Traffic is evacuated within 45 minutes; and between 50 and 60 minutes after water diversion to the holding pond, it spills over the bellmouth and into the tunnel. If the heavy rainstorm is prolonged, the tunnel will continue to be used for the flood passage. The highway is reopened within 48 hours after the closure to restart serving as a traffic route in the city of Kuala Lumpur.

Nastaran Pour Ebrahim

See also: Channel Tunnel (France and United Kingdom); Seikan Tunnel (Japan); Tünel (Turkey).

Further Reading

Darby, Arthur. 2007. "A Dual-Purpose Tunnel." *Ingenia Online* 30. http://www.ingenia.org
    .uk/ingenia/articles.aspx?Index=411. Accessed August 26, 2015.
SMART Motorway Tunnel. http://smarttunnel.com.my/. Accessed August 26, 2015.

STREETCARS (UNITED STATES)

A streetcar or trolley car is an electrically powered passenger vehicle that runs on rails. They are distinguished from other urban rail systems by being powered by an overhead electrical wire and by passengers stepping up from a street or sidewalk into the trolley car, as they would on a modern bus. As the name suggests, streetcars usually ran in the street, either in a central median or in a regular travel lane. They run alongside of other street traffic at similar speeds and must stop for stoplights.

Streetcars once were one of the most common forms of urban transport in the United States. Montgomery, Alabama, became the first American city to have an entirely electric streetcar system, and nearby Birmingham eventually possessed the second largest such system, serving to connect the city's far-flung industrial districts. Los Angeles had the biggest streetcar system, which helped create the low-density dispersed pattern of the metropolitan area. Almost every large town and city in America once had at least one streetcar line. These systems comprised hundreds of miles of routes and dozens of lines, running along nearly every main street downtown. After World War II, most streetcar systems were eliminated and replaced by buses. In a few cities streetcars survived and became tourist attractions.

Starting in the 1970s, when construction of new streetcar systems began, the term "light rail" came into use to describe the systems. This term has nothing to do with the weight of the rails but simply is a new term used to describe what otherwise would seem like an old-fashioned transport system. These versions of streetcars might run in streets, central medians, below ground, and elevated above ground level. The Portland MAX system has a long subway section and includes the deepest rail station in North America. They can be operated in short trains, making the distinction between streetcars and rapid transit less clear. The term "modern streetcar" also is becoming common. Unlike light rail, these systems still operate as single vehicles, just as historic streetcar systems did.

Many cities have built new streetcar systems, including Phoenix (with light rail trains) and Tucson (modern streetcar) in Arizona. These current systems often comprise just one or a few routes connecting downtown to a nearby area. These are very different from early streetcar systems, in which a city might have dozens of lines serving almost all the city, much as a bus system might today. They also differ from past systems in that they usually are built by a local government with public money, rather than being privately owned and operated, as in the past. Critics of streetcar and light rail point out that these systems could be designed and

Streetcar on St. Charles Avenue, New Orleans, Louisiana. (Library of Congress)

promoted as much for real estate development or urban renewal purposes as for actual transportation. They might be considered vital to promoting a city's image as prosperous and fun. Additionally, they almost never meet the ridership numbers originally predicted.

Even though few Americans can remember a time when streetcars were common, they remain a part of American culture. Tennessee Williams's play and movie *A Streetcar Named Desire* referred to a streetcar route in New Orleans that operated from 1920 to 1948. Desire Street was the end of the line and so Desire was listed on the front of the car. The Brooklyn Dodgers baseball team got their name from the practice of dodging streetcars in the street.

Joe Weber

See also: Cable Cars; Light Rail Transit Systems; San Diego Trolleys (United States); San Francisco Cable Cars (United States).

Further Reading

Crump, Spencer. 1970. *Ride the Big Red Cars: How Trolleys Helped Build Southern California.* Corona Del Mar, CA: Trans-Anglo Books.

Hudson, Alvin W., and Harold E. Cox. 1976. *Street Railways of Birmingham*. Forty Fort, PA: Harold E. Cox.

STRING TRANSPORTATION SYSTEM

Developed in 1977 by Russian scientist, Anatoly Yunitskiy, string transport system (STS) is an alternative rail infrastructure technology sometimes also known as Yunitskiy String Transport (UST). Unlike conventional steel track, this novel technology utilizes strings built with extremely high-tensioned steel wires and cable that are inserted within concrete columns elevated from the ground. Wheeled vehicles can ride on them with 50 passengers and up to 5 tons of freight capacity, and vehicles travel at up to 220 miles per hour. Because the STS is elevated between 30 and 300 feet above the ground with supporting towers approximately 90 feet apart, it looks more like a pre-stressed concrete bridge. Of all the possible train technologies that exist today, string technology promises to build string rail track at a fraction of the cost of a ground rail system or even a motorway. It could cost about US $50,000 to build 0.62 mile of string rail track; whereas the same length of low-speed surface rail might cost up to US $345 million. Most importantly, it would not require building overpasses, tunnels, or other expensive infrastructure because the system already is elevated. The system claims to be earthquake, hurricane, and terrorist-proof and also capable of running straight on rough terrain and providing the smoothest transport system a person can experience.

Yet, there still is no active railway built using the concept of string transport. A pilot project was planned in 2008 in Khabarovsk, but it was not implemented because the project received negative feedback from railway experts from Moscow State University. Presently, there is a feasibility study for building a passenger rail using STS between Sydney and Kingsford-Smith airport in Australia. This study found that the STS would be able to handle current traffic demand of 12,300 passengers per day; the speed would be much greater than current public transport options, and it would be feasible for highly urbanized areas.

Selima Sultana

See also: Autonomous Vehicles; Flying Cars; Solar Roads; Vacuumed Trains.

Further Reading

Hargraves, A. J. 2014. "A Feasibility Study into the Use of String Transport Systems for Passenger Rail in New South Wales. Presented at the 32nd Conference of Australian Institutes of Transport Research, 17th and 18th February 2014." http://www.sidrasolutions.com/cms_data/contents/sidra/media/articles/caitr2014_hargraves.pdf. Accessed July 24, 2015.

Unitsky String Transport Co. Ltd. 2005. High-Speed String Transportation Route: Abu Dhabi–Dubai–Sharjah." http://www.yunitskiy.com/author/2005/2005_10.pdf. Accessed August 8, 2015.

SUNSET LIMITED, THE (UNITED STATES)

The Sunset Limited is an Amtrak passenger train running 1,995 miles between New Orleans, Louisiana, and Los Angeles, California. The train departs New Orleans Union Passenger Terminal, completed in 1954 and one of the newest union stations in the United States. The train crosses the Mississippi River on the massive Huey Long Bridge. The railroad tracks are more than 130 feet above the river and must rise and descend gradually from this height. The total length of the viaducts and bridge is 4.5 miles, making it one of the longest railroad bridges in the world. Heading west from the Mississippi River, the Sunset Limited serves Lake Charles, Texas; Houston, Texas; San Antonio, Texas; El Paso, Texas; Tucson, Arizona; Palm Springs, California; Los Angeles, California; and many smaller places. Each stop has three westbound and three eastbound departures each week. The train features sleeper cars, coach seating, dining cars, and upper-level observation windows. Riding the full distance takes 48 hours.

The Sunset Limited is notable for several reasons. Although Amtrak has operated the train since 1971, the name first was used in 1894 by the Southern Pacific Railroad, making it the oldest Amtrak train name. Although the name has not changed, its route has. In 1993, the eastern end of the line was moved to Florida, making it the first and only transcontinental passenger train to operate in the United States. Jacksonville, Miami, and Orlando all served as endpoints of the line at various times. In 1996 Phoenix, Arizona, was bypassed because the line west of the city was a remote desert line scheduled for abandonment. After Hurricane Katrina devastated the Mississippi and Louisiana coast in late August 2005 the eastern endpoint was moved back to New Orleans. The topic of once again extending the Sunset Limited to Florida has been continually discussed, but with no plans in place 10 years after Hurricane Katrina. The Sunset Limited also is noteworthy due to several serious accidents. In 1993, the Sunset Limited wrecked at a damaged bridge at Big Bayou Canot north of Mobile, Alabama, killing 47 people and injuring 103. It was Amtrak's worst accident in history. In 1995, the track west of Phoenix, Arizona, was sabotaged by an unknown group and caused the train to derail, killing 1 person and injuring 78. The crime never was solved.

Joe Weber

See also: Amtrak (United States); Coast Starlight, The (United States); Crescent, The (United States); Railroads.

Further Reading

Amtrak. "Sunset Limited." http://www.amtrak.com/sunset-limited-train. Accessed August 27, 2015.

Amtrak. "Sunset Limited Route Guide." 2015. http://www.amtrak.com/ccurl/57/168 /Amtrak-Sunset-Train-Route-Guide.pdf. Accessed December 16, 2015.

SUSPENSION BRIDGES

Suspension bridges are among the most visible and dramatic components of modern transport infrastructure. They are found on roads around the world and their towers are visible for miles and often dominate the skylines of their communities. This type of bridge dates to antiquity, originally made of vines and wood, but it was in the early 19th century that it became a noteworthy bridge design. There are many variations, but the basic idea is a flexible roadway deck hanging from ropes, chains, or cables that run from one end of the bridge to the other. The cables run over the tops of towers—which can be more than 500 feet tall—and are anchored at each end. The tall towers required of suspension bridges often become landmarks. The stone towers and arched openings of the Brooklyn Bridge, the heavy truss design of the George Washington Bridge, and the red Art Deco towers of the Golden Gate Bridge are examples.

The oldest suspension bridge in the United States is the 1849 Delaware Aqueduct (now called the Roebling Bridge), spanning the Delaware River between New York and Pennsylvania. It was designed by John Roebling (1806–1869), who built several important early suspension bridges. Among his designs were the Cincinnati Bridge opened in 1865 over the Ohio River, followed by the Brooklyn Bridge between Manhattan and Brooklyn. In the United States this type of bridge is famous for the huge structures found in large coastal cities: the Brooklyn Bridge, the George

Golden Gate Bridge in the fog in San Francisco, 2013. (Juliengrondin/Dreamstime.com)

Washington Bridge, and Verrazano Narrows Bridge in New York City, and the Golden Gate Bridge and Oakland Bridge in San Francisco, California. Many smaller suspension bridges exist in the Midwest and elsewhere.

Since the 1840s, the longest bridge in the world has been a suspension bridge. The suspension bridge over the Ohio River at Wheeling, West Virginia, was the first bridge in the world to have a span of more than 1,000 feet. Today, this has been surpassed many times. The longest bridge currently is the Akashi Kaikyo Bridge in Japan, which has a 6,532-foot main span. The bridge opened in 1998 as part of a highway between the islands of Honshu and Shikoku. It is likely that longer bridges will be built, but suspension bridges will continue to be the longest. A project that has been discussed for many years is to bridge the Straits of Messina between Italy and the island of Sicily. This would require a main span of more than 10,000 feet.

Joe Weber

See also: Cable-Stayed Bridges; Great Belt Fixed Link (Denmark).

Further Reading

Van Der Zee, John. 1986. *Gate: The True Story of the Design and Construction of the Golden Gate Bridge*. New York: Simon and Schuster.

Whitney, Charles S. 2003. *Bridges of the World: Their Design and Construction*. New York: Dover.

SUV

See Sport Utility Vehicles.

T

TAIPEI METRO (TAIWAN)

The Taipei Metro system is the first and the largest subway system in Taiwan, and serves the capital city of Taipei and its metropolitan area. An initial Taipei MRT network was approved in 1986 by the Central Government of Taiwan and construction started in 1988. The Taipei MRT has been built line-by-line and section-by-section over the years. As new sections and lines were added to the network, names of Taipei MRT lines were changed accordingly to reflect the new line configurations. The first operational line of Taipei MRT, the Muzha Line, began its service in 1996, for example, and became part of the Wenhu Line in 2009.

As of January 2015, there were five main lines and two branch lines in operation. The five main lines are Wenhu Line (Line 1/Brown Line, 15.66 miles and 24 stations), Tamsui-Xinyi Line (Line 2/Red Line, 18.21 miles and 28 stations), Songshan-Xindian Line (Line 3/Green Line, 12.86 miles and 20 stations), Zhongshan-Xinlu Line (Line 4/Orange Line, 18.21 miles and 26 stations), and Bannan Line (Line 5/Blue Line, 15.35 miles and 22 stations) (TRTC 2015). The Xinbeitou branch line—which connects Xinbeitou to the Beitou station on the Tamsui-Xinyi Line—is only 0.75 miles long. Xiaobitan branch line has a length of 1.18 miles and connects Xiaobitan to the Qizhang station on the Songshan-Xindian line.

Taipei MRT operates both medium-capacity and high-capacity trains. The Wenhu line uses medium-capacity trains that are fully automated and driverless trains that have rubber tires. Each train consists of two two-car units that can accommodate approximately 450 passengers. The other four major lines use high-capacity trains that have steel wheels and are controlled by computers. Each high-capacity train, which has an onboard operator who can override the computer-controlled system in case of an emergency, consists of two three-car units that can accommodate roughly 2,000 passengers. Both medium-capacity and high-capacity trains can operate at a maximum speed of 50 miles per hour, but their average operating speed is approximately 22 miles per hour.

Fares for the Taipei MRT are NT$20 for the first three miles and NT$5 for each additional three miles. Riders can buy an RFID IC token or use a smartcard to ride Taipei MRT trains. Discount fares are available for senior citizens, students, physically challenged people, smart card users, and groups of 10 or more people. Normal operation hours of Taipei MRT trains are from 6 a.m. to midnight. The annual passenger volume of Taipei MRT in 1997, the first full year of operation, was slightly more than 31 million passengers. Network expansions over the years have enabled the annual passenger volume to increase to approximately 680 million passengers by 2014. On the New Year's Eve 2014—in the 24-hour period from 6 a.m. December 31, 2014, to 6 a.m. January 1, 2015—the Taipei MRT carried a total of 2.96 million passengers.

Due to strong economic growth in Taiwan since the 1970s and high population concentration in the Taipei metropolitan area, traffic conditions in Taipei were quite bad by the 1980s. Construction of multiple Taipei MRT lines during the late 1980s and the 1990s caused further traffic challenges in Taipei. Additionally, negative publicity of some major incidents of the Muzha Line—such as system malfunctions, train fires, and change of contractor—led to harsh criticisms and questions about the value of Taipei MRT. Fortunately, when a network of MRT lines began to provide better accessibility among different locations in Taipei and the increasing MRT passenger volume started to alleviate the surface traffic congestion, the public began to see many direct and indirect benefits of the Taipei MRT. The Taipei MRT has played a significant role in the redeveloping of old urban areas, land development of the outskirts served by MRT lines, and the shaping of land values in the Taipei metropolitan area. Further, the Taipei MRT has successfully prompted etiquette that encourages orderly, courteous, and safe behaviors when riding the MRT that also has changed rider behaviors in other public transit systems. It is fair to say that the Taipei MRT now is considered to have introduced many positive changes to the Taipei metropolitan area and that people have positive expectations of its future expansions.

Shih-Lung Shaw

See also: Light Rail Transit Systems; Mass Rapid Transit (Singapore).

Further Reading

Department of Rapid Transit Systems (DORTS). Taipei Municipal Government. 2015. http://www.dorts.gov.tw/. Accessed January 10, 2015.

Taipei Rapid Transit Corporation (TRTC). 2015. http://english.trtc.com.tw/. Accessed January 10, 2015.

TOKYO SUBWAY SYSTEM (JAPAN)

The Tokyo subway system is a part of the urban railway network in the Greater Tokyo area and consists of 13 lines of 193 miles and 285 stations operated by Tokyo Metro and Toei Subway. Tokyo Metro, privatized in 2004, operates 121 miles of 9 lines—Ginza, Marunouchi, Hibiya, Tozai, Chiyoda, Yurakucho, Hanzomon, Namboku, and Fukutoshin—with 179 stations. Toei Subway is run by the Tokyo Metropolitan Bureau of Transportation under the Tokyo Metropolitan Government and serves four lines—Asakusa, Mita, Shinjuku, and Ōedo—with 106 stations.

The Tokyo subway network mainly covers the city center. Tokyo consists of 23 central special wards (formerly the city of Tokyo), with prefectures in the west and outlying islands. Tokyo and surrounding cities—such as Yokohama, Kawasaki,

Rush hour at the Shinjuku train station, Tokyo, Japan, 2013. Shinjuku is one of the biggest train stations in Japan. (Siraanamwong/Dreamstime.com)

Urawa, Sagamihara, Saitama, and Chiba—comprise Greater Tokyo, the national capital region of Japan. After postrecovery from World War II, the fast growth of Tokyo as a center of business, commerce, and information in Japan has promoted continuous residential development in suburban areas. As a result, 13.2 million people live in the Tokyo metropolis and 37.9 million in Greater Tokyo, whereas 9 million were concentrated in the 23 wards of Tokyo in 2012. Considerable business activity still concentrates in central Tokyo, however, which creates overcrowded commuting traffic between suburban areas and the central business district (CBD) of Tokyo—up to 37 miles from the center of Tokyo.

To provide for heavy commuting traffic, Tokyo's government focused on expanding the railway network as well as the road network and the capacity of five major national railways connecting suburban areas to central Tokyo was improved. At the same time, private railway lines played a major role in serving commuters of Greater Tokyo. Private railway lines were not allowed to enter the CBD, however, hence most of the private lines terminate at the stations of the Yamanote Line, an important circular line connecting the urban centers of Tokyo. To cover heavy commuting flow to the CBD, Tokyo's government has been expanding the subway network to connect with private railways and East Japan Railway Company (JR East) with through service for improving the efficiency of passenger travel. Through service means running trains between different railway operators, so suburban residents can access Tokyo's CBD without changing trains. The Asakusa Line provided the first through service between Tokyo's subway network and private railways and JR East in 1960. Since then, the 9 lines of Tokyo's subway system provide through service with the 32 lines of private railways and JR East.

The dense network of Tokyo subway lines provides high accessibility to important urban centers. Most railway corporations cover their network maintenance depending on revenue; thus, there is a different fare system between railway operators, and passengers have to pay additional charges when they change railway operators even though they use the same train.

The Tokyo subway system plays a central role in traveling within Tokyo, attracting passengers by convenient through service, with trains arriving and departing frequently, and accessible stations. The Tokyo Metro and Toei Subway carried 9.32 million daily passengers in 2013, which is 24 percent of the 38.6 million daily rail passengers of Greater Tokyo. As a result, the level of dependence on the railway within the 23 wards of Tokyo clearly is high. The significant role of the Tokyo subway system is shown by the fact that 74 percent of people in the 23 wards choose railways for commuting, as do 52 percent of people in Greater Tokyo.

Hyojin Kim

See also: Light Rail Transit Systems; Shinkansen (Japan).

Further Reading

Hirooka, Haruya. 2000. "The Development of Tokyo's Rail Network." *Japan Railway and Transport Review* 23: 22–30.

Ito, Makoto. 2014. "Through Service between Railway Operations in Greater Tokyo." *Japan Railway and Transport Review* 63: 14–21.

Kurosaki, Fumio. 2014. "Through-Train Services: A Comparison between Japan and Europe." *Japan Railway and Transport Review* 63: 22–25.

Nakamura, Hideo. 1995. "Transportation Problems in Tokyo." *Japan Railway and Transport Review* 4: 2–7.

Train-Media.net. 2013. "Tokyo Subway Ridership." http://www.train-media.net /report/1410/index.html. Accessed June 5, 2015.

Wakuda, Yasuo. 1997. "Improvement of Urban Railways." *Japan Railway and Transport Review* 11: 46–49.

TOLL ROADS

See Turnpikes (United States).

TRANS-AFRICAN HIGHWAY

The Trans-African Highway network (TAH) is a collection of nine separate highways, totaling 35,221 miles in length, which interconnects each corner of the continent. Although the endpoints are in large cities, the routes are constructed in such a way that smaller, developing countries also can access and benefit from the

TAH. Due to the lack of a major highway network, interior African countries have long endured steep transportation and trade costs to port cities and to other trade areas within the continent. The Trans-African Highway system consists of nine intercontinental routes: Highway 1, Cairo-Dakar; Highway 2, Algiers-Lagos; Highway 3, Tripoli-Windhoek; Highway 4, Cairo-Gaborone; Highway 5, Dakar-N'Djamena; Highway 6, N'Djamena-Djibouti; Highway 7, Dakar-Lagos; Highway 8, Lagos-Mombasa; and Highway 9, Beira-Lobito. The longest proposed route is Highway 3 at 6,715 miles, and the shortest is Highway 9 at 2,189 miles. Many of these connections are incomplete, however, and lack significant links.

Initial ideas and planning for the TAH began in the early 1970s with funding by the African Development bank, the United Nations Economic Commission for Africa (UNECA), and the New Partnership for Africa's Development fund (NEPAD). The main purpose for this large development project is to help African countries grow economically and reduce poverty levels. According to a 2006 World Bank report, "isolation from regional and international markets" was found to be contributing "significantly to poverty in many sub-Saharan African countries." Once the roads are completed, "the continental network would yield an extra $250bn in overland trade within fifteen years, with substantial benefits for the rural poor," according to the same report. An example on the United Nation's website cites the owner of a shipping company who stated that before the new Trans-African Highway, it used to take his drivers all day to make the 236-mile journey from Ghana to Nigeria. Roads were narrow, hard to navigate, filled with potholes, and required continual stopping at the many checkpoints along the way. With the improvements of the new road, it currently takes the drivers only eight hours to complete the same trip. By helping decrease travel time and fuel costs, the TAH has greatly improved and increased trade throughout Africa.

Many are skeptical of its completion, however, as the Trans-African Highway must overcome significant obstacles before and after being constructed. Some of the impediments that engineers and builders face with the development of the TAH include varying climate and geography where flooding is common or vegetation is extremely thick (such as the Congo), local warfare that stalls new progress and damages completed sections of road, and inadequate building materials that do not hold up well under Africa's intense environmental conditions. It is important to keep in mind that although these roads might be labeled highways, they are much different than what would constitute a highway in the United States. Although many portions of the TAH are paved, other sections are covered with gravel and some are simply dirt (track) roads.

For many of the interior countries of Africa, the TAH does not currently hold a high enough priority to secure funding, which has left many broken links throughout the different routes. Although these countries recognize the benefits of having an intercontinental highway to connect them with other areas, the primary focus continues to remain on building and maintaining roads to the larger port cities where trade is most extensive. Because many of these countries do not have adequate funding for infrastructure purposes, development of the Trans-African Highway in these areas has been placed on hold until other "more important" projects are completed.

Selima Sultana

See also: Alaska Highway (Canada and United States); National Highways Development Project (India); National Trunk Highway System (China).

Further Reading

African Development Bank. 2015. "Review of the Implementation Status of the Trans African Highways and the Missing Links." http://www.afdb.org/fileadmin/uploads/afdb /Documents/Project-and-Operations/00473227-EN-TAH-FINAL-VOL2.PDF. Accessed June 4, 2015.

Whitehead, Eleanor. 2015. "Trans-African Highway Remains a Road to Nowhere." *How We Made It in Africa. DHL.* http://www.howwemadeitinafrica.com/trans-african-highway -remains-a-road-to-nowhere/39863/2/. Accessed June 4, 2015.

World Bank Development Report. 2006. "Equity and Development." http://www.world bank.org/servlet/WDSContentServer/WDSP/IB/2005/09/20/000112742_200509201 10826/Rendered/PDF/322040World0Development0Report02006.pdf. Accessed June 4, 2015.

TRANS-AMAZON HIGHWAY (BRAZIL)

The Trans-Amazon Highway, locally known as *Rodovia TransAmazônica* (BR-230), is a Brazilian interstate road that was inaugurated on September 27, 1972, and extends more than 2,500 miles across the northern part of the country. The goal of the highway is to link the Atlantic Coast through the isolated villages of the Amazon Rainforest to the border with Colombia and Peru, in the hope of improving the living conditions of the local villages and population, and to encourage commerce and development to the isolated Brazil's hinterland.

The highway begins in the coastal city of João Pessoa in state of Paraíba. The city is known for its beautiful white sand beaches and historical architecture. Further to the west the highway crosses the states of Ceará, Piauí, Maranhão, and Tocantins, Pará, and finally ends in the village of Lábrea in the heart of the Amazon Rainforest in the state of Amazonas. The highway provides easy access from densely populated Atlantic coast cities (e.g., Maceió, Recife, Natal, João Pessoa), to less densely populated western interior parts of Brazil. The highway is a mixture of both paved and unpaved segments. Unpaved areas are located in low-population density areas of the states of Amazonas and Pará, where the primary economic activities are cattle ranching, logging, and soybean agriculture. In contrast, paved areas are located in more densely populated states where the people rely heavily on the highway for commerce and transportation of goods.

The construction of the highway began during the presidency of Emílio Garrastazu Médici in 1970. Médici, who was the army general for Brazil prior to his

Trans-Amazon Highway, crossing in the middle of the jungle. (Alvaro Pantoja/Dreamstime .com)

presidency, in early 1970 visited the state of Pernambuco—an area known for its consistent droughts—and was disturbed by the extreme poverty and terrible living conditions of the local dwellers. Médici proposed a plan to connect the impoverished regions of the interior to the more affluent coastal cities. Through a series of reforms he increased access by the development of a road system into the interior part of the country and encouraged people to relocate and settle in the interior areas to farm the lands. The Brazilian government provided financial incentives for farmers to settle along the highway, but only was able to influence people that could acclimate easily to extreme heat and seasonally wet conditions. The financial incentives included a loan program that many of the citizens accessed, creating significant amounts of personal debt that they have not been able to pay off due to the lack of sustainable agriculture. As a result, the country experienced a devaluation of its currency ("cruzeiro") and a rapid rise in inflation of the costs of products and services. The resulting economic disruption over the next two decades was the basis for the fragmented condition of the highway, and the lengthy delay in its construction.

The Trans-Amazon Highway crosses a variety of tropical climates that include Koppen's classifications: Tropical rainforest (*Af*) near the town of Lábrea, tropical monsoon (*Am*), tropical with dry winter (*Aw*), tropical with dry summer (*As*), and the semi-arid hot and dry climate (*BSh*) along the north region of the Brazilian Highlands. In wetter parts of the country the unpaved highway becomes unusable as heavy rains inundate the roads causing muddy and wet conditions, and the roads close for extended periods. This affects the local agricultural economy greatly during rainy season, as produce does not reach its destination on time.

Today, accessibility to isolated villages in the so-called frontier forest of the Amazon is raising many questions for conservation groups regarding the environmental impact of this project on the Amazon's ecosystem. Slash and burn is common along sections of the highway, turning forests into soybean plantations and pasture fields for cattle ranching. This is the main problem of land degradation in the Amazon. An observation of an aerial map from Google Earth along the stretch between the city of Marabá and Itaituba (more than 600 miles) in the state of Pará reveals the degree of deforestation and the direct relationship that the highway has on the Amazonian rainforest.

Land cover changes as well as land degradation will continue to occur as more people migrate into the area of the Amazon wilderness. This could lead to the disappearance of the vast forest. A paved road connecting the borders will increase development as more trade between countries will stimulate the booming Brazilian economy and establish the groundwork for modernization for the region. More importantly the improvement of the Brazilian infrastructure and corresponding technology could lead to strategies for recovering the forest and reversing the ill-effects of current degradation.

Alejandro Molina

See also: Alaska Highway (Canada and United States); National Highways Development Project (India); National Trunk Highway System (China).

Further Reading

Ahrens, C. Donald. 2013. "Global Climate." In *Meteorology Today: An Introduction to Weather, Climate, and the Environment*, 10th ed. Belmont, CA: Brooks/Cole.

Cooper, James M., and Christine Hunefeldt. 2013. *Amazonia: National Politics and Interests, Laws, and the Environment*, 98–101. Brighton: Sussex Academic Press.

Dangerousroads: The World's Most Spectacular Roads. "Rodovia Transamazônica (BR-230)." http://www.dangerousroads.org/south-america/brazil/431-rodovia-transamazonica -brasil.html. Accessed December 16, 2015.

Google Earth. Northern Brazil. www.google.com/earth. Accessed December 6, 2014.

Soares-Filho, Britaldo, Ane Alencar, Daniel Nepstad, Gustavo Cerqueira, Maria del Carmen Vera Diaz, Sérgio Rivero, and Eliane Voll. 2004. "Simulating the Response of Land-Cover Changes to Road Paving and Governance Along a Major Amazon Highway: The Santarem-Cuiaba Corridor." *Global Change Biology* 10(5): 745–764.

TRANSMILENIO, BOGOTÁ (COLOMBIA)

The TransMilenio of Bogotá, literally meaning "transcending the millennium," is the highest-capacity bus-rapid transit (BRT) system in the world, which has been serving the city of Bogotá, Colombia since 2000. Bus-rapid transit essentially up-grades traditional bus service by adding features befitting many light rail systems, but with reduced costs. This is accomplished by allowing the bus to operate for a majority of time on dedicated bus lanes that usually are separated from traffic by curbs or guiderails. Increased efficiency and speed are also improved by including stations with electronic fare collection, platforms that align with the bus floor, and bus priority at intersections.

The Brazilian city of Curitiba was where BRT first was established and although the pioneering system provided the source of inspiration for its Colombian coun-terpart, it now is Bogotá that boasts the greatest carrying capacity at 1.98 million passengers per day. The system covers more than 50 miles with nearly 1,300 buses traveling along 9 corridors. The highest peak loads on a single corridor are seen on Avenida Caracas, which carries 45,000 passengers per hour in each direction. Buses travel at an average speed of 10 miles per hour but increase to 18 miles per hour along some corridors. Traveling along neighborhood streets is a network of much smaller feeder buses that transport people to the central stations. TransMile-nio also is heavily supported by bicycle infrastructure. There are three terminals and one station with bicycle parking facilities.

TransMilenio was an integral part of a wider program to improve the quality of life in Bogotá, a city that was afflicted by violence and fear stemming from the narco-culture of the 1990s. The initiative began with the 1995 election of Mayor Antanas Mockus, an academic who had no previous political experience, but who none-theless became known for his eccentric endeavors at improving civic life, one of which included hiring a team of 420 mimes to publicly shame traffic violators. Dur-ing his *Formar Ciudad* (Educate the City) administration, traffic fatalities dropped by more than 50 percent and the homicide rate fell by 70 percent. Mayor Mockus educated the public through avenues such as the environment, public space, urban productivity, and a "culture of citizenship," which he defined as "the sum of habits, behaviors, actions and minimum common rules that generate a sense of belonging, facilitate harmony among citizens, and lead to respect for shared property and heritage and the recognition of citizens' rights and duties." This heritage of citizenry changed the attitudes of many *Bogotanos* and laid the foundation for the subsequent mayor, Enrique Penalosa, who would be responsible for transforming the shape of public space with the implementation of TransMilenio.

When Enrique Penalosa became mayor in 1997, he was responsible for a num-ber of urban renewal projects that essentially involved reclaiming public space for the citizens of Bogotá. This was accomplished by improving previously ill-maintained parks, raising sidewalks and installing bollards in areas where motorists previously (selfishly) parked, and most importantly beginning the construction of TransMilenio, the brainchild of his election. As a way to mitigate traffic congestion he also instated the controversial policy of *pico y placa*. This system helps to regulate traffic during rush hours by restricting vehicles with license plates that end in cer-tain digits from traveling on certain days. This strategy has since been embraced by other Colombian cities. Equally important was the construction of the city's *cicloruta* (bike route), which provided more incentive for leaving automobiles at home. Under Penalosa's leadership, the integration of bicycle infrastructure, mass transit, and pedestrian-friendly spaces led to the recognition of Bogotá as the first winner of the Sustainable Transportation Award in 2005.

The impacts of TransMilenio might be best appreciated when compared with the antiquated bus system of the past, which was a confusing array of independent, uncoordinated minibuses numbering in the thousands. Drivers were paid on a per-passenger basis, which led to competitive—and usually unsafe—practices. The *guerra de centavos* (penny wars, as they had come to be known) were fought with a penchant for dangerous driving done to maximize profits. Drivers also toiled under long shifts—up to 15 hours—which not only exacerbated the unsafe conditions, but also caused the maintenance of the vehicles to decline. Yet even more absurd was the fact that although 240 routes reached the city center, they were scarcely served throughout the poorer districts along the periphery

In 1998, an 18-mile trip by bus took 2 hours and 15 minutes. Today, that same trip takes only 55 minutes when using TransMilenio. Further, with the new regulation of fuel quality, the TransMilenio will eliminate nearly 1 million tons of carbon dioxide per year in Bogotá.

Other Colombian cities have requested TransMilenio buses to alleviate their own transit issues. The founders of TransMilenio have taken their expertise across the globe to educate other developing cities on proper implementation of BRT. TransMilenio has become a world-renowned model for similar systems in Guangzhou, China, and Istanbul, Turkey, among others.

Despite these positives, the current state of TransMilenio is in continued disinvestment. The public's impression of the system began to degrade in 2004, when passengers protested poor service. Stirs of discontent continued throughout the years until violence erupted in 2012, when riots led to the destruction of five bus stations. Penalosa's policies attempted to alleviate traffic congestion; ironically, however, commuters now cram into buses during rush hour, often leaving no space for others who are entering and exiting. The crowding also has created a lucrative supply of victims for pickpockets. The high fare of US $1 is another concern for low-income riders, whose daily income is only US $3.

Bogotá is one of the few major South American cities without a metro-rail system. Yet this could cease to be true in the near future. When Enrique Penalosa stepped down as mayor in 2001, he also allowed the future of TransMilenio to be decided by others. Subsequent mayors have not supported or invested in the system nearly enough. Politicians instead have focused their efforts on an underground subway system, projected to be in operation by 2021. In the past 25 years, the city of Bogotá has made astonishing progress in increasing the mobility of its citizens. A subway might be the next destination in an ever-expanding metropolis, but it was the humble bus, the TransMilenio, that transformed the mobility of a city.

Selima Sultana

See also: Buses; Mexico City Metro (Mexico); Santiago Metro (Chile).

Further Reading

Cain, Alasdair, Georges Darido, Michael R. Baltes, Pilar Rodriguez, and John C. Barrios. 2007. "Applicability of Bogota's TransMilenio BRT System to the United States." *Transportation Research Record: Journal of the Transportation Research Board* 2034: 45–54.

Jaffe, Eric. 2012 (March 20). "Why Are People Rioting Over Bogota's Public Transit System?" *CityLab.* http://www.citylab.com/commute/2012/03/why-are-people-rioting-over-bogotas-public-transit-system/1537/. Accessed July 8, 2015.

TRANS-SIBERIAN RAILWAY (RUSSIA)

The Trans-Siberian Railway (*Transsibirskaya Zheleznodorozhnaya Magistral*) links the Russian capital of Moscow in Western Russia to the Pacific port of Vladivostok in the Russian Far East. At 5,772 miles, it is the longest railway in the world and connects with other major rail systems in Russia itself, Western Europe, and Central and East Asia. Upon its completion in 1916 it was acclaimed one of the greatest engineering accomplishments of its time.

Russia began its expansion eastward into Siberia in the 16th century, and by the mid-17th century had extended its control to the Pacific Ocean. Siberia is rich in natural resources, but its immense size and extreme climatic conditions made transportation very difficult. Those who wished to travel through it by coach had one primitive route available: the Great Siberian Post Road. It was obvious that better means of crossing were needed if Russia were to tap the region's resources and compete with other world powers. Earlier plans for a railway had been discarded, but proposals finally were approved by Tsar Alexander III. Progress stalled, however, after construction began at Vladivostok in 1891. At the suggestion of Count Sergei Witte, who became minister of finance in 1892, Alexander appointed a Committee of the Siberian Railway to oversee the project. Although the czar's own son, Tsarevich Nicholas, headed the committee, it was Witte who controlled it.

Rail lines already existed between Moscow and the city of Chelyabinsk in westernmost Russian Asia. The remainder of the route was conceived and built in six sections—the Western Siberian, the Mid-Siberian, the Circum-Baikal, the Transbaikal, the Amur, and the Ussuri. Four of these had been completed and one reconfigured by 1902, but the Circum-Baikal section—which involved blasting a difficult route along high cliffs around the southern shore of Lake Baikal—was not finished until 1905. That year marked the "first" completion of the Trans-Siberian Railway, as the Russians had abandoned their plans for the Amur section and, with Chinese permission, followed a shorter route to Vladivostok across the Chinese territory of Manchuria. After its defeat in the Russo-Japanese War of 1904–1905, however, Russia feared Japanese designs on Manchuria and completed the Amur section entirely within Russian territory in 1916.

Journey through steppes of Mongolia on the Trans-Siberian Railway, 2012. (Photonphotos /Dreamstime.com)

The construction of the railway was carried out under almost inconceivably difficult conditions. In the coldest parts of Siberia it only was possible to work four months a year. Swaths of dense taiga forest had to be cut and hundreds of bridges built. Much of the route lay across permafrost, which turned to mud when it thawed. There was very little stone or quality timber available locally (the coniferous trees of the taiga being generally unsuitable), so such materials had to be transported great distances to the construction site, usually by river.

Many of the project's other problems involved personnel. Surveyors and contractors were notoriously corrupt, and the workers themselves generally were unskilled. There were seldom enough local men available, so foreign workers were hired, soldiers assigned, and convicts recruited with offers of reduced sentences. The tools and equipment available were primitive, and almost all tasks were performed with the use of horses or by hand. Anthrax and mosquito-borne malaria were common.

Russia adopted its current system of time zones after the completion of the Trans-Siberian Railway, and in the early 21st century the route crosses seven zones and connects nearly 90 cities, including Omsk and Irkutsk. The gauge of the railway's track is nearly 5 feet—several inches wider than the standard gauge more familiar to Westerners. The most important passenger train on the route, the *Rossiya*, runs from Moscow to Vladivostok and vice versa on alternate days. It offers three classes of accommodation, the most comfortable being two-berth sleeping compartments. Trains with shorter runs typically are older and less comfortable.

Passage between Moscow and Vladivostok takes at least six days, and although such lengthy trips are popular with foreign tourists, Russians themselves tend to use the line for shorter trips. The railway's primary function is the transportation of freight, however; much of it in the form of cargo containers. In 2013, Russia appropriated $17 billion to improve and expand service along the route with a goal of moving freight at the same speed as passengers are transported. Such an accomplishment would give overland transport of freight between Western Europe and China an even greater economic advantage over seaborne transport than it enjoys already, and would enhance Russia's importance as a link between East and West.

Grove Koger

See also: Alaska Railroad (United States); Railroads; Trans-African Highway.

Further Reading

Haywood, Anthony. 2012. *Trans-Siberian Railway*, 4th ed. Footscray, Australia: Lonely Planet.

Manley, Deborah. 2009. *The Trans-Siberian Railway: A Traveller's Anthology*. Oxford: Signal Books.

Wolmar, Christian. 2013. *To the Edge of the World: The Story of the Trans-Siberian Railway.* New York: Atlantic Books.

TRIMET METROPOLITAN AREA EXPRESS (MAX), PORTLAND (UNITED STATES)

The TriMet Metropolitan Area Express (MAX) is the light rail system of transportation in the city of Portland, Oregon. Operated by TriMet, the Tri-County Metropolitan Transportation District of Oregon, which provides bus, light rail, and commuter rail transit services throughout the city and surrounding communities, the MAX Light Rail line began service in 1986 and consists of four separate lines serving 87 stations situated along 52 miles of track. It is unique because in the 1970s—a time when most major metropolitan areas were using federal funds to build freeways—Portland chose to invest its funds in a public light rail system instead.

In 1943, the city of Portland hired Robert Moses, the mastermind behind New York City's massive grid of roads and bridges, to help plan the Portland's future roads, just in time for the 22 percent increase in population experienced by Portland immediately following World War II. Moses laid out a blueprint of freeways running through the city and to outlying neighborhoods. A result of Moses's plan was the Interstate 405 link to Interstate 5, the West's major north-south corridor running from Southern California into Canada. To add to Moses's ideas and as a way to allow commuters direct access to downtown, local city planners in the 1960s proposed freeways running throughout the city, the largest of the roads being the Mount Hood Freeway, an eight-lane thoroughfare passing through Southeast Portland.

The State of Oregon prepared for the proposed freeway by buying up property in the Southeast neighborhoods using federal money that already was approved for the project. More than 1,700 homes would have to be razed to make way for the freeway. Grassroots opposition was strong, however, as residents looked at the freeway not as a link to downtown but as a way to divide the city even further. Still, the project seemed destined since both the federal and state governments were on board.

Neil Goldschmidt, a legal aid lawyer, was opposed to the Mount Hood Freeway and in 1970—with public sentiment on his side—he ran for a seat on the City Council and won. Two years later he was elected mayor while using anti–freeway expansion as the centerpiece of his campaign platform. A 1973 environmental impact study supported the anti-freeway movement by declaring that the road would send more traffic downtown than surface streets could support and that the need for the freeway would be obsolete nearly as soon as it was completed. With $500 million of federal funds about to be lost if plans for the freeway fizzled, Goldschmidt chose to push public transit as a way to use those funds for the public good. He proposed a bus system along dedicated roadways. His staff, however, convinced him that a light rail system running from downtown southeast to Gresham was a better use of funds.

With opposing voices emanating from the city's major newspaper, *The Oregonian*, and others who thought Portlanders—like most Americans—would rather commute by automobile than ride on mass transit, the mayor was met with stiff opposition. On February 4, 1974, however, U.S. district judge James M. Burns ruled that the freeway route was decided before the state went through the proper channels. Burns's decision delayed the freeway plan and opened up the possibility that it would be scrapped—especially after the Multnomah County Commission also adopted a resolution in favor of stopping the freeway. Finally, in July 1974, the city council voted to kill the freeway plan.

Stopping the Mount Hood Freeway construction reflected Portland's history of opposition to urban sprawl and its belief in linking land use with transportation options. In 1976, the city replaced a four-lane freeway along the Willamette River to build Tom McCall Waterfront Park; and in December 1977, the downtown Transit Mall was opened, with one-way streets intended solely for use by mass transit. The Transit Mall was among the first of its kind in the United States and led to downtown redevelopment. Another incident occurred in 1981, when the city built Pioneer Courthouse Square, affectionately referred to as "Portland's living room," in place of a proposed 10-story parking garage.

Before building of the MAX light rail system began, voters approved the development of Metro, a regional government board responsible for overseeing the region's future transit plans. Metro adopted the urban growth boundary, or UGB, to manage regional land use and development. In 1980, the federal government approved construction of the first of four MAX lines, the Banfield Light Rail Project on Portland's east side. It would be built with funds from the Urban Mass Transportation Act of 1964. Although the first of the second-generation, or modern, light rail systems had just been built in 1981 in San Diego, some Portland officials questioned its fit in their city. But Metro chose to focus growth around light rail.

Opening on September 5, 1986, the Banfield Light Rail line (later referred to as the Eastside MAX line when a line on the west side of Portland was proposed) consisted of 15 miles of track between the suburb of Gresham and Portland City Center. The Westside line, traveling from downtown to Beaverton and Hillsboro, began construction in 1993 and was complete in 1998. With this addition, the MAX totaled 33 miles from end to end.

The entire MAX system was built as five separate projects, the Eastside, Westside, Airport, Interstate, and Green Line, with each segment connecting to the others. The use of color names for the individual lines was adopted in 2000, with the Eastside and Westside lines becoming the Blue Line. The Red Line, connecting the city of Beaverton to the west of Portland with the Airport and City Center, was opened in September 2001; followed by the Yellow Line connecting North and Northeast Portland and City Center in May 2004. The newest link, the Green Line, spans to Southeast Portland and Clackamas County. It was opened in September 2009. Construction on a fifth line, the 7.3-mile Orange Line, from Portland to the cities of Milwaukie and Oak Grove in Clackamas County, was proposed in 2008 and opened in September 2015.

All MAX projects were completed on time or ahead of schedule and on or under budget, a feat practically unheard of for a major transit system of its size. An estimated $6 billion in development has occurred along the MAX lines since its inception in 1978.

Rosemarie Boucher Leenerts

See also: Light Rail Transit Systems; Schwebebahn-Wuppertal (Germany).

Further Reading

Mesh, Aaron. 2014 (November 5). "Feb. 4, 1974: Portland Kills the Mount Hood Freeway. . . ." *Willamette Week*. http://www.wweek.com/portland/article-23466-feb-4 -1974-portland-kills-the-mount-hood-freeway.html. Accessed November 5, 2014.

Portland-Milwaukie Light Rail Transit Project. "Route and Stations." http://www.trimet.org /pm/routeandstations/index.htm. Accessed August 26, 2015.

TÜNEL (TURKEY)

The two tunnels (*tünel* in Turkish) covered in this entry are both located in İstanbul, the largest city and the commercial heart of the modern Republic of Turkey. Uniquely situated, İstanbul straddles Europe and Asia, separated by the natural strait of Bosphorus that connects the Black Sea to the Sea of Marmara. Nearly

Tünel underground railway car waiting for its passengers, 2015. Tünel is a historic underground railway line with two stations, connecting the quarters of Karakoy and Beyoglu in Istanbul, Turkey. (Özgör Gövenç/Dreamstime.com)

14.2 million people call İstanbul home (2013), making it the largest city in Europe, followed by Moscow (11.5 million) and London (8.4 million). About two-thirds of all İstanbulites live on the European side. Due to its prized location, İstanbul always has been a desirable settlement, but it has grown exponentially in recent decades. Since 1970, İstanbul gained more than 11 million new residents, many migrating from rural communities in search of economic opportunity.

Crossing the Bosphorus has been one of İstanbul's main transportation challenges from its earliest days, and one of the reasons life in the city historically has been concentrated on the European side. For centuries, people and goods crossed the Bosphorus via water. As population and economic activity increased after World War II, and motorized road transport became widespread, developing fixed-link alternatives to ferry crossing of the Bosphorus emerged as a necessity. Such a link had implications far beyond serving İstanbul: it would seamlessly connect Europe with Anatolia (which constitutes the majority of Turkey's land mass), the Middle East, and the rest of Asia.

Dreams of either a bridge or tunnel connection between the two continents were realized only recently. The first suspension bridge (the Boğaziçi Bridge) entered into service in 1973. During the 1980s, rising population and rapid expansion of the urban periphery necessitated construction of another suspension bridge, three miles north of the first. By the time the second bridge (the Fatih

Sultan Mehmet Bridge) opened in 1988, vehicle crossings on the first bridge already had quadrupled—far exceeding initial expectations. In recent years, hours-long traffic jams have made driving over these bridges—now the primary means of getting from one side of the city to the other—an ordeal. In essence, the two bridges might have accelerated urbanization and encouraged vehicle dependence.

In 1999, İstanbul suffered a catastrophic earthquake, a jarring reminder of the seismic instability of the city. The demands of an ever growing metropolis and the realities of it being situated near an active fault, have since shaped the transportation infrastructure decisions. This entry reviews two of these: the Marmaray railway tunnel, and the Eurasia Tunnel highway.

The Marmaray project is an 8.5-mile-long railway tunnel that allows crossing of the Bosphorus via rail transit. The subsea portion consists of a 87-mile-long "immersed tube" tunnel running 180 feet below sea level that is the deepest of its kind in the world. It was built in sections, submerged into place below the seabed and then assembled. About 6 miles of the remaining tunnel were dug using tunnel-boring machines, and the 1.5 miles closest to the surface was built using the cut-and-cover method. The subsea portion has been designed to withstand a 7.5-magnitude earthquake. The tunnel walls have two layers, each independently watertight: a steel envelope with a concrete layer inside. Engineers believe that the long and narrow tunnel will flex "like a straw in gel" during an earthquake. Should the walls fail, floodgates at the end of each section would close to isolate the flooding.

Construction of Marmaray started in 2004, but work was suspended after an 8,000-year-old archeological site was discovered at the European-side terminus. After four years of delay, Marmaray's first phase, which links up with the city's relatively young rail transit system (light rail and subway), opened in 2013. It can carry up to 75,000 passengers per hour on two bidirectional tracks. The second phase of the project includes upgrading 39 miles of commuter railway. Once it is done freight trains also will be able to utilize the tunnel during off-peak periods. Ultimately, Marmaray could be part of a planned transcontinental railway extending from London to Beijing.

The Eurasia Tunnel is a highway tunnel that is 8 miles long. It currently is under construction and is expected to open in 2016. The subsea section will be a 3.4-mile-long, two-deck tunnel, buried in the seabed at the south entrance of the Bosphorus. Custom tunnel-boring machines are used to dig the subsea portion. The project will provide an alternative to the bridges for vehicle crossings between the European and Asian sides. It also will offer a more direct road link between İstanbul's two airports. Like the Marmaray project, the Eurasia Tunnel is earthquake-resistant up to a magnitude of 7.5.

Although these tunnels will bring relief to those traversing the Bosphorus to and from the southern and central districts of İstanbul, residents of the northern suburbs will continue to rely on the bridges. Inevitably, construction of a third, northernmost, suspension bridge is under way. This will be the longest and widest of all three bridges, with both road and rail components. It will link to a new highway that circumvents the city from the north, aimed at diverting heavy truck traffic away from the city.

Nazli Uludere Aragon

See also: Channel Tunnel (France and United Kingdom); Railroads.

Further Reading

Avrasya Tüneli (Eurasia Tunnel) Project website. 2015. http://www.avrasyatuneli.com.tr /en. Accessed August 27, 2015.
Harris, William. 2015. "How Tunnels Work." HowStuffWorks.com. http://science.howstuff works.com/engineering/structural/tunnel.htm. Accessed August 27, 2015.
Marmaray Railway website (in Turkish). 2015. http://www.marmaray.gov.tr. Accessed August 27, 2015.
Railyway-Technology.com. 2015. "Marmaray Railway Engineering Project, Turkey." http:// www.railway-technology.com/projects/marmaray/. Accessed August 27, 2015.
RoadTraffic-Technology.com. 2015. "Eurasia Tunnel Project, Istanbul, Turkey." http://www .roadtraffic-technology.com/projects/eurasia-tunnel-project-istanbul/. Accessed August 27, 2015.
World Heritage Encyclopedia. 2015. "Marmaray Project." http://www.worldheritage.org /articles/Marmaray_project#cite_note-19. Accessed August 27, 2015.

TURNPIKES (UNITED STATES)

Turnpikes or "toll roads" are those that directly charge drivers to use the road. In the United States all drivers pay for roads through a tax on gasoline, but this usually is not counted, nor is the frequent need to pay for parking. Toll roads exist around the world and have been used since antiquity. The term "turnpike" refers to an early practice of placing a physical barrier—or pike—across a road. Many toll roads are freeways, and some even are part of the U.S. Interstate Highway System.

The Turnpike Boom of 1800–1845 saw a proliferation in the number of privately owned toll roads in the northeastern and midwestern United States, with as many as 3,000 different companies in operation. As many as 50,000 miles of turnpikes could have been built during the boom. Though rarely profitable they were important in providing improved roads at a time when settlement was rapidly moving away from coasts and waterways. Many of these roads were built with wooden-plank surfaces, an early form of paving. The turnpike boom faded in the late 19th century when railroads emerged as the principal long-distance mode of transport. Changing sentiments caused state governments to take action to eliminate toll roads in the early 20th century.

Turnpikes became popular again in the years after World War II, when demand for new highways was growing rapidly but little money was available for construction. Building a turnpike enabled those using the road to pay for construction costs. These toll roads usually were freeways and were built by state turnpike commissions—state agencies separate from state highway departments. The first major postwar turnpike built was the New Jersey Turnpike, which opened in 1952. The Ohio and Indiana turnpikes soon followed, as did others in the Northeast.

The creation of the Interstate Highway System in 1956 ended the popularity of toll roads in most states, because the Highway Trust Fund provided plenty of money. Turnpikes continued to be built in Oklahoma and Florida. In Oklahoma they provide a set of freeways connecting to the main Interstate routes. In Florida turnpikes often provide for urban freeways, such as that in Orlando. Although tolls were being charged to pay construction costs, the tolls rarely were ended. The Boulder-Denver Turnpike in Colorado was one of the few roads for which tolls were lifted. In most cases, tolls remain in place to finance maintenance or the construction of new toll roads.

In the 1990s, toll roads again became popular with highway builders due to the lack of sufficient funds available to build roads to keep up with traffic growth. Charging users once again seemed the best way to pay for the roads. Instead of building a new highway, tolled lanes can be added to a freeway. These sometimes are called "Lexus lanes" because they are popularly associated with wealthy drivers. New toll roads and lanes have been built in the Los Angeles, California, area—which also is the birthplace of the "freeway" or untolled roads.

Not all toll roads are part of a freeway; tolls also may be collected on a bridge or ferry crossing. The Federal Highway Administration provides statistics on the number of toll roads, bridges, tunnels, and ferries in the United States. In the United States, as of 2013 there were 142 toll roads in 27 states, with Texas and Florida having the most, and 153 toll bridges or tunnels, with Texas and New York having the most. There also are 120 toll ferries in 28 states and territories, providing service across rivers, lakes, and in coastal areas.

The price paid to drive on a toll highway or bridge varies tremendously. Some tolls might be less than a dollar. The most expensive tolls in the United States are for the Verrazano Narrows Bridge in New York City and the Chesapeake Bay Bridge/Tunnel, both of which cost $15 one way ($30 round trip). Tolls charged in other parts of the world could be even greater. Some Japanese toll bridges charge the equivalent of $20 or more each way. Rates for passenger cars almost always are less than rates for trucks. Toll lanes on freeways often make use of variable pricing, in which the cost varies by time of day or traffic levels. The heavier the traffic, the greater the cost to bypass it by using a toll lane.

To make payment easier and reduce congestion at toll booths, the use of electronic payment systems has become common. Vehicles equipped with electronic transponders do not stop at toll booths and are charged automatically. The E-ZPass system used in 14 states in the Northeast and Midwest allows vehicles to use a variety of toll roads and bridges in different states. In the United States, drivers who dislike toll roads might go out of their way to use different routes and avoid tolls; this practice is called "shunpiking."

Joe Weber

See also: Beltways; Freeways; Interstate Highway System (United States).

Further Reading

Cupper, Dan. 1990. *The Pennsylvania Turnpike: A History*. Lebanon, PA: Applied Arts Publishers.

Lewis, Tom. 1997. *Divided Highways: Building the Interstate Highways, Transforming American Life*. New York: Penguin.

Radde, Bruce. 1993. *The Merritt Parkway*. New Haven, CT: Yale University Press.

U

UBER

Headquartered in San Francisco, California, Uber (originally called UberCab) is an American transportation company officially launched in 2010 and operating worldwide as a type of on-call taxicab or ride-on-demand service. Unlike other cab companies (e.g., Yellow Cab), any car owner passing Uber's required driving background checks can pick up passengers who have requested rides using Uber's apps. These on-call drivers are not under any obligation to Uber, so some work during their free time and earn extra money. Users who are interested in finding an Uber ride/cab must create an account on the Uber website using Uber apps and provide their personal information, such as name, cell phone numbers, and credit card number and expiration date. A valid Uber account enables riders to use the Uber app to check for the nearest available Uber driver and book a ride. Uber's pricing is similar to metered taxis, but all payment is handled exclusively through Uber and automatically billed to the customer's credit card. Uber does not require riders to pay tips and drivers cannot collect tips personally. There is no exchange of cash between drivers and passengers. Uber's fare is calculated on a time basis—but in some cities the price is based on a distance travelled.

As the company continues to grow and expand geographically, Uber faces opposition from the taxi industry. The major concern is safety, as it has been claimed that Uber's on-demand ridesharing program is more dangerous than taking a traditional taxicab. There is no evidence to support this statement, however. In fact, Uber's screening of drivers' background is superior to the screening of taxi drivers, and Uber's business model offers greater safety advantages for the drivers than found in the traditional taxi industry. In support of the latter argument, Uber claims that its users (passengers) are not anonymous, as they must create an account linked to a credit card to receive Uber service. As a result, if an Uber passenger commits a crime during a ride then the passenger is easier to trace. Additionally, Uber's drivers and users do not have to carry or exchange cash. These important differences from traditional taxi services ensure that Uber services are safe. Nonetheless, there remain legitimate concerns regarding passengers' safety, how they are insured, and how their privacy is protected by Uber. Because many Uber drivers do not hold commercial insurance, Uber has started purchasing insurance to cover accidents involving drivers. Despite these concerns, it is expected

Uber application startup page on an Apple iPhone display, 2015. (Aleksey Boldin/Dreamstime.com)

Smartphone Ridesharing Apps

Smartphone ridesharing apps are a revolutionary solution for the urban transport industry worldwide. Ridesharing apps such as Uber, Lyft, and Sidecar can be downloaded onto a smartphone. The apps enable passengers to find rides in a more convenient, flexible, and cost-efficient way. Ridesharing apps also offer more choices and pricing options for riders. These services also benefit the urban environment in various ways, such as by lessening traffic congestion and reducing trips made by solo drivers, hence there is less pollution. The taxi industry and the obtaining of a New York City taxi medallion, for example, are huge barriers to wider adoption of new technologies. The cab companies have raised concerns about the safety of passengers who use ridesharing apps and impose rules that must be followed to hold a taxi medallion. This means that drivers have to acquire a license, which costs $200,000. The taxi medallion company has claimed that any car owner can pick up passengers who have requested rides using ridesharing apps, and this raises serious safety issues for both drivers and passengers that are not faced when taking a traditional taxicab. Nonetheless, ridesharing has increased, and the value of the taxi permit has declined by 57 percent over the past few years. There is tremendous potential for ridesharing apps to entirely change the future of the taxi industry worldwide.

Selima Sultana

that Uber will continue to provide its on-demand ridesharing service anywhere in the world.

Selima Sultana

See also: Lyft; Moto-Taxis; Slugging (United States); Yellow Cab, New York City (United States).

Further Reading

Feeney, Matthew. 2015. "Is Ridesharing Safe?" *Policy Analysis at the Cato Institute* 767: 1–15. http://www.memphistn.gov/Portals/0/pdf_forms/CATO.pdf. Accessed August 27, 2015.

UNITED STATES NUMBERED HIGHWAY SYSTEM

The United States Numbered Highway System was America's first national highway system. It was created in 1926, at a time when long-distance travel by automobile was increasing. Navigating across the unmarked roads of the era was very difficult. One answer to this had been the naming of highways, beginning in 1912 with the Lincoln Highway. This was created by a private association to mark a New York City to San Francisco, California, highway. Many other associations were formed to create other routes, such as the Dixie Highway between Michigan and Chicago and Florida, and dozens of others. The marking of a road did not mean it was anything but a rough dirt road, and poor marking also created confusion. Another method of marking roads was needed.

In 1918, Wisconsin became the first state to number its highways. This idea soon spread to other states, and a national set of numbered highways was discussed. In 1926, the 48 state highway departments and the American Association of State Highway Officials came up with a 78,500-mile highway network. All highways in the system were numbered using a consistent system. Routes running east-west have even numbers, with smaller numbers in the north and larger numbers farther south; U.S. 80 being the southernmost highway running coast to coast, and U.S. 2 the northernmost. North-south running highways have odd numbers, with smaller numbers on the east coast and larger numbers farther west; U.S. 101 being the westernmost main north-south highway, and U.S. 1 the easternmost.

In addition to the main network, many additional routes were part of the numbered system. These roads have three-digit numbers, the second and third correspond with the main highway from which the road branches off. Other routes were lettered to show routes for traffic to bypass the congestion of big cities. Bypass or truck routes indicate detour routes around congested points. The original route would have a B (for business) or C (for city) suffix. An alternate route would be shown with an "A" suffix.

Since 1926, the system has changed considerably, with many new highways being added and many rerouted; a number of roads even have disappeared from the map. This especially was the case from the 1960s through the 1980s when the Interstate Highway System was constructed. This national freeway network—which also is numbered—paralleled or overlaid many U.S. highway routes, making them redundant. Other highways were shortened; for example, U.S. 80 once was a transcontinental highway running from Savannah, Georgia, to San Diego, California. It now runs only as far west as central Texas; from Texas westward it has been replaced by I-20, I-10, and I-8. The highway, however, remains important. In Alabama part of it is the Selma to Montgomery National Scenic Byway, which commemorates the 1965 Selma to Montgomery civil rights march.

The most famous of all the numbered highways was U.S. 66, or Route 66. This ran from Chicago, Illinois, to Los Angeles, California, via St. Louis, Missouri; Tulsa, Oklahoma; Oklahoma City, Oklahoma; Amarillo, Texas; and Albuquerque, New Mexico. It appeared in Steinbeck's *The Grapes of Wrath* as a route used by the Okies to escape poverty in the Oklahoma Dust Bowl and go search for a better life in California. After World War II, U.S. 66 was the subject of a song written by Bobby Troup and first recorded by Nat King Cole in 1946. As the Interstate routes were completed along Route 66, the extra number was not needed. Removing the U.S. 66 signs began in the 1970s, with the last coming down in Williams, Arizona, in 1985. Since then, the highway has become an important tourist attraction and has been resigned as Historic U.S. 66 in many areas. The story has become familiar to a new generation thanks to the 2006 Disney movie *Cars*.

The U.S. Numbered Highway System continues to serve as the main roads of the country and will not disappear. There are more than 150,000 miles of highway in this network, which provide the main roads away from the Interstate routes.

Joe Weber

See also: Alaska Highway (Canada and United States); Interstate Highway System (United States); Pacific Coast Highway, California (United States).

Further Reading

Stewart, George R. 1953. *U.S. 40: Cross Section of the United States of America.* Boston: Houghton Mifflin.

V

VACUUMED TRAINS

Humans continually seek faster modes of transport to make the entire globe within their reach in the shortest time possible. Unfortunately, the very air we breathe is a powerful obstacle to accelerating speed. That is why airplanes must ascend up to 40,000 feet above the ground before they can fly 570 miles per hour.

Vacuumed trains are extreme high-speed rail transport that uses evacuated tubes. Theoretically, the vacuumed trains could travel up to 5,000 miles per hour—10 times faster than today's airplanes and 6 times faster than speed of sound. The concept is that by building a tunnel underground, evacuating all the air, and operating a rail line inside it using maglev (magnetic levitation) technology, people could travel all the way around world almost soundlessly and within a few hours. In fact, a trip between New York and Beijing would take less than two hours and a commute between Europe and North America would take only an hour. It's more akin to what Ray Bradbury described as a "silent air-propelled train" that "slid soundlessly down its lubricated flute in the earth" in his most famous science fiction book, *Fahrenheit 451*.

Although the concept of vacuumed trains is not new (the idea of building a tube shuttle was proposed by Robert M. Salter of RAND in 1972), in recent years there has been a push for widespread adoption of vacuum-tube technology. A team from the Massachusetts Institute of Technology's (MIT) mechanical and ocean engineering department conducted experiments in the 1990s by building a half-mile-long air-evacuated tube. The team found that a nearly evacuated tube can enable a maximum speed of 580 miles per hour, which led to the proposing of a rail route between Boston and New York using vacuumed technology. Many concerns exist regarding the technological limitations of the vacuum and maglev systems, however, as well as the cost of the two systems. Major technological breakthroughs in tunneling as well as in maglev are necessary to reduce the cost of these technologies, however; otherwise this system is too expensive to build. There also are concerns about safety and maintenance costs as well as land-acquisition costs. Conversely, proponents of the vacuumed train believe it would not cost as much as some estimate. One study calculated $2 million per mile of construction cost with a speed 350 miles per hour, which is very similar to the cost of a proposed high-speed rail project in the United States.

Selima Sultana

See also: Autonomous Vehicles; Flying Cars; Solar Roads; String Transportation System.

Further Reading

Stewart, Jon. 2014. "Vacuum Trains: A High-Speed Pipe Dream?" BBC. http://www.bbc
.com/future/story/20120601-high-speed-pipedreams. Accessed December 17, 2015.

VELOMOBILES

The velomobile, in general, is a derivative of bicycles and automobiles. It looks like a racing car with seats for back support and a cover to protect riders and passengers from weather. Mostly propelled by human power, this vehicle—also referred to as a "cyclecar," "pedal cars," or "light cars," requires three to four times less energy than does pedaling a bicycle, and thus allows greater speeds (25 miles per hour) and longer pedaling time than a "standard" bicycle. Velomobiles are clean, efficient vehicles and earned praise for their environmental friendliness. Despite these advantages, they remain a very marginal mode of transportation in the United States. The designers of the velomobile, however, are working hard to make a vehicle that will play an important role as a future modern mode of individual transportation in urban areas as well as for medium-distance transportation.

Velomobile development began in Europe in the late 19th century—a period when a variety of technological innovations created new types of transportation, such as the bicycle, the motorcycle, and the automobile. The first cycling technology that resembled an automobile emerged in 1885, but the first velomobile was not produced until the 1920s. The original idea of building the velomobile came from Charles Mochet—the largest manufacturer of velomobiles—when his young son George wished for a bicycle. George's mother did not grant his wish because she considered bicycles too dangerous to use in Paris traffic. Mochet came up with a different idea—he built a lightweight "pedal car" for George, which was faster than a cycle. In 1925, Mochet started producing human-powered velocars commercially. Some velocars were fitted with small motors but were never made without pedals. Velocars came with several variations; some were built with one seat but most had two seats, and some velocars even had room for luggage. The purchase price of velocars was similar to that of motorcycles, but no fuel costs were incurred and the velocar could be maintained by standard bike mechanics. The velocar also received attention for the health benefits of its manual pedaling as well as its affordability, particularly for people who could not purchase motorized vehicles. As a result, before and during World War II, velocars became a great success for practical reasons in France.

Velocars did not sell well in the postwar market, however, due to the financial devastation of the war. This also was a period when Europe began experiencing an automobile boom and saw a massive decline in bicycle use. The production of velocars stopped when a law was passed in France that limited the speed of small cars without licensed drivers to 25 miles per hour.

Interest in velomobiles increased again after the oil crisis of the 1970s, with the organization of the International Human Powered Vehicle Association (IHPVA). This

A velomobile, or bicycle car, travelling down the road, 2012. (Modfos/Dreamstime.com)

group originated in the United States with collaborative members from the United Kingdom and Germany. Many top universities got involved with IHPVA to design human-powered vehicles, focusing solely on increasing efficiency and speed. The first new-generation velomobile, "Leitra," was built by hand in the 1980s. It is much faster than a regular bicycle and has proven to be a functional mode of transportation. Unfortunately, the Leitra was so expensive that only 200 vehicles were sold.

A more recent design, "The Alleweder," has gained popularity and generated enthusiasm in the Netherlands, Germany, and Belgium. Buyers of Alleweders are pleased with its efficiency and speed. There now are about nine types of velomobiles commercially available.

There are many challenges that velomobiles must face, however, before they are accepted widely as a form of transport. In addition to engineering challenges such as crash protection and good visibility for the rider, the greatest challenge is the cost of velomobiles. Makers of velomobiles must set a price that is acceptable for those who cannot afford cars. Reducing the costs could be accomplished by using mass production, but that requires significant consumer demand. It seems that velomobile designers also are car designers, and the existing velomobiles already have shown great promise in speed and efficiency, therefore the vehicle already is a functional mode of transportation and not some futuristic dream. The velomobile designers and bicycle advocates strongly believe that, in the future, velomobiles will be recognized as a valuable mode of sustainable transportation, especially in urban areas.

Selima Sultana

See also: Bicycles and Tricycles; Minicars; Mopeds.

Further Reading

Horton, Dave, Paul Rosen, and Peter Cox (eds.). 2007. *Cycling and Society.* Burlington, VT: Ashgate.

Van De Walle, Frederik. 2004. "The Velomobile As a Vehicle for More Sustainable Transportation: Reshaping the Social Construction of Cycling Technology." Master's thesis. Department of Infrastructure, Royal Institute of Technology, Stockholm. http://users .telenet.be/fietser/fotos/VM4SD-FVDWsm.pdf. Accessed July 20, 2015.

W

WALKING SCHOOL BUSES (UNITED STATES)

A walking school bus can be as informal as a parent walking with a group of children or as formal as a planned walking route with "bus stops" where children are collected at a particular time and with a regular schedule of volunteers. Although the type of program varies, the desired result of increased physical activity and healthier children remains the same.

Rapidly increasing rates of childhood obesity and related health concerns coupled with decreasing levels of physical activity among children have resulted in concerted attempts to encourage more children to walk to school. These efforts have led to development of the walking school bus program, whereby a group of children walk to school with one or more adults. Supported by the 2005 national Safe Routes to School (SRTS) program, federal funding has been allocated to efforts designed to enable and encourage children to walk and bicycle to school. A majority of the funding has been allocated to infrastructure and related improvements, such as sidewalk and crosswalk improvements, traffic calming and diversion, and new or improved bicycle facilities. Additional funding has been allocated to public awareness campaigns, safety programs for children, training for SRTS volunteers, and community outreach. The principal objective of these efforts is to reverse the decades-long downward trend in the rates of children who walk or bike to school. By using foot or pedal power, children not only experience the health benefits associated with increased physical activity, but can accrue such additional benefits as further developing an appreciation for nature and building a sense of community. The most effective walking school bus programs have been found to be those that can boast of strong parental involvement along with active support from principals and teachers.

Development of the walking school bus program has been necessitated by a dramatic downturn in the rates by which children walk or bicycle to school. Between 1969 and 2009, the rate by which children actively traveled to school dropped from nearly 50 percent to only 13 percent. During the same timeframe, overall activity rates among children also have declined and youth obesity has more than tripled. Concerns over these statistics are coupled with awareness that children who walk to school display reduced Body Mass Indices (BMIs) and increased cardiorespiratory fitness. Yet walking school bus programs face many structural and societal obstacles that work to limit their success. Among the most difficult issues to address is the fact that settlement patterns over the past half-century have been characterized by extensive suburban sprawl. The associated pattern of consolidation into fewer but larger schools is revealing, with the number of schools declining by nearly 70 percent between 1940 and 1990. Large schools requiring large areas of land frequently are built on the outskirts of town, thereby limiting the ability of children to reach the school without motorized transport.

Suburban sprawl also is characterized by automobile dominance and significant volumes of traffic along major thoroughfares, and many studies demonstrate that parental concerns are tied to the built environment. Improvements to density, diversity, and design (the "3 Ds" of walkability) therefore have garnered much attention recently. Density of population allows for efficiency in walking and can reduce automobile traffic, speed, and congestion. Diversity of land use and building type results in the proximity of many walkable destinations. Design frequently is associated with a pedestrian-friendly interconnected street network. Yet even the provision of a walkable environment does not guarantee increased rates of pedestrian activity among children if other parental and societal concerns remain unaddressed. Fear of crime and associated perceptions that walking to school is not a safe activity for children permeate society and could serve as a deterrent. Time and convenience also are considerations that have been identified as potentially limiting factors to children walking to school. Thus, a walking school bus program that entails more than alterations to the built environment is needed. Coordination between and among parents and schoolteachers and administrators can provide the requisite support for the walking school bus to succeed.

The rate of children who walk or bike to school is only a fraction of what it was a generation ago, therefore habits must be changed and cultural attitudes must be shifted for a walking renaissance to take place. Fortunately, the walking school bus program affords tremendous opportunity initiate such changes in habits and attitudes. Successful walking school bus initiatives have been associated with the "five Es": encouragement, education, enforcement, engineering, and evaluation. Parents and schools can be encouraged and educated about the benefits of walking. Enforcement involves collaboration with police departments to enforce speed limits, and engineering refers to the construction of necessary crosswalks, sidewalks, and traffic-calming devices. Evaluation of the program's success allows for identification of any necessary changes or improvements.

Christopher Cusack

See also: Bicycle Lanes; Bicycles and Tricycles.

Further Reading

Chillón, Palma, Kelly R. Evenson, Amber Vaughn, and Dianne S. Ward. 2011. "A Systematic Review of Interventions for Promoting Active Transportation to School." *International Journal of Behavioral Nutrition and Physical Activity* 8(10).

McDonald, Noreen C., Ruth L. Steiner, Chanam Lee, Tori Rhoulac Smith, Xuemei Zhu, and Yizhao Yang. 2014. "Impact of Safe Routes to School Program on Walking and Bicycling." *Journal of the American Planning Association* 80(2): 153–167.

WASHINGTON METRO (UNITED STATES)

The Washington Metro is the rapid transit system for Washington, DC. It is a modern system; the first portion opened in 1976. The current network includes five lines identified by color, 91 stations, and 117 route miles; and 50 miles of the system is underground. It is the second-busiest urban rapid transit system in the United States (after New York City), with an annual ridership more than 200 million and an average weekday ridership of 750,000. The Washington Metropolitan Area Transit Authority (WMATA) operates the rail network (as well as the bus network in DC). The rail network cuts across Washington, DC; Montgomery County and Prince George's County in Maryland; and in Fairfax County, Arlington County, and the City of Alexandria in Virginia.

In 1956, the Federal-Aid Highway Act passed, and plans for a national interstate highway system began. At the time, the National Capital Planning Commission favored the highway network to other options, including the construction of a rapid rail system. The commission thought that a rail could not be self-sufficient, and in the 1950s transit ridership was in decline as residents moved to the suburbs and favored commuting by automobile. By 1960, however, the U.S. federal government created the National Capital Transportation Agency (NCTA). Its role was to develop a rapid rail system for Washington, DC. In 1966, a bill passed with support of the federal government, the District of Columbia, Virginia, and Maryland created the WMATA. NCTA's planning power shifted to the WMATA. As the rapid transit plans began, growing opposition to the federal highway plan mounted. Initially (and despite the plan for a new rapid transit system), new interstate highways were being considered that would go through Washington DC, including plans for an inner beltway that would link them. The opposition continued until 1977, when the original plans were cancelled in favor of an outer beltway (which was opened) and funds from the cancelled portion went toward rapid transit.

In 1968, the WMATA approved plans for a 98-mile system and construction began in 1969. Metro construction required several billion dollars, and Congress appropriated the funds under the authority of the National Capital Transportation Act of 1969. Two-thirds of the funding came from the federal government and a third came from local sources. Federal funding allocated during the 1980s allowed for the entire to be completed. The first portion of the system opened in 1976, a mere 4.6-mile section on the Red Line with only five stations, all of it in Washington, DC. The system quickly expanded into Arlington County, Virginia, by 1977, including a station at National Airport (thus linking the airport to Washington, DC, by rapid transit). Extensions into Maryland first occurred by 1978, and into Fairfax County, and the city of Alexandria, Virginia, by 1983. Tall concrete arches that resemble honeycombs and indirect lighting highlight the design of underground stations. The architecture drew praise from critics and offered a stark contrast to other rapid transit systems built earlier in the century. In general, the relatively fast, clean, and safe Metro drew riders as an alternative to commuting by auto.

The construction of the Metro was not without controversy or ridicule. For the first time in decades, a new rapid transit was being built through a major metropolitan area in the United States. Residents and businesses had to endure years of construction along the planned routes; local groups opposed the location of a route or a planned station if they did not agree with the plan. Although the Red, Blue, Orange, and Yellow Lines faced issues as they were constructed, it was the Green Line that faced the most opposition from communities that would be impacted by the construction. The Green Line also faced opposition from riders and suburban communities that feared that the new extensions into unsafe neighborhoods would affect the safety of the network. It took until 2001 for the Green Line to be finished, a period much longer than that for the other routes. Unfortunately, the Green Line does have the highest crime rate within the system.

The opening of the Green Line in 2001 completed the originally planned metro network, the next expansion to the system opened in 2004 on the Blue Line. The Dulles (Airport) Corridor Metrorail Project has been under way since 2009, and the first phase opened in 2014 with five new stations to the network. The extension marked the addition of exclusive stations for the Silver Line, which previously operated entirely on other lines in the system. It is expected that the second phase (to Dulles Airport) will open in 2018.

Jason Greenberg

See also: London Underground/The "Tube" (United Kingdom); New York City Subway (United States).

Further Reading

Ozer, Mark. 2013. *Washington Metroland.* CreateSpace Independent Publishing Platform.
Schrag, Zachary, 2006. *The Great Society Subway: A History of the Washington Metro.* Baltimore: Johns Hopkins University Press.
Thompson, R. Wayne. 2001. *Metro at 25: Celebrating the Past, Building the Future.* Washington DC: Washington Metropolitan Area Transit Authority.

WHEELCHAIRS

Wheelchairs are a mode of transportation in which a chair is fitted with wheels for use by people who otherwise would be immobile. There are many forms of wheelchairs, ranging from manual to electric-powered propulsion. China is perhaps the first to have created a device for disabled people, as depicted on a stone slate from the fifth century. The first records of wheeled seats being used, however, are found during the 15th and 16th centuries in Europe. The first wheelchair was introduced in America in 1887 in Atlantic City, New Jersey. Over the years, wheelchairs have evolved with technological and legal changes.

Currently, the Americans with Disability Act (ADA) of 1990 has increased the mobility of people who have a disability. The key campaign of ADA is to design buildings

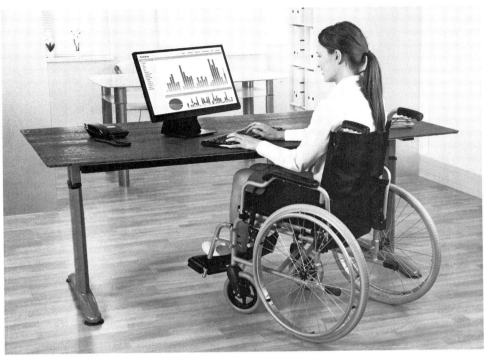

Wheelchairs allow those with limited physical functionality to participate in everyday activities, as seen in this photo of a young businesswoman in a wheelchair working on a computer, 2015. (Andrey Popov/Dreamstime.com)

and transportation systems that increase the accessibility for those individuals who use a wheelchair. Along with Canada, Europe also is developing human rights acts intended to improve the well-being of disabled people along with improving their forms of transportation. Germany, for instance, has added newer buses that have lower floors, making the buses more wheelchair accessible. Some European countries started using taxis or taxi buses to pick up individuals who are confined to a wheelchair. Over the years, the adaptations for people with disabilities have increased tremendously in the developed world. There now are accessible buildings, designated parking, and curbs that are easy to use by people in wheelchairs.

Personal transportation systems also include powered wheelchairs and mobility scooters. Mobility scooters have become popular choices for the elderly and those with significant arthritis or respiratory issues and who are capable of mobility but benefit from assistance or tire quickly, particularly for outings such as shopping or visiting parks. Many supermarkets and malls now provide mobility scooters of three-, four-, or five-wheel designs. This service is especially beneficial because Medicare does not pay for powered wheelchairs (of which scooters are considered a subset) for individuals who do not need them inside the home, thus limiting some individuals' activities.

Despite all these changes that make it possible for every individual with a mobile disability to participate in society, they are still not enough. A survey in 2004 of individuals who have a disability shows that more than half of the respondents stated that transportation was a major problem and that it needed to be improved.

Selima Sultana

See also: Ambulances; Segway.

Further Reading

Rosenbloom, Sandra. 2007. *Transportation Patterns and Problems of People with Disabilities.* Washington, DC: National Academic Press. National Center for Biotechnology Information. http://www.ncbi.nlm.nih.gov/books/NBK11420/. Accessed August 27, 2015.

Y

YELLOW CAB, NEW YORK CITY (UNITED STATES)

New York City's iconic yellow taxi cabs are among a fleet of 10,000 cars for hire throughout the five boroughs of New York. Known as "medallion cabs" because of the official medallions on the cars' hoods, the canary-yellow cars' drivers are required to pick up passengers when hailed and deliver them throughout New York's five boroughs—the Bronx, Brooklyn, Manhattan, Queens, and Staten Island. The Yellow Cab Company also runs "livery" or "boro" cabs, which are painted apple green and carry passengers only in boroughs outside Manhattan and in northern Manhattan. Taxicabs in New York are operated by private companies, all of which are licensed by the New York City Taxi and Limousine Commission (TLC).

Horse-drawn carriages had been a popular means of transportation in New York City since the early 19th century, but horses were considered unpredictable, expensive to keep, and dangerous, and they produced 1 million pounds of manure each day. Alternatives were sought and the first paid transport service by an electric vehicle in New York City occurred in 1897. Operated by the Samuel's Electric Carriage and Wagon Company, it ran 12 electric hansom cabs. The fleet reached 62 and its financiers changed the company name to the Electric Vehicle Company, which soon began building the Electrobat car. There were 100 Electrobats on city streets by 1899, and as many as 1,000 were in use just a few years later. A fire in January 1907 destroyed 300 of the cars; the company folded soon after this disaster.

Gasoline-powered taxicabs have existed in New York City since October 1, 1907, when a fleet of 65 metered cars paraded down Fifth Avenue. These vehicles replaced electric-and steam-powered cars. The idea for the service came from Harry N. Allen after he experienced what he believed to be extortion when he and his companion were charged $5 (equivalent to about $125 in today's terms) for a three-quarter-mile ride by hansom cab. Allen vowed to start a cab service that charged a flat rate per mile. With European and U.S. financial backing, Allen purchased 600 cars from France and equipped them with taximeters to gauge fares. With them he began the New York Taxicab Company. Drivers wore crisp uniforms and were instructed to act courteously toward customers. Pleased with their work, Allen gave faithful drivers a gold watch and tempted them with the possibility of a pension plan, but by October 8, 1908, 500 drivers had walked out in a wage dispute. Labor unrest would mar the taxi industry, but industrialists still put money into the business, which proved to be lucrative.

In fact, by the 1920s, automobile manufacturers such as General Motors and Ford Motor Company not only built but operated fleets of cabs. The largest

New York City taxicab passing by the Chrysler Building in Manhattan, New York, 2009. Medallion taxis, the familiar yellow cabs, are the only vehicles in the city permitted to pick up passengers in response to a street hail. (Rafael Ben-ari/Dreamstime.com)

manufacturer was the Checker Cab Manufacturing Company of Illinois, owned by Russian-born Morris Markin. He led a partnership that purchased the Chicago-based Yellow Cab Company from immigrant John D. Hertz, who had left the taxi business to concentrate on his car-rental venture. After taking over Yellow Cab, Markin painted his fleet of New York Checker Taxi cars yellow and kept the iconic checkerboard pattern running along the body. These large, roomy cabs became a familiar symbol of New York City.

Taxis became a common mode of transportation throughout the city, as the majority of New York's residents did not own a car. The increase in the number of taxis in the 1930s gave rise to the need for enforceable regulations to protect drivers and passengers. Corruption in the form of unfair labor practices and the overcharging of passengers was widespread and the Taxicab Commission could not keep it in check. Competition became fierce during the Great Depression and the Checker Cab Company was accused of bribing New York mayor James J. Walker. In New York's Times Square in 1934, more than 2,000 drivers staged what was considered to be the biggest strike in history. To quell the unrest, the new mayor, Fiorello LaGuardia, signed the Haas Act of 1937, which issued a limited number of official taxi licenses and ordered each taxi to display a medallion certifying it for passenger use. The act attempted to increase wages for drivers, many of whom were immigrants, but it also increased the power of the few companies awarded the medallions.

During World War II, from 1939 to 1945, when fuel and auto parts were rationed, many big-city dwellers turned to taxis as their primary mode of transportation. Profits soared and, in turn, the value of the $10 medallions grew, motivating some licensed companies to sell them for as much as $5,000 apiece. The demographic of taxi drivers also changed during the war, as more women and African Americans took to driving taxis for a living.

In the 1960s, to decrease the confusion between licensed and unlicensed cabs, city lawmakers initiated a requirement that all medallion cabs be painted yellow. Not only was more consistent regulation of medallion cabs put in place, but the commission also began acknowledging and legitimizing livery cabs—those not carrying an official medallion and, therefore, unable to legally pick up people from the street. They concentrated their business in poorer neighborhoods that many regulated cabs illegally avoided. More regulations designed to keep both drivers and passengers safe followed; for instance, many cabs of the time were equipped with bulletproof glass.

By the 1980s, a medallion cost $125,000—making owning a car nearly impossible for drivers and causing companies to lease their cabs for days and weeks at a time. Trading for as much as $1 million each, medallion prices decreased in 2014 to $872,000, reflecting an increase in competition from private companies such as Uber (founded by Travis Kalanick), which offers rides in privately owned vehicles that people book using a mobile-phone application. By 2014, there were still more than 51,000 men and women licensed to drive the nearly 13,500 medallion cabs in New York. Additionally, 8,000 of the cabs are hybrids, representing the majority of taxis in service, which is the most of any city in North America.

Rosemarie Boucher Leenerts

See also: Lyft; Slugging (United States); Uber.

Further Reading

Taxi Dreams. "Taxi History Part 1: 1890–1930s." PBS. http://www.pbs.org/wnet/taxi dreams/history/. Accessed June 5, 2015.

Wilson, Mark. R., Stephen R. Porter, and Janice L. Reiff. "Yellow Cab Co. (of Chicago)." *Encyclopedia of Chicago*. http://www.encyclopedia.chicagohistory.org/pages/2912.html. Accessed June 5, 2015.

Bibliography

Advani, M., and G Tiwari. 2005. "Evaluation of Public Transport Systems: Case Study of Delhi Metro." Proceeding in START-2005 Conference, Kharagpur, India. http://tripp.iitd.ernet.in/publications/paper/planning/mukti_metro_kharag pur_05.pdf. Accessed July 28, 2015.

Albalate, D., and G. Bel. 2012. "High-Speed Rail: Lessons for Policy Makers from Experiences Abroad." *Public Administration Review* 72(3): 336–349.

Allen, R. S. 1970. *Covered Bridges of the South*. Brattleboro, VT: Stephen Greene Press.

Alshalalfah, B., A. Shalaby, S. Dale, and F. Othman. 2012. "Aerial Ropeway Transportation Systems in the Urban Environment: State of the Art." *Journal of Transport Engineering* 138(3): 253–262.

Alshalalfah, B., A. Shalaby, S. Dale, and F. Othman. 2013. "Improvements and Innovations in Aerial Ropeway Transportation Technologies: Observations from Recent Implementations." *Journal of Transport Engineering* 139(8): 814–821.

Anderson, James M., Nidhi Kalra, Karlyn D. Stanley, Paul Sorenson, Constantine Samaras, and Oluwatobi A. Oluwatola. 2014. "Autonomous Vehicles: A Guide for Policymakers." Santa Monica: RAND Corporation. http://www.rand.org /content/dam/rand/pubs/research_reports/RR400/RR443-1/RAND_RR443-1 .pdf. Accessed April 1, 2015.

Avci, Buket, Karan Girotra, and Serguei Netessine. 2012. "Electric Vehicles with a Battery Switching Station: Adoption and Environmental Impact." Faculty and Research Working Paper, INSEAD, Fontainebleau, France. http://www.insead .edu/facultyresearch/research/doc.cfm?did=49338. Accessed November 10, 2014.

Bachleda, F. Lynne. 2011. *Guide to the Natchez Trace Parkway*. 2nd ed. Birmingham: Menasha Ridge Press.

Bale, John. 2004. *Running Cultures: Racing in Time and Space*. London: Routledge.

Beaucage-Gauvreau, E., G. A. Dumas, and M. Lawani. 2011. "Head Load Carriage and Pregnancy in West Africa." *Clinical Biomechanics* 26: 889–894.

Beijing Subway Official website [in Chinese]. 2015. http://www.bjsubway.com /corporate/dtdsj/. Accessed August 2015.

Bel Geddes, N. 1940. *Magic Motorways*. New York: Random House.

Bergner, A. 2013. The Bamboo Train in Battambang: The Good and the Ugly. http:// thatbackpacker.com/2013/04/18/bamboo-train-battambang-cambodia-the -good-and-the-ugly/. Accessed June 9, 2014.

Bertaud, A. 2003. *Metropolitan Structures around the World: What Is Common? What Is Different? What Relevance to Marikina in the Context of Metro Manila?* http://alainbertaud.com/wp-content/uploads/2013/07/Metropolitan_Structures _around_the_World.pdf. Accessed August 2015.

Black, William R. 2010. *Sustainable Transportation: Problems and Solutions*. New York: Guilford Press.

Blodgett, Peter J. (ed.). 2015. *Motoring West*. Vol. I: Automobile Pioneers, 1900–1909. Norman, OK: Arthur H. Clark.

Blotevogel, Hans H. 2004. "Aktuelles zur Zentralitätsdiskussion in Deutschland." Presented at Conference of OEGR and ÖROK, Vienna, Germany, September 11, 2004. http://www.oerok.gv.at/fileadmin/Bilder/2.Reiter-Raum_u._Region/3. Themen_und_Forschungsbereiche/9.Zentralitaet_u._Raumentwicklung /Zentralitaet_u._Raumentwicklung/Fachtagung/Blotevogel.pdf. Accessed January 11, 2016.

Bonsor, Kevin. 2000. "How Flying Cars Will Work." HowStuffWorks.com. December 1, 2000. http://auto.howstuffworks.com/flying-car.htm. Accessed February 5, 2015.

Brodsly, David. 1981. *L.A. Freeway: An Appreciative Essay*. Berkeley: University of California Press.

Busan Transportation Corporation. 2015. http://www.humetro.busan.kr/english /main/. Accessed June 5, 2015.

Button, K. 2012. "Is There Any Economic Justification for High-Speed Railways in the United States?" *Journal of Transport Geography* 22: 300–302.

Buzi, Miriam, and Daniel Shefer. 1978. "The Carmelit: Transit Characteristics and Mode Choice Determinants." *Traffic Quarterly* 32(1): 145–167.

Cain, Alasdair, D. Georges, R. B. Michael, R. Pilar, and C. B. Johan. 2007. "Applicability of Bogota's TransMilenio BRT System to the United States." *Transportation Research Record: Journal of the Transportation Research Board* 2034: 45–54. doi: 10.3141/2034-06.

Car2Go. 2015. https://www.car2go.com/. Accessed June 5, 2015.

Cervero, R. 1998. *The Transit Metropolis: A Global Inquiry*. Washington, DC: Island Press.

Chen, X., and R. Greene. 2012. "Spatial-Temporal Dynamics of China's Changing Urban Hierarchy (1950–2005)." *Urban Studies Research* (2012): 1–13. doi:10.1155/2012/162965. http://www.hindawi.com/journals/usr/2012/162 965/. Accessed August 4, 2015.

Chester, M., and A. Horvath. 2012. "High-Speed Rail with Emerging Automobiles and Aircraft Can Reduce Environmental Impacts in California's Future." *Environmental Research Letters* 7(3): 034012. http://iopscience.iop.org/1748-9326 /7/3/034012/pdf/1748-9326_7_3_034012.pdf. doi:10.1088/1748-9326/7/3/0 34012.

Chou, Michael. 2010 (November 3). "A Railway under the Ocean: The Channel Tunnel Linking Britain and France." *Illumin: A Review of Engineering in Everyday Life*. http://illumin.usc.edu/printer/172/a-railway-under-the-ocean-the-channel -tunnel-linking-britain-and-france/. Accessed July 2, 2015.

City of Chicago. 2015. Sustainable Chicago 2015. http://www.cityofchicago.org/city/en/progs/env/sustainable_chicago2015.html. Accessed June 5, 2015.

Cookridge, E. H. 1978. *Orient Express: The Life and Times of the World's Most Famous Train.* New York: Random House.

Cooper, James M., and Christine Hunefeldt. 2013. *Amazonia: National Politics and Interests, Laws, and the Environment.* Brighton: Sussex Academic Press.

Curtis, P. 2007 (January 26). Living Near a Motorway Damages Children's Lungs, Research Reveals. http://www.theguardian.com/environment/2007/jan/26/pollution.transportintheuk. Accessed June 5, 2015.

Cycle Super Highways of Greater Copenhagen. 2015. Cycle Super Highways: Capital Regions of Denmark. http://www.supercykelstier.dk/sites/default/files/Cycle%20Superhighways_UK_maj%202014.pdf. Accessed July 22, 2015.

Department of Rapid Transit Systems (DORTS), Taipei Municipal Government. 2015. http://www.dorts.gov.tw/. Accessed January 10, 2015.

Dick, W., and A. Lichtenberg. 2012. The Myth of Hitler's Role in Building the Autobahn. http://www.dw.de/the-myth-of-hitlers-role-in-building-the-autobahn/a-16144981. Accessed June 6, 2014.

Docherty, I., and J. Shaw. 2003. *A New Deal for Transport?* Malden, MA: Blackwell Publishing.

Fagnant, Daniel J., and Kara M. Kockelman. 2013. *Preparing a Nation for Autonomous Vehicles: Opportunities, Barriers and Policy Recommendations.* Eno Center for Transportation, Washington, DC. http://www.enotrans.org/wp-content/uploads/wpsc/downloadables/AV-paper.pdf. Accessed April 1, 2015.

Fink, Charles, and Robert Searns. 1993. *Greenways: A Guide to Planning, Design, and Development.* Washington, DC: Island Press.

Frey, T. 2015. 2050 and the Future of Transportation. http://www.futuristspeaker.com/2008/05/2050-and-the-future-of-transportation/. Accessed July 24, 2015.

Garratt, Collin. 2002. *The World Encyclopedia of Locomotives.* London: Lorenz.

Garrett, M. (ed.). 2014. *Encyclopedia of Transportation: Social Science and Policy.* Thousand Oaks: Sage Publication. doi: http://dx.doi.org/10.4135/9781483346526.

Givon, M., and D. Banister (eds.). 2013. *Moving Towards Low Carbon Mobility.* Northampton, MA: Edward Elgar.

Glover, John. 2010. *London's Underground.* 11th ed. Shepperton, UK: Ian Allan Publishing.

Grava, S. 2003. *Urban Transportation Systems: Choice for Communities.* New York: McGraw-Hill.

Haller, John S. 1992. *Battlefield Medicine: A History of the Military Ambulance from the Napoleonic Wars through World War I.* Carbondale: Southern Illinois University Press.

Hargraves, A. J. 2014. A Feasibility Study into the Use of String Transport Systems for Passenger Rail in New South Wales. Presented at the 32nd Conference of Australian Institutes of Transport Research, 17th and 18th February 2014. http://www.sidrasolutions.com/cms_data/contents/sidra/media/articles/caitr2014_hargraves.pdf. Accessed July 24, 2015.

Hasan, Raqibul MD. 2009. "Problems and Prospects of a Railway: A Case Study of Bangladesh Railway." *Journal of Service Marketing* 4: 124–136. http://www.slide share.net/sohagmal/a-case-study-of-bangladesh-railway-33311962. Accessed March 15, 2015.

Hass-Klau, C. 2015. *The Pedestrian and the City*. New York: Routledge.

Hays, K. J. 2002. *An American Cycle Odyssey*. Lincoln and London: University of Nebraska Press.

History of Motorways. 2014. http://www.theaa.com/public_affairs/reports/history -of-motorways.html. Accessed June 5, 2015.

Hood, C. 2006. *Shinkansen: From Bullet Train to Symbol of Modern Japan*. London: Routledge.

Hood, Clifton. 2004. *722 Miles: The Building of the Subways and How They Trans- formed New York*. Baltimore: Johns Hopkins University Press.

Hoton, D., P. Rosen, and P. Cox (eds.). 2007. *Cycle and Society*. Burlington, VT: Ashgate Publishing Company.

Hoyle, B., and R. Knowles. 1999. *Modern Transport Geography*. New York: Wiley and Sons.

Ioannou, P. A. (ed.). 1997. *Automated Highway Systems*. New York: Plennum Press.

Ioannou, P. A., and C. C. Chien. 1993. "Autonomous Intelligent Cruise Control." *IEEE Transactions on Vehicular Technology* (42)4: 657–672.

Jacobs, Allan B., Elizabeth MacDonald, and Yodan Rofé. 2003. *The Boulevard Book*, Cambridge, MA: MIT Press.

Jakle, J. A., and K. A. Sculle. 2008. *Motoring: The Highway Experience in America*. Athens: The University of Georgia Press.

Jang, S. Y., H. Y. Jung, and S. K. Baik. 2013. A Study on the Improving the Services for Users of LRT (Light Rail Transit) by Structural Equation Model-Focus on Busan Gimhae Light Rail Transit and Busan Subway Line 4 (Bansong Route). http://www.researchgate.net/publication/264134438. Accessed August 8, 2015.

Johnson, B. E. 2012. "American Intercity Passenger Rail Must Be Truly High-Speed and Transit-Oriented." *Journal of Transport Geography* 22: 295–296.

Johnson, C., and D. Hettinger. 2014. *Geography of Existing and Potential Alternative Fuel Markets in the United States*. National Renewable Energy Laboratory. http:// www.afdc.energy.gov/uploads/publication/geography_alt_fuel_markets.pdf. Accessed June 5, 2015.

Johnson, Stephen, and Roberto T. Leon. 2002. *Encyclopedia of Bridges and Tunnels*. New York: Checkmark Books.

Kaushik. 2012. "Chicken Buses of Guatemala." Amusing Planet. http://www. amusingplanet.com/2012/10/chicken-buses-of-guatemala.html. Accessed August 8, 2015.

Kerr, Ian J. 2006. *Engines of Change: The Railroads That Made India*. Westport, CT: Praeger.

Kim, G., and J. Rim. 2000. "Seoul's Urban Transportation Policy and Rail Transit Plan: Present and Future." *Japan Railway & Transport Review* 25: 25–31.

Kim, H., and S. Sultana. 2015. The Impacts of High-Speed Rail Extensions on Ac- cessibility and Spatial Equity Changes in South Korea from 2004 to 2018. *Jour- nal of Transport Geography* 45: 48–61. doi:10.1016/j.jtrangeo.2015.04.007.

Kingsbury, K. T., M. B. Lowry, and M. P. Dixon. 2011. "What Makes a 'Complete Street' Complete? A Robust Definition, Given Context and Public Input." *Transportation Research Record: Journal of the Transportation Research Board* 2245. Washington, DC: Transportation Research Board of the National Academies. 103–110. doi: 10.3141/2245-13.

Krivit, S. B. (ed.). 2010. *Nuclear Energy Encyclopedia: Science, Technology, and Applications.* Hoboken, NJ: John Wiley & Sons.

Lane, B. W. 2012. "On the Utility and Challenges of High-Speed Rail in the United States." *Journal of Transport Geography* 22: 282–284.

League of American Cyclists. 2014. Where We Ride: Analysis of Bicycle Commuting in American Cities. http://bikeleague.org/sites/default/files/ACS_report_2014_forweb.pdf. Accessed February 25, 2015.

Lee, K. 2013 (June 10). Students Install the World's First Solar Pavement Panels in Virginia [blog post]. http://inhabitat.com/students-install-the-worlds-first-solar-pavement-panels-in-virginia/solar-walk-up-jmc-2013-9954-460x260/. Accessed May 2015.

Lee, S., S. Lee, and Y. I. Lee. 2006. "Innovative Public Transport Oriented Policies in Seoul." *Transportation* 33(2): 189–204.

Lewis, Tom. 1997. *Divided Highways: Building the Interstate Highways, Transforming American Life.* New York: Penguin.

Light Rail Transit Authority. Department of Transportation and Communications. 2014. http://lrta.gov.ph/. Accessed August 8, 2015.

Lloyd, R., B. Parr, S. Davies, T. Partridge, and C. Cooke. 2010. "A Comparison of the Physiological Consequences of Head-Loading and Back-Loading for African and European Women." *European Journal of Applied Physiology* 109: 607–616. doi: 10.1007/s00421-0101395-9.

Longhurst, James. 2015. *Bike Battles: A History of Sharing the American Road.* Seattle: University of Washington Press.

Loukaitou-Sideris, A., and R Ehrenfeucht. 2009. *Sidewalks: Conflict and Negotiation over Public Space.* Cambridge, MA: The MIT Press.

Marchand, Jean-Pierre, Pierre Riquet, and Roger Brunet. 1996. "Europe du Nord, Europe Médiane." *Géographie Universelle* 9. Paris: Belin-Reclus.

Marin, G. D., G. F. Naterer, and K. Gabriel. 2010. Rail Transportation by Hydrogen vs. Electrification, a Case Study for Ontario, Canada, II: Energy Supply and Distribution. *International Journal of Hydrogen Energy* 35: 6097–6107.

Masters, Nathan. 2012. From Roosevelt Highway to the 1: A Brief History of Pacific Coast Highway. http://www.kcet.org/updaily/socal_focus/history/la-as-subject/from-the-roosevelt-highway-to-the-one-a-brief-history-of-pacific-coast-highway.html. Accessed August 8, 2015.

Matsuo, Shogo. 1986. "An Overview of the Seikan Tunnel Project." *Tunneling and Underground Space Technology* 1: 323–331.

McDonald, Noreen C., Ruth L. Steiner, Chanam Lee, Tori Rhoulac Smith, Xuemei Zhu, and Yizhao Yang. 2014. "Impact of Safe Routes to School Program on Walking and Bicycling." *Journal of the American Planning Association* 80(2): 153–167.

McQueen, B., and J. McQueen. 1999. *Intelligent Transportation Systems Architecture.* Norwood, MA: Artech House.

Mielenz, J. R. 2001. "Ethanol Production from Biomass: Technology and Commercialization Status." *Current Opinion in Microbiology* 4: 324–329.

MRTA Annual Report 2013. 2013. *Connecting Today for the Future.* http://www.mrta.co.th/en/aboutMRTA/annualReport/All2556eng.pdf. Accessed August 8, 2015.

Mun, J., and D. Kim. 2012. *Construction of High-Speed Rail in Korea.* The KDI School of Public Policy and Management.

National Association of City Transportation Officials. 2015. NACTO Urban Bikeway Design Guide. http://nacto.org/cities-for-cycling/design-guide/. Accessed August 8, 2015.

National Bicycle Dealers Association. 2013. Industry Overview. http://nbda.com/articles/industry-overview-2013-pg34.htm. Accessed August 8, 2015.

National Society for the Preservation of Covered Bridges. 2009. *World Guide to Covered Bridges.*

National Society for the Preservation of Covered Bridges. 2015. http://www.coveredbridgesociety.org/. Accessed August 8, 2015.

National Transportation Safety Board. 2014. Mandate Motor Vehicle Collision Avoidance Technologies. http://www.ntsb.gov/safety/mwl/Pages/mwl10_2012.aspx. Accessed January 8, 2015.

O'Garra, T., S. Mourato, and P. Pearson. 2005. "Analyzing Awareness and Acceptability of Hydrogen Vehicles: A London Case Study." *International Journal of Hydrogen* 30(6): 649–659.

Ogden, J. M. 1999. "Developing an Infrastructure for Hydrogen Vehicles: A Southern California Case Study." *International Journal of Hydrogen Energy* 24: 709–730.

Pacione, M. 1974. "Italian Motorway." *Geographical Analysis* 59(1): 35–41.

Pavan, G. 2015. Green Cars Adoption and the Geography of Supply of Alternative Fuels. Center for Economic Policy Research. http://www.cepr.org/sites/default/files/events/Pavan_Aprile2015.pdf. Accessed August 8, 2015.

Pedestrian and Bicycle Information Center. http://www.pedbikeinfo.org/. Accessed August 8, 2015.

People for Bikes. 2015. U. S. Bicycling Participation Benchmarking Study Report. http://www.peopleforbikes.org/resources/entry/u.s.-bicycling-participation-benchmarking-report. Accessed August 8, 2015.

Perl, A., and A. Goetz. 2015. "Getting Up to Speed: Assessing the Usable Knowledge from Global High-Speed Rail Experience for the United States." *Transportation Research Board 94th Annual Meeting* (No. 15-0761).

Perrier, D. 2009. *Onramps and Overpasses: A Cultural History of Interstate Travel.* Gainesville: University Press of Florida.

Peterman, D. R., J. Frittelli, and W. J. Mallett. 2009 (December). "High Speed Rail (HSR) in the United States." Library of Congress, Washington, DC: Congressional Research Service.

Pilkington, John. 2006 (October 21). "Dying Trade of the Sahara Camel Train." *BBC NEWS*. http://news.bbc.co.uk/2/hi/programmes/from_our_own_corre spondent/6070400.stm. Accessed August 8, 2015.

Pimentel, D., and T. W. Patzek. 2005. Ethanol Production Using Corn, Switch-grass, and Wood; Biodiesel Production Using Soybean and Sunflower. *Natural Resources Research* 14(1): 65–76. http://www.precaution.org/lib/ethanol_and _biodiesel.050301.pdf. doi: 10.1007/s11053-005-4679-8. Accessed August 8, 2015.

Plug In America. 2015. "Plug-In Vehicle Tracker: What's Coming, When." http:// www.pluginamerica.org/vehicles. Accessed August 8, 2015.

PlugInCars.com. 2015. Nissan Leaf Review: Affordable All-Electric Car. http:// www.plugincars.com/nissan-leaf. Accessed August 8, 2015.

Porter, G. 2002. "Living in a Walking World: Rural Mobility and Social Equity Is-sues in Sub-Saharan Africa." *World Development* 302: 285–300.

Porter, G. 2007. "Transport Planning in Sub-Saharan Africa." *Progress in Develop-ment Studies* 7(3): 251–257.

Probáld, F., and P. Szabó (eds.). 2007. *Európa Regionális Földrajza. Társadalomföl-drajz Egyetemi Tankönyv*. Budapest: ELTE Eötvös Kiadó.

Pucher, J., and R. Buehler (eds.). 2012. *City Cycling*. Cambridge, MA: MIT Press.

Railway-technology.com. 2015. The Website for the Railway Industry. http://www. railway-technology.com/. Accessed August 8, 2015.

Reference Note. 2013. *National Highways Development Project—An Overview*. No. 23/RN/Ref./August/2013. http://164.100.47.134/intranet/NHDP.pdf. Accessed August 8, 2015.

Rehmeyer, J. 2007. "Road Bumps: Why Dirt Roads Develop a Washboard Surface." *Science News*. https://www.sciencenews.org/article/road-bumps-why-dirt-roads -develop-washboard-surface.

Reid, C. 2015. *Roads Were Not Built for Cars: How Cyclists Were the First to Push for Good Roads & Became the Pioneers of Motoring*. Washington, DC: Island Press.

Reidel, H-U. 2014 (November). "Chinese Metro Boom Shows No Sign of Abating." *International Railway Journal*. http://www.railjournal.com/index.php/metros /chinese-metro-boom-shows-no-sign-of-abating.html?channel=525. Accessed August 8, 2015.

Renewable Fuels Association. 2015. World Fuel Ethanol Production. http://ethan olrfa.org/pages/World-Fuel-Ethanol-Production. Accessed January 5, 2015.

Riverson, J. D. N., and S. Carapetis. 1991. "Intermediate Means of Transport in Sub-Saharan Africa: Its Potential for Improving Rural Travel and Transport." *World Bank Technical Paper* No. 161. Africa Technical Department Series. http:// www4.worldbank.org/afr/ssatp/Resources/WorldBank-TechnicalPapers/TP161 .pdf. Accessed May 13, 2014.

Rodrigue, J-P, C. Comtois, and B. Slack. 1998. *The Geography of Transport Systems*. Hempstead, NY: Hofstra University.

Rodrigue, J-P., T. Notteboom, and J. Shaw. 2013. *The Sage Handbook of Transport Studies*. Los Angeles, CA: Sage.

Roney, J. Matthew. 2008. Bicycles Pedaling into the Spotlight. Earth Policy Institute Eco-Economy Indicators. http://www.earth-policy.org/indicators/C48.

Rosen, D. A., F. J. Mammano, and R. Favout. 1970. "An Electronic Route-Guidance System for Highway Vehicles." *IEEE Transactions on Vehicular Technology* (19)1: 143–152.

Rowley, R. J. 2013. *Everyday Las Vegas: Local Life in a Tourist Town.* Reno: University of Nevada Press.

Roy, S., and K. Hannam. 2013. "Embodying the Mobilities of the Darjeeling Himalayan Railway." *Mobilities* 8(4): 580–594.

Savage, N. 2011 (June 23). "The Ideal Biofuel." *Nature* 474: s9–s11.

Schivelbusch, W. 2006. *Three New Deals: Reflections of Roosevelt's America, Mussolini's Italy, and Hitler's Germany, 1933–1939.* Translated by Jefferson Chase. New York: Metropolitan Books Henry Holt and Company, LLC.

Schwandl, Robert. 2015. UrbanRail.Net. http://www.urbanrail.net/. Accessed August 8, 2015.

Serna, Joseph. 2015. The Ins and Outs and U.S.-Mexico Border Tunnels. http://www.latimes.com/local/lanow/la-me-ln-border-tunnels-20150501-htmlstory.html.

Shaheen, Susan A., Adam P. Cohen, and Melissa S. Chung. 2009. "North American Carsharing." *Transportation Research Record: Journal of the Transportation Research Board* (2010)1: 35–44.

Shapouri, H., J. Duffield, and M. Wang. 2002. *The Energy Balance of Corn-Ethanol: An Update.* Washington, DC: U.S. Department of Agriculture.

Shaw, S. L., Fang, Z., Lu, S., and Tao, R. 2014. "Impacts of High Speed Rail on Railroad Network Accessibility in China." *Journal of Transport Geography* 40: 112–122.

Shenzhen News. 2010 (April 14). "Shenzhen Metro: Ten-Year's Dream Came True." http://sztqb.sznews.com/html/2010-04/14/content_1035002.htm [in Chinese]. Accessed November 13, 2014.

Shenzhen News. 2014. Free WiFi to be Provided at the End of the Month on Three Lines of the Shenzhen Metro. http://travel.people.com.cn/n/2014/0411/c41570-24880120.html [in Chinese]. Accessed November 14, 2014.

Siler, W. 2011. "Motorcycle Taxis Come to America." *RideApart.* http://rideapart.com/articles/motorcycle-taxis-come-to-america. Accessed August 8, 2015.

Stille, William M. 2014. "Mobility on the Move: Rickshaws in Asia." *Transfers* 4(4): 88–107.

Suh, N. P., D. H. Cho, and C. T. Rim. 2011. "Design of On-Line Electric Vehicle (OLEV)." In *Global Product Development* 3–8. Springer.

Sulkin, M. 1999. "Personal Rapid Transit Déjà Vu." *Transportation Research Record* 1677: 58–63.

Sutton, John C. 2015. *Gridlock: Congested Cities, Contested Policies, Unsustainable Mobility.* Florence, KY: Routledge.

Taplin, Michael. 1998. The History of Tramways and Evolution of Light Rail. Light Rail Transit Association. http://www.lrta.org/mrthistory.html. Accessed November 14, 2014.

Tesla Motors. 2015. Supercharger. http://www.teslamotors.com/supercharger. Accessed February 25, 2015.

Tesla Motors Blog. 2014 (November 5). "Model S Achieves Euro NCAP 5-Star Safety Rating." http://www.teslamotors.com/blog/model-s-achieves-euro-ncap-5star-safety-rating. Accessed February 25, 2015.

Thompson, L. S. 1994. "High-Speed Rail (HSR) in the United States—Why Isn't There More?" *Japan Railway & Transport Review* 32–39.

Thong, Melvyn, and Adrian Cheong. 2012. "Energy Efficiency in Singapore's Rapid Transit System." *Journeys* 38–47.

Todorovich, P., D. Schned, and R. Lane. 2011. *High Speed Rail: International Lessons for US Policy Makers* (No. Policy Focus Report/Code PF029).

Tollefson, J. 2010. "Hydrogen Vehicles: Fuel of the Future?" *Nature News* 464: 1262–1264 http://www.nature.com/news/2010/100428/pdf/4641262a.pdf. doi: 10.1038/4641262a. Accessed August 8, 2015.

Tryler, N. 2010. History of Transportation Systems in the United States. http://tyler topics.com/TranspBookJan2010.pdf. Accessed December 18, 2015.

Twitchell, Heath. 1992. *Northwest Epic: The Building of the Alaska Highway*. New York: St. Martin's Press.

United States Congress. (1995). *High-Tech Highways: Intelligent Transportation Systems and Policy*. Washington, DC: Congressional Budget Office.

United States Department of Agriculture. Economic Research Service. 2015. http://www.ers.usda.gov/data-products/us-bioenergy-statistics.aspx. Accessed January 6, 2015.

United States Department of Energy. Alternative Fuels Data Center website. 2014. http://www.afdc.energy.gov/fuels/emerging_biobutanol.html. Accessed January 10, 2015.

United States Department of Transportation. Federal Highway Administration. 2015. http://www.fhwa.dot.gov/. Accessed August 5, 2015.

Unitsky String Transport Co. Ltd. 2005. Pre-Project Proposal High-Speed String Transportation Route "Abu Dhabi—Dubai—Sharjah." http://www.yunitskiy .com/author/2005/2005_10.pdf. Accessed July 24, 2015.

Universiade Shenzhen. 2013 (August 10). Shenzhen Metro. http://www.sz2011 .org/szbk/bmfw/36222.shtml [in Chinese]. Accessed February 25, 2015.

Urban Rail. 2015. Urban Rail website. http://www.urbanrail.net/. Accessed August 5, 2015.

Van De Walle, Frederik. 2004. "The Velomobile As a Vehicle for More Sustainable Transportation: Reshaping the Social Construction of Cycling Technology." MSc thesis. Royal Institute of Technology, Department of Infrastructure, Stockholm. http://users.telenet.be/fietser/fotos/VM4SD-FVDWsm.pdf. Accessed July 20, 2015.

Voice of America. 2014 (April 4). Vehicles May Soon Be Talking to Each Other. http://www.voanews.com/content/vehicles-may-soon-be-talking-to-each -other-/1886895.html. Accessed January 8, 2015.

Voth, H., and N. Voigtländer. 2014 (May 22). "Nazi Pork and Popularity: How Hitler's Roads Won German Hearts and Minds." http://www.voxeu.org/article

/nazi-pork-and-popularity-how-hitler-s-roads-won-german-hearts-and-minds. Accessed June 12, 2014.

Vyas, A., D. Santini, and L. Johnson. 2009. "Plug-In Hybrid Electric Vehicles' Potential for Petroleum Use Reduction: Issues Involved in Developing Reliable Estimates." Presented at 88th Annual Meeting of the *Transportation Research Board*. Washington, DC. January 11–15, 2009.

Walker, D. A. 2009. "Biofuels, Facts, Fantasy, and Feasibility." *Journal of Applied Phycology* 21(5): 509–517.

Walker, J. 2012. *Human Transit: How Clearer Thinking about Public Transit Can Enrich Our Communities and Lives*. Washington, DC: Island Press.

Washington Metropolitan Area Transit Authority. 2001. *Metro at 25: Celebrating the Past, Building the Future*. Washington, DC: WMATA.

Weinberger, H. 2014 (November 19). Netherlands Unveils "Starry Night" Solar Bike Path [blog post]. http://www.outsideonline.com/1927506/netherlands -unveils-starry-night-solar-bike-path. Accessed December 18, 2015.

"What Happened to the Flying Car?" 2012 (March 3). *The Economist* 402 (8774): S.3–S.4.

Whisnant, Anne Mitchell. 2006. *Super-Scenic Motorway: A Blue Ridge Parkway History*. Chapel Hill: University of North Carolina Press.

Whitney, Charles S. 2003. *Bridges of the World: Their Design and Construction*. New York: Dover.

Williams, A. 1992. "Transport and the Future." In *Modern Transport Geography*. Edited by B. S. Hoyle, and R. D. Knowles. 257–70. London: Belhaven Press.

Williams, Carrie. [n.d.]. "Railroads in New Mexico Tags: Libguides." *Narrow Gauge Railroads*. Friends for the Public Library. Accessed November 3, 2014.

Wilner, Frank N. 2012. *Amtrak: Past, Present, Future*. Omaha, NE: Simmons-Boardman Books.

World Highways. 2015. World Highways website. http://www.worldhighways .com/. Accessed August 5, 2015.

Ye, X., Z. Lian, C. Jiang, Z. Zhou, and H. Chen. 2010. "Investigation of Indoor Environmental Quality in Shanghai Metro Stations, China." *Environmental Monitoring and Assessment* 167: 643–651. doi: 10.1007/s10661-009-1080-9.

Yenne, Bill. 2005. *Great Northern Empire Builder*. Minneapolis, MN: MBI.

Younger, Emily. KRQE.com. 2014 (October 1). Route 66 'Singing Road' Debuts in New Mexico. http://krqe.com/2014/10/01/route-66-singing-road-debuts-in -new-mexico/. Accessed June 5, 2015.

Zusman, E., A. Srinivasan, and S. Dhakal. 2012. *Low Carbon Transport in Asia: Strategies for Optimizing Co-Benefits*. London: Earthscan from Routledge.

About the Editors and Contributors

Editors

Selima Sultana (PhD, 2000, Geography, University of Georgia) is a professor and the director of Graduate Studies of Geography at the University of North Carolina, Greensboro (UNCG). Her research foci are in the area of urban and transportation geography. Very specifically Dr. Sultana studies the commuting patterns of individuals, households, and among different race/ethnic groups, focusing on how people negotiate the conflicting demands of household responsibilities and the changing urban settings of their lives. Dr. Sultana has authored more than 40 scholarly articles and her work has appeared in leading geographical journals such as the *Annals of the Association of American Geographers, The Professional Geographer, Journal of Transport Geography, Transport Policy, Tourism Geographies, Urban Geography, Urban Studies, Growth & Change, Southeastern Geographer,* and on the London School of Economics and Political Science's (LSE) blog website. Dr. Sultana's research has been supported by the Department of Transportation through the University Transportation Center of Alabama, the Center for Sustainability at Auburn University, the University of North Carolina at Greensboro, and Guilford County Child Development. Prior to appointment at UNCG, Dr. Sultana was an assistant professor at Auburn University. She currently is working on several projects including a book (coauthored with Dr. Joe Weber) on the U.S. National Parks.

Joe Weber (PhD, 2001, Geography, Ohio State University) is a professor of geography at the University of Alabama. Dr. Weber's research focuses on individual and household mobility, accessibility within changing urban environments, and the transportation networks that sustain and limit this mobility. He has long been fascinated by the changing geography of the American highway system and has written about the U.S. Interstate Highway System, roads in U.S. national parks, changing regional accessibility patterns, space-time convergence, and the cultural geography of highways and the roadside landscape. His favorite activities include searching out abandoned roads and bridges by car, on foot, and on Google Earth. Dr. Weber has published numerous articles in leading geographical journals such as *Annals of the Association of American Geographers, Professional Geographers, Geographical Analysis, Journal of Geographical System, Urban Studies, Urban Geography, Journal of Historical Geography, Geographical Review, Cultural Geography,* and *Journal of Transport Geography*. His research has been supported by the Department of Transportation through the University Transportation Center of Alabama, and National Institutes of Health (NIH). Dr. Weber currently is working on several projects including a book (coauthored with Dr. Selima Sultana) on the U.S. National Parks. Weber teaches classes on transportation geography, national parks, and Geographic Information Systems.

Contributors

Adri, Neelopal (PhD). Dr. Adri is a lecturer of Urban and Regional Planning in Bangladesh Institute of Technology (BUET), Dhaka, Bangladesh.

Akter, Taslima. Taslima is a PhD student in the Department of Geography and Planning in University of Toledo, Ohio, USA.

Alam, Bhuiyan Monwar (PhD). Dr. Alam is an associate professor of geography and planning in University of Toledo, Ohio, USA.

Aragon, Nazli Uludere. Nazlil is a PhD student in geography at Arizona State University, Arizona, USA.

Bjelland, Mark (PhD). Dr. Bjelland is a faculty member for geology, geography, and environment planning at Calvin College, Grand Rapids, Michigan, USA.

Burke, Charles M. Charles is a PhD candidate in the Department of Geography and Earth Science at McMaster University, Hamilton, Ontario, Canada.

Chen, Xuwei (PhD). Dr. Chen is an associate professor of geography at Northern Illinois University, DeKalb, Illinois, USA.

Cook, Simon (PhD). Dr. Cook is a postdoctoral research associate in the Department of Geography at Royal Holloway, University of London, United Kingdom.

Coolbaugh, Dylan. Dylan is a graduate student in geography at the University of North Carolina, Greensboro, North Carolina, USA.

Cusack, Christopher (PhD). Dr. Cusack is a professor of geography at Keene State College, Keene, New Hampshire, USA.

Dede-Bamfo, Nathaniel (PhD). Dr. Dede-Bamfo is an adjunct faculty member in geography at Texas State University, San Marcos, Texas, USA.

Di Gianni, Joe. Joe is a PhD student in the Department of Earth and Environment Studies at Montclair State University, Montclair, New Jersey, USA.

Greenberg, Jason (PhD). Dr. Greenberg is a professor of geography at Sullivan University, Louisville, Kentucky, USA.

Hagge, Patrick D. (PhD). Dr. Hagge is an assistant professor of geography at Arkansas Tech University, Russellville, Arkansas, USA.

Jiao, Jingjuan (PhD). Dr. Jiao is a researcher at the Chinese Institute of Geographic Sciences and Natural Resources Research.

Kelley, Scott. Scott is a PhD student in geography at Arizona State University, Phoenix, Arizona, USA.

Kim, Hyojin. Hyojin is a PhD candidate in geography at the University of North Carolina-Greensboro, Greensboro, North Carolina, USA.

Kim, Hyun (PhD). Dr. Kim is an assistant professor of geography at University of Tennessee, Knoxville, Tennessee, USA.

Kim, Jong-Geun (PhD). Dr. Kim is a lecturer in geography at Seoul National University, Seoul, Korea.

Koger, Grove. Grove is an independent freelance writer.

Kuby, Michael (PhD). Dr. Kuby is a professor of geography at Arizona State University, Phoenix, Arizona, USA.

Lane, Bradley W. (PhD). Dr. Lane is an assistant professor of geography at the University of Kansas, Lawrence, Kansas, USA.

Le Vine, Scott E. (PhD). Dr. Le Vine is an assistant professor at the State University of New York, New Paltz, New York, USA, and a research associate at the Imperial College, London, UK.

Leenerts, Rosemarie Boucher. Rosemarie is an independent freelance writer.

Minn, Michael (PhD). Dr. Minn is a PhD student in geography at the University of Illinois-Urbana in Champaign, Illinois, USA.

Molina, Alejandro. Alejandro is a graduate student in geography at the University of North Carolina at Greensboro, North Carolina, USA.

Nichols, Terri. Terri is an independent freelance researcher and writer.

Oetter, Doug R. (PhD). Dr. Oetter is a professor of geography at Georgia College in Milledgeville, Georgia, USA.

Pour Ebrahim, Nastaran. Nastaran is a PhD student in geography at the University of North Carolina-Greensboro, Greensboro, North Carolina, USA.

Richard, Amanda. Amanda is a PhD student in geography at Florida State University, Tallahassee, Florida, USA.

Rivera, Hector Agredano. Hector is a PhD student in geography at City College of New York, New York, USA.

Rowley, Rex J. (PhD). Dr. Rowley is an assistant professor of geography at Illinois State University, in Normal, Illinois, USA.

Rupe, Rachel. Rachel is a graduate student in geography at the University of North Carolina, in Greensboro, North Carolina, USA.

Sharma, Purva. Purva is a PhD student in geography at the University of North Carolina-Greensboro, in Greensboro, North Carolina, USA.

Shaw, Shih-Lung (PhD). Dr. Shaw is a professor of geography at University of Tennessee, in Knoxville, Tennessee, USA.

Sorrensen, Cynthia (PhD). Dr. Sorrensen is an assistant professor of geography at Texas Tech University, Lubbock, Texas, USA.

Index

Page numbers in **boldface** indicate main entries in the volume.

Abt, Carl Roman, xxiv, 120, 247
Adri, Neelopal, 350
aerial tramways, **1–3**; cost of, 2; future of, 3; iconic aerial tramways, 1; Jackson Hole ski lift, 2; lifespan of, 1; lifespan of aerial tram systems, 2; photograph of, 2; Pipestem tram system, 2; Portland, Oregon, 3; replaced or discontinued, 1–2; Roosevelt Island Tramway, 1; Sandia Peak Tramway, 1; Schmittenhöhebahn tramway, 1; in Tatev, Armenia, 1; as tourist attractions, 1; tramway systems, 2–3
Afghan Express, the, 123
Agassi, Shai, 35
Akashi Kaikyo Bridge, 299
Akter, Taslima, 350
Alam, Bhuiyan, 350
Alaska Marine Ferry, 4
Alaska or Alaska-Canada (ALCAN) Highway, xxiv, **3–4**; alternatives to, 4; beginning of, 4; best guide to, 4; date completed, 3; date started, 3; Dawson Creek, 4; endpoint of, 4; highest elevation of, 4; initial road, 3; length and location of, 4; *The Milepost*, 4; military use of, 3; original purpose of, 3; reconstruction of, 3; use of, 4
Alaska Railroad (United States), xxiv, **4–6**; Anchorage to Fairbanks train, 6; construction of, 5; cruise ship passengers and, 6; current status of, 6; date completed, 5; Denali National Park and, 6; the Hurricane Turn train, 6; length of, 4; ownership of, 5; passenger service of, 5–6; photograph of President

Harding driving the last spike of, 5; uniqueness of, 5; White Pass and Yukon Route railroad, 5
Allen, Harry N., 335
Allen, Richard Sanders, 87
alpine tunnels (Europe), xxiv, **6–8**; AlpTransit or New Rail Link, 7; Arlberg Railway Tunnel, 7; Arlberg Road Tunnel, 7; Ceneri Base Tunnel, 7; Fréjus Rail Tunnel (Mont Cenis Tunnel),, 7; Fréjus Tunnel, 7; Gotthard Base Tunnel, 7; Gotthard Tunnel, 7; Great St. Bernard, 7; human cost of, 8; Lötschberg Base Tunnel, 7; Lötschberg Tunnel, 7; Mont Blanc Tunnel, 7; Simplon tunnels, 7; St. Gotthard Tunnel, 7; technological advancements, 7; Urnerloch tunnel, 6; vehicular tunnels, 7
alternative fuels, **8–10**; alternative fuel options, 8; biodiesel fuel, 8, 9; charging stations, 8; corporate interests, 9; cultural factor involving, 9; economic factors involving, 9; environmental benefits, 9; ethanol, 8, 9; hybrid electric vehicles, 8; natural gas, 9; plug-in electric vehicle, 8; political factors involving, 9; propane, 9; refineries and, 8–9; renewable natural gas, 9; significance of, 10
Amazon.com, 118
ambulances, xxiv, **10–12**; Ambulance Corps Act (U.S.), 10; ambulance designs, 11; Bellevue Hospital, NYC, 11; Commercial Hospital of Cincinnati, 11; definition of, 10; Dominique-Jean, Larry, 10; emergency transport system (U.S.),